A STRUCTURED INTRODUCTION
TO

NUMERICAL MATHEMATICS

A Structured Introduction to
Numerical Mathematics

P. J. HARTLEY   MA MSc PhD AFIMA

and

A. WYNN-EVANS MA

Department of Mathematics, Lanchester Polytechnic

Stanley Thornes (Publishers) Ltd.

First published in 1979 by

Stanley Thornes (Publishers) Ltd.
EDUCA House
Liddington Estate
Lekhampton Road
CHELTENHAM GL53 0DN
England

ISBN 0 85950 426 3 student flexiback edition

Typeset at The Alden Press, Oxford, London and Northampton and printed and bound by Redwood Burn Ltd, Trowbridge and Esher.

Contents

Preface

This text is primarily for first-year undergraduates. It differs from many other first courses in the subject by not assuming any knowledge of mathematical analysis, and yet not being simply a list of methods. The prerequisites for the course are a sound knowledge of elementary differentiation and integration and a mathematical maturity such as would be obtained by studying any A-level course in mathematics.

The book comprises essentially the set of course notes with exercises which the authors have used with reasonable success at Lanchester Polytechnic. The courses which have used the material are the first year of each of:

(a) the Modular Degree Courses in Science scheme, which attracts students of all scientific disciplines, interests and abilities;

(b) the Part-Time Degree and Diploma Courses in Mathematics; and

(c) the HND in Mathematics, Statistics and Computing.

Hence the text aims to satisfy a wide spectrum of students. Nevertheless a serious attempt is made to be analytical where this is feasible with such a mixed audience.

It also attempts to be sufficiently wide-ranging to be a reasonably complete course for students who are unlikely to proceed further with the subject, but also to lay the foundations for further study at a higher level for others.

We believe that undergraduates need to be encouraged to read mathematics books, and to study with discipline. To aid the former we have attempted an informal style and have given rather fuller explanations than are usual in a text book. To encourage discipline the book is carefully structured, being divided into 24 lessons of approximately equal weight, each in a set format. The format is: objectives; material, with concurrent examples and exercises; answers to exercises; further reading; comprehension test; supplementary exercises. In practice each lesson has been associated with three hours classroom time, with the assumption that the average student will also require one or two hours of his own time to complete the lesson. The 'harder' sections and exercises within the book are indicated with an asterisk.

We would like to acknowledge the help we have received in the writing and publication of this book. Firstly we thank our colleagues at Lanchester Polytechnic for not pouring too much scorn on our efforts and for making useful suggestions. Secondly we thank the many students who have patiently pointed out a host of mistakes in the original notes. (We look forward to hearing about any that still remain!) Lastly we note our gratitude for the tolerance and skill shown by the staff of Stanley Thornes (Publishers) Ltd.

P. J. Hartley
A. Wynn-Evans

PART A **GETTING STARTED**

Lesson 1 **Introduction**

OBJECTIVES

At the end of this lesson you should

(a) have an appreciation of the need for numerical mathematics;

(b) have an appreciation of the sort of problems studied in numerical mathematics;

(c) understand the idea of an algorithm and of a flow diagram;

(d) have an appreciation of the idea of a *direct* method of solution and of an *iterative* method of solution to a problem.

CONTENTS

1.1 METHODS OF STUDY AND HOW TO USE THESE LESSONS!

Many people find studying mathematics difficult. To start this lesson it may be appropriate to give some advice on studying mathematics in general and these lessons in particular. In this course, as in any study of mathematics, numerical or otherwise, you should not necessarily expect to understand the contents completely at the first reading. The mathematical arguments and explanations should be read at least twice for reasonable understanding. In order to help your comprehension of the processes and ideas various examples are given in the text. Normally you should work through these examples yourself. Exercises are set to test and reinforce your understanding, and it may well be that only in doing these exercises will you really get to grips with the essence of the material covered in the text. Answers to the exercises appear at the end of each lesson, together with suggestions for further reading, a comprehension test, and, usually, some supplementary exercises. This course is intended for readers with a wide variety of backgrounds. The only assumptions made are that you should be able to perform elementary algebraic operations with reasonable facility and have a working knowledge of elementary calculus.

Where results are used that may not be familiar to you appendices are referred to and are provided separately. This approach is intended to enable you to work through the material of the lessons without losing sight of the main ideas. You may wish to work through the appendices in detail. On the other hand, you may well be best advised simply to note the results of the work contained in the relevant appendix and then to continue with the work of the lesson.

Some sections of the lessons and some exercises are marked with an asterisk. These are intended for students who are not experiencing much difficulty with the main work. We would expect that they *will* be read by students intending to study mathematics as a main subject.

Generally speaking for any course that you do it is a good idea to imagine that *you* are going to give a course of lectures on the subject. If you put together a set of notes in your own words which is suitable for that purpose, you will have done really well and you should know a good deal about it. You will then be able to extract from your notes a shorter set of 'signposts' which should remind you of the main points of the course. It might also be a good idea to keep a clear and complete set of solutions to the exercises set.

Some of the notation used in the lessons and the appendices may prove difficult or unusual by the standards of your previous experience. Where this is the case attempts have been made to give exercises to help with the familiarization process and you should realise that becoming familiar with and learning how to use a new notation is part of your mathematical education. The study of mathematics consists of the learning of a language as well as the learning of concepts, techniques and methods for solving problems.

Within each lesson the following numbering system has been adopted:

Main sections (with titles) are numbered 1.1, 1.2, 1.3, etc., so that 2.3 is the third section of Lesson 2.

Examples are also numbered 1.1, 1.2, 1.3, etc., so that Example 3.4 is the fourth example of Lesson 3. Example 3.4 will not necessarily be in Section 3.4 of Lesson 3. **Exercises** are similarly numbered.

Equations are numbered 1-1, 1-2, 1-3, etc., so that Equation 3-5 is the fifth equation of Lesson 3.

Diagrams are labelled Fig. 1.1, Fig. 1.2, etc., and again are not necessarily in the same-numbered section of their lesson.

1.2   WHY NUMERICAL MATHEMATICS?

As the name implies, numerical mathematics is concerned with calculating numerical solutions of problems which can be expressed in mathematical form. Such problems arise in many different disciplines and the use of numerical mathematics is frequently the only way in which useable answers can be obtained. Pure mathematics, on the other hand, gives considerable insight into the structure and properties of mathematical problems but often supplies methods of solutions which work only for very restricted cases.

Here are some examples of practical problems that can be solved by numerical methods.

EXAMPLE 1.1   In civil engineering an approximation to the rise time for a reservoir turned out to be

$$\int_{0.25}^{0.5} \frac{4000}{10 - 20\,h^{3/2}}\, dh.$$

Such an integral cannot obviously be evaluated by finding an indefinite integral so that numerical techniques must be employed. These are essentially geometrically-based techniques for estimating the area under a graph.

EXAMPLE 1.2   The positions of planets are given at various fixed times in tables. If the position of a planet is required at a time not given in the tables, a numerical process called interpolation may be used to find the required position.

EXAMPLE 1.3   The following equations arise from an approximation to the equations describing heat flow in a bar:

$$\begin{aligned}
5u_1 - 2u_2 \qquad\;\; &= \; 4 \\
-2u_1 + 5u_2 - 2u_3 &= \; 3 \\
-2u_2 + 5u_3 &= 14.
\end{aligned}$$

[1-1]

We can solve these without too much trouble by an elementary extension of the techniques we all learn at school for solving a pair of simultaneous linear equations.

Normally however we have rather more than three equations to solve — in some cases as many as 2000! Here is an example of a system of eight linear equations:

$$
\begin{aligned}
-4x_1 + x_2 \quad &\quad + x_5 &&= a_1 \\
x_1 - 4x_2 + x_3 \quad &\quad + x_6 &&= a_2 \\
x_2 - 4x_3 + x_4 \quad &\quad + x_7 &&= a_3 \\
x_3 - 4x_4 \quad &\quad + x_8 &&= a_4 \\
x_1 \quad &\quad -4x_5 + x_6 &&= a_5 \\
x_2 \quad &\quad + x_5 - 4x_6 + x_7 &&= a_6 \\
x_3 \quad &\quad + x_6 - 4x_7 + x_8 &&= a_7 \\
x_4 \quad &\quad + x_7 - 4x_8 &&= a_8.
\end{aligned}
\tag{1-2}
$$

So we have eight equations in the eight unknowns $x_1, x_2, x_3, x_4, x_5, x_6, x_7, x_8$ (which we normally write $x_1, x_2, \ldots, x_8$ for brevity), in which $a_1, a_2, \ldots, a_8$ would be given numbers in practice. Clearly the solution of this set of equations is not so straightforward as the previous set of three. The questions then arise as to the best method of solution and the accuracy of the answers obtained. Answering those two questions is the province of numerical mathematics.

Numerical mathematics has, in the past, suffered from having to concern itself to a very great extent with the drudgery of numerical calculations. In consequence little interest was shown in the subject and little progress was made. However, with the advent of reliable mechanical desk calculators and later electronic desk calculators, and especially computers, the range of problems that can be tackled by numerical means has expanded enormously and in all manner of fields, for example, in pure science, applied science, engineering, economics, and geography.

1.3   WHAT IS NUMERICAL MATHEMATICS?

As we have already said, numerical mathematics is mainly concerned with finding numerical answers to problems stated in mathematical form. We can express this problem-solving activity in a simple diagrammatical form as shown in Fig. 1.1.

In this diagram the word *data* refers to all the information we have about the problem. This will usually be in two parts: first there will be an equation or equations to be solved; and second there will be some numbers, which may be obtained by experiment or observation, or may be physical or mathematical constants, or may be numbers chosen to test the effectiveness

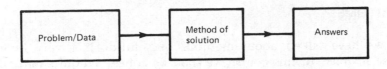

Fig. 1.1

of the method as compared to some other method. Sometimes we shall use the word *data* to refer to just the numbers.

In most cases the method of solution can be chosen from many possible methods. One of the most important aspects of numerical mathematics is the consideration of which are the better methods. Various criteria are used for judging which are the better methods, among which we might mention

(a) their speed (or economy) of calculation;

(b) the accuracy of the resulting answers;

(c) their safety (i.e., whether they go badly wrong in some cases).

By an accurate answer we mean, of course, one which is 'near' to the true answer. The difference between a true answer and a calculated answer is called the *error*. The consideration of the sources and the effects of errors in calculations is the other important part of numerical mathematics. We shall consider the term error more precisely in Lesson 2. For the moment we shall be content to say that when we talk of errors we do *not* include the effects of mistakes in the method used or in the calculations.

The most common sources of error are the data for the problem (experimental error for example), the method employed (which may make an approximation in order to make the problem solvable) and any machine used in the calculation (because numbers tend to be rounded in machines).

To summarise, in numerical mathematics we study, construct and compare numerical methods of solution on the one hand, and analyse error production and propagation on the other.

Finally we should say that the activity described by Fig. 1.1 may be just the numerical-mathematical core of a wider process which starts with devising a model to describe or simulate a physical situation and ends with the interpretation of the results as applied to that physical situation. Example 1.3 is typical of this sort of process. A great deal of work has gone into converting the physical problem into a set of linear equations. When the equations have been solved there would remain the interpretation of the answers in the context of the original problem.

In general we shall be primarily concerned with just the numerical solution of the problem stated in mathematical terms — i.e., with the activity given by Fig. 1.1.

1.4 ALGORITHMS

So far we have talked about methods of solution in a very general and imprecise manner. In this course we shall be concerned quite naturally with *numerical* methods for solving mathematical problems. When we talk of a particular *method* we shall mean the method as it might be described without going into the precise detail of the calculation.

On the other hand, if we give precise instructions as to how a solution may be obtained, then we shall use the term *algorithm* to describe such an implementation of a method.

In consequence we will take the term *algorithm* to mean a complete and unambiguous description of a method of constructing a solution of a mathematical problem. In doing so precise mention must be made of what operations are allowed. For example, in geometrical algorithms, rulers, pencils and compasses might be specified. In this book only numerical algorithms will be discussed and for these algorithms the basic arithmetic operations of addition, subtraction, multiplication, division and raising to a power will be required, together with certain logical operations based on comparing numbers for equality or inequality.

A numerical algorithm must describe the method, be such that the solution, if it exists, is obtained in a finite number of steps, and contain checks to ensure that, when a solution does not exist or when the process appears to be taking an excessive number of steps, suitable action is taken. If these properties hold then an algorithm can be implemented using a computer in an automatic fashion.

EXAMPLE 1.4   (a) Let us consider the equation

$$ax = b$$

where a and b are given and x is unknown, and try to describe an algorithm for finding x whatever a and b are.

"Divide b by a" is not a proper algorithm because it would fail if $a = 0$.

A proper algorithm might be described by Fig. 1.2 called a *flow chart* or *flow diagram*.

(b) Problem — to find the largest of a set of positive numbers.

Possible algorithm — compare each number in turn with all others until one is found to be larger than all others. This is a valid algorithm, but clearly slower (in general) than one which compares the first number with the second and rejects the smaller, compares the retained number with the third similarly, and so on to the end of the list of numbers.

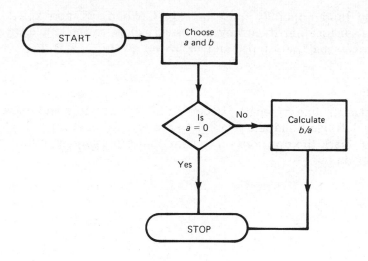

Fig. 1.2

An algorithm may be *direct* or *iterative*. An example of a direct algorithm is that in which a set of linear algebraic equations is solved by eliminating one variable at a time.

EXAMPLE 1.5   To find y given that

$$2x + y = 4$$
$$x - y = -1.$$

We eliminate x in the usual way

$$2x + y = 4$$
$$2x - 2y = -2$$
$$\therefore \qquad 3y = 6$$
$$\therefore \qquad y = 2.$$

Clearly such an algorithm takes a fixed number of steps.

In an iterative algorithm an estimate is made of the solution, say x_0. Then, using the properties of the problem, a second estimate, x_1, is calculated from x_0. Similarly from x_1 an x_2 is found, and so a sequence of estimates is formed:

$$x_0, x_1, x_2, x_3, \ldots .$$

If the iterative algorithm is well-chosen the estimates of the answer (the numbers $x_0, x_1, x_2, x_3, \ldots$) should get nearer and nearer to the true answer. This means that they will get nearer and nearer to each other eventually (and hopefully quite quickly!). When x_n and x_{n+1}, say, are close enough together then they are usually close to the true answer as well, so the process may be stopped and an answer given.

EXERCISE 1.1

Draw a straight line of length 16 cm on a piece of paper and mark an origin O at the left-hand end and a point to represent the number 1 at the right-hand end. Mark the positions on the line that represent the numbers $x_0, x_1, x_2, \ldots$ given by

$$x_0 = 1$$
$$x_1 = x_0/2$$
$$x_2 = x_1/2$$
$$\cdots \cdots \cdots$$
$$x_n = x_{n-1}/2$$
$$\cdots \cdots \cdots$$

continuing until you cannot distinguish two successive points.

Given that $x_0, x_1, x_2, \ldots$ is a sequence of numbers that is supposed to be getting closer and closer to the solution of a problem, what do you think the solution is? If we stop the process as soon as two successive values differ by less than 0.001 what will the last value be?

We shall see many examples of iterative methods in this course. It is interesting to note that many of the day-to-day activities of life are iterative in nature.

EXAMPLE 1.6   A designer will make a first attempt at a design, analyse its implications, and then successively improve it until it meets the design objectives sufficiently closely — see Fig. 1.3.

1.5   EXERCISES

To conclude the main part of Lesson 1 here are some "warming-up" exercises. They are designed to give you a small insight into the kind of calculations that will occur in the lessons to come.

EXERCISE 1.2

Use a calculator to do these calculations. Each is designed so that you need write down no intermediate results: the complete calculation can be done

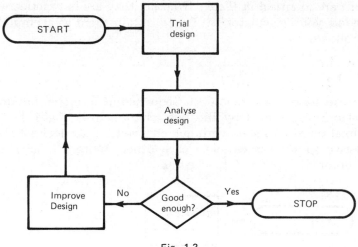

Fig. 1.3

inside the calculator, provided that it has at least one memory and a square root facility.

(a) Evaluate the two roots of the quadratic equation

$$x^2 + 20x + 1 = 0$$

giving the answers to 3 decimal places.

(b) Evaluate the solutions

$$\begin{cases} x = (ce - bf)/(ae - bd) \\ y = (af - cd)/(ae - bd) \end{cases}$$

of the equations

$$\begin{cases} ax + by = c \\ dx + ey = f \end{cases}$$

(i) when $a = 2$, $b = 1$, $c = 4$, $d = 1$, $e = -1$, $f = -1$
(ii) when $a = 2$, $b = 1$, $c = 4$, $d = -2$, $e = -1$, $f = -1$.

(c) Evaluate $5x^4 - 2x^3 + x^2 - x + 3$, for $x = 1.7$, using the "nested" form

$$(((5x - 2)x + 1)x - 1)x + 3.$$

(d) Evaluate
(i) $\sqrt[8]{(1.092)^9}$
(ii) $\sqrt{40.001} - \sqrt{40.000}$
(iii) $0.001/(\sqrt{40.001} + \sqrt{40.000})$.

(ii) and (iii) are identical *in theory*. Whether they are in practice will depend on how good your calculator is. Try to obtain them as accurately as your calculator allows.

EXERCISE 1.3

(a) Use a calculator to calculate the sequence of numbers obtained by the iterative algorithm defined by the flow chart below (Fig. 1.4). Work with three decimal places to start with and change to four decimal places when the successive values get close to each other. Write the successive values down in a column.

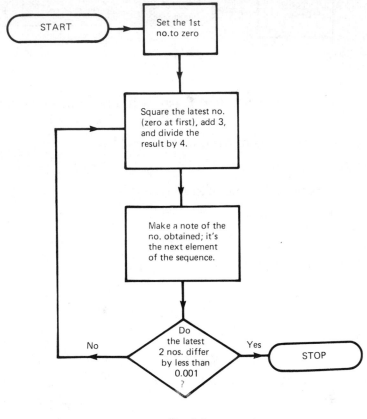

Fig. 1.4

(b) Write down the formula which gives each new number x_{n+1} in terms of the previous number x_n.

(c) Try to deduce what problem this algorithm has solved.

(d) Try to redraw the flow chart using mathematical language instead of

English. For example, the first box could be

```
Set x_0 = 0
```

and the second box must be something to do with (b).

Solve these pairs of linear equations:

(a) $x + y = 2$
 $x + 1.01y = 2.01$

(b) $x + y = 2$
 $x + 1.01y = 2$

(c) $x + y = 2$
 $x + y = 2.01$

(d) $x + y = 2$
 $x + y = 2.$

Any comments?

ANSWERS TO EXERCISES

1.1 The sequence is $1, \frac{1}{2}, \frac{1}{4}, \frac{1}{8}, \ldots, (\frac{1}{2})^n, \ldots$ and the numbers get as close as you like to zero. The first two successive numbers that differ by less than 0.001 are 1/512 and 1/1024, so that the last value is 1/1024.

1.2 (a) $-0.050, -19.950$
(b) (i) $x = 1, y = 2$ (ii) No solutions
(c) 36.1245
(d) (i) 1.10408 to 5 decimal places
   (ii) 0.000 079 056 457 $\Big\}$ using a Texas SR50.
   (iii) 0.000 079 056 447 4

1.3 (a)

i	x_i	Differences
0	0	
1	0.75	
2	0.891	
3	0.948	
4	0.975	
5	0.988	
6	0.994	
7	0.997	0.003
8	0.9985	0.0015
9	0.9993	0.0008

(b) $x_{n+1} = (x_n^2 + 3)/4$.

(c) It has found an approximation to the solution $x = 1$ of $x^2 - 4x + 3 = 0$.

*1.4   (a) $x = 1, y = 1$.

(b) $x = 2, y = 0$.

(c) No pair of values x, y, will satisfy both equations.

(d) Any values $x = \alpha$, $y = 2 - \alpha$, where α is any real number, will satisfy this pair of equations as they are, of course, the same equation.

FURTHER READING

References in these Further Reading sections will be to the author's name only, unless this would be ambiguous. Titles and publishers are given in the Bibliography at the end of the book.

We encourage you to consider at least one of the references given for two reasons:

(a) if you are having difficulty with the work you may like to read a simpler version;

(b) it is important not to think that our version is the only one or in any sense the absolute truth; reading another version broadens your view of the subject.

Here are some references for Lesson 1.

(a) Wilkes, Chapter 1 — a short, descriptive introduction.

(b) Walsh, Chapter 1, Sections 1 and 2 — also descriptive, but more advanced in concept.

(c) Henrici, Chapter 1 — a much fuller treatment with a historical perspective and some specific examples. Worth reading for the flavour even if you do not understand the details.

(d) Hosking, Joyce and Turner, *Prologue* — another short introduction.

(e) Dixon, Chapter 1 — yet another short introduction!

Lesson 2 Sets, Numbers and Errors

OBJECTIVES

At the end of this lesson you should

(a)   understand the concepts of and the notations for sets, open intervals, and closed intervals;

(b)   know what decimal places and significant figures are and be able to round a number correctly;

(c)   know the definitions of the various measures of error;

(d)   be able to analyse the effect of rounding errors in addition, subtraction, multiplication and division;

(e)   appreciate the way numbers are used in computers.

CONTENTS

2.1   SETS

A *set* is a collection of objects (usually called *elements*).

The number of objects (or elements) in a particular set may be finite or infinite. The elements are not to be regarded in any particular order. It is usual to denote a set S in which the elements are $a, b, c, \ldots$ by the notation

$$S = \{a, b, c, \ldots\}$$

unless there is some specially-designed alternative notation.

If the element a belongs to the set S we write

$$a \in S.$$

Sometimes we say a is a member of S.

EXAMPLE 2.1   Let M be the set of all men. (It is a finite set!) Let R be the set of all red-headed men. Clearly a redheaded man is a man. If r represents a red-headed man we write

$$r \in R$$

and hence

$$r \in M.$$

In Example 2.1 we have "$r \in R$" implies that "$r \in M$". It is easier to use the symbol $\Rightarrow$ for "implies that". Thus we have

$$r \in R \Rightarrow r \in M.$$

It can also be read as "if r is a red-headed man then r is a man". This kind of abbreviation is common in mathematics, avoiding as it does the use of long, wordy arguments.

2.2   SETS OF NUMBERS

We are all familiar with what are usually called the *natural numbers*, 1, 2, 3, ... They form a set $\mathbb{N}$ and we can write

$$\mathbb{N} = \{1, 2, 3, \ldots\}.$$

The natural numbers can be used for example for counting the amount of money in a bank balance or for ordering the finishers in the Derby.

If a bank balance is empty we need another number to denote this state of affairs and of course we use *zero*, 0, for this purpose.

When a bank balance is overdrawn (and money is owed to the bank) we can introduce a minus sign to denote this state of affairs and produce the numbers

$$-1, -2, -3, \ldots.$$

It is usual to denote the infinite set $\{0, \pm 1, \pm 2, \ldots\}$ by $\mathbb{Z}$ and these numbers are called the *integers*.

Thus $n \in \mathbb{Z}$ means that n is one of the numbers $0, \pm 1, \pm 2, \pm 3, \ldots.$ This idea is often used when a formula is given in which a general integer appears. For example,

$$\sin(\theta + 2k\pi) = \sin\theta, \quad k \in \mathbb{Z}.$$

This means that the formula is true for any integer k.

A convenient way of representing the integers ($\mathbb{Z}$) is to draw a straight line that is notionally of infinite length. We can take any point on this line and use it to represent zero. If we mark off equal lengths on this line, starting at 0 and proceeding in both directions, we gain an infinite set of equally-spaced points and these can be used to represent the rest of $\mathbb{Z}$. The points nearest to 0 represent ± 1, the next nearest ± 2, and so on. All the points on one side of 0 will represent positive numbers and those on the other side negative numbers.

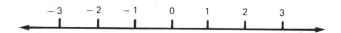

The arrows at the end of this line are used to show that the line is supposed to be infinite in length.

So far we have a set of isolated points on the line, which we will call the *real line*. It may be asked "if we take a point on the real line which does not represent a member of $\mathbb{Z}$ how can we interpret this point?" Such a point represents a *real number* and we denote the (again infinite) set of real numbers by $\mathbb{R}$.

$\mathbb{R}$ contains an infinite set of numbers called the *rational numbers*, so-called because each is the *ratio* of two integers ($1/2$, $-3/4$, $107/5$ are rational numbers for example).

Not all the points of the real line represent rational numbers. Those that do not are called the *irrational numbers*. The irrational numbers are so called because they cannot be written as the ratio of two integers. (Well known examples of irrational numbers are $\sqrt{2}$, π and e.) Taken together the set of rational numbers and the set of irrational numbers form the set of real numbers. A point on the real line represents either a rational number or an irrational number.

The fact that the real numbers contain all the rationals and all the irrationals, and a rigorous definition of what an irrational number is, are part of the proper study of pure mathematics and outside our objectives. One way of distinguishing rationals and irrationals is to use the fact that the decimal representations of the former are always repeating decimals. For example,

$$1/3 = 0.333\ldots.$$
$$1/5 = 0.200\ldots.$$
$$1/11 = 0.090\,909\ldots.$$
$$19/13 = 1.461\,538\,461\,538\ldots.$$

EXERCISE 2.1

Write out the following statements in words.

(a) $n \in \mathbb{N} \Rightarrow n \in \mathbb{Z}$ and $n \in \mathbb{R}$

(b) $z \in \mathbb{Z} \Rightarrow z \in \mathbb{R}$.

EXERCISE 2.2

Give an example (a so-called counter-example) to prove that

$$r \in \mathbb{R} \Rightarrow r \in \mathbb{Z}$$

is not necessarily true.

Sometimes it is useful to be able to refer to an unbroken part, P, of the real line. Such a part is usually called an *interval*.

A *closed interval* is the set of points on the real line which represents the numbers x where $a \leqslant x \leqslant b$ and a and b are real numbers. Notice that a and b are members of the set, as well as the numbers between them. This closed interval is denoted by $[a, b]$.

An *open interval* is the set of points on the real line which represents the numbers x where $a < x < b$ and a and b are real numbers. Notice that a and b are *not* members of the set this time. This open interval is denoted by (a, b).

Open and closed intervals can be shown diagrammatically. Here, as examples, are $(-\infty, -2)$, $[-1, 1]$, $(3, 4)$.

EXERCISE 2.3

Give the relation between the points x and the intervals I given below, e.g., if $x = 3$ and $I = (2, 4)$ then $x \in I$, but if $x = 6$ and $I = (2, 4)$ then $x \notin I$ (x is not a member of I).

	x	I
(a)	4	$(6, 8)$
(b)	5	$(2, 7)$
(c)	0	$[-1, 2]$
(d)	2	$[-1, 2]$
(e)	2	$(-1, 2)$
(f)	2	$(-1, 0)$

Hint: Draw diagrams in each case.

2.3 ERRORS

The answers to the practical problems tackled by numerical mathematics are almost always in the form of numbers. To be meaningful it is essential that these answers are accompanied by a description of their accuracy. Sometimes this is very difficult to do, but it is always necessary to attempt it.

The commonest way of describing the accuracy of a number is to state its number of correct decimal places or significant figures. For example, suppose that the true answer to a problem is (unknown to us) $X = \sqrt{2}$, then we might find the answer $x = 1.414$ (correct to three decimal places).

In the following Sections (2.4 to 2.9) we shall go into some detail about errors in numbers. In Section 2.10 we shall study the errors in sums, differences, products and quotients of numbers, and in Sections 2.11 and 2.12 we shall give a brief treatment of the way numbers are stored and processed in computers.

2.4 DECIMAL PLACES

If a number X is quoted as x correct to n decimal places (written nD) the value of X lies between

$$x - 0.5 \times 10^{-n} \quad \text{and} \quad x + 0.5 \times 10^{-n}.$$

It is usual to write $X = x$ (to nD).

EXAMPLE 2.2   If $x = 1.414$ (to 3D) then the true value X lies between 1.4135 and 1.4145, i.e. $X \in [1.4135, 1.4145]$.

(X *could* be $\sqrt{2}$ for example but we cannot deduce that it is from the information given.)

EXERCISE 2.4

Give the limits between which X lies when x is given as

(a) 2.467 (to 3D)   (b) 5.4387 (to 4D)   (c) 0.002178 (to 6D)

(d) 1.5 (to 1D)   (e) 1.99 (to 2D)   (f) 13 (to 0D).

2.5 SIGNIFICANT FIGURES

If a number X is quoted as x correct to n significant figures (written nS), the first n digits of x are to be taken as correct, counting from the left and starting with the first non-zero digit. It is usual to write $X = x$ (to nS).

EXAMPLE 2.3   If $x = 1.414$ (to 4S) then X lies between 1.4135 and 1.4145 as before. (Once again X could be $\sqrt{2}$, but we cannot deduce that it is from $x = 1.414$ (to 4S) alone.)

EXERCISE 2.5

Give the limits between which X lies when x is given as

(a) 1.23 (to 3S)   (b) 4.2785 (to 5S)   (c) 0.002 76 (to 3S)

(d) 11.5 (to 3S)   (e) 1.99 (to 3S)   (f) 130 (to 3S)

(g) 130 (to 2S).

2.6 ROUNDING DECIMAL NUMBERS

In numerical processes, it is possible for errors to come into the calculations because only a finite number of decimal places or significant figures can be retained in the calculation. This is true whether pencil and paper, a calculating machine or a computer is used. It is usually assumed that if a number is rounded to nS or nD the first nS or nD are retained and the rest are discarded according to the following rule:

if the amount discarded is less than $\frac{1}{2}$ in the nth place, the nth place is unchanged; if it is greater than $\frac{1}{2}$ in the nth place, 1 is added to the nth place; if it is precisely equal to $\frac{1}{2}$ in the nth place, the number is rounded up or down to make the nth place an even digit.

EXAMPLE 2.4

(a) $1.414 = 1.41$ (to 2D)

(b) $5.318 = 5.32$ (to 3S)

(c) $6.715 = 6.72$ (to 2D)

(d) $0.020\,45 = 0.0204$ (to 3S)

(e) $0.020\,450\,1 = 0.0205$ (to 3S).

EXERCISE 2.6

Round the following numbers to (i) 4S (ii) 3S (iii) 2S (iv) 3D (v) 2D.

(a) 5.98724   (b) 11.55   (c) 0.0172   (d) 7.8854.

2.7   ERRORS AND ABSOLUTE ERRORS

In numerical mathematics there are some common terms which you should know. In Lesson 1 we said that an error is the difference between the true answer and the calculated answer. Let us now look at the precise definition. (You may find the words complicated in this section. Don't let yourself be put off by them; they are part of the 'language' of the subject that you have to learn in order to study it.)

Let X be defined or given exactly and let x be an approximation to X.

The *error* ϵ (the Greek letter 'epsilon') in this approximation can be defined by the equation

$$X = x + \epsilon.$$

(In words: the true value is the approximate value plus the error. ϵ can be positive or negative.)

EXAMPLE 2.5   Let $X = 5.342$. If X is rounded to 3S (or 2D) the approximation x for X is then given by $x = 5.34$. The error $\epsilon = X - x = 0.002$.

The *absolute error* is the absolute value of the error (which means the magnitude of the error regardless of its sign ($\pm$)) and is thus denoted by $|\epsilon|$ (read as 'the modulus of ϵ' or just 'mod ϵ').

EXAMPLE 2.6 Let $X = 4.387$. If X is rounded to 2S the approximation x for X is then given by $x = 4.4$. The error is $\epsilon = X - x = 4.387 - 4.4 = -0.013$. The absolute error is $|\epsilon| = |-0.013| = 0.013$.

EXERCISE 2.7

Calculate the error and the absolute error in the approximations

(a) 2.718 to e $= 2.718\,28 \ldots$.
(b) 2.24 to $\sqrt{5} = 2.236\,06 \ldots$.

2.8 RELATIVE ERRORS AND ABSOLUTE RELATIVE ERRORS

In some calculations we have to compare the accuracies of quantities of very different sizes. In such cases we need a measure of error that is independent of the magnitudes or dimensions of the quantities involved. In fact we use the *relative error*, which is defined to be ϵ/X using the notation of Section 2.7.

In practice we rarely know the true value X and have to use ϵ/x as an approximate relative error. This will be a good approximation provided the relative error is reasonably small. (The difference $\epsilon/X - \epsilon/x$ is $-\epsilon^2/Xx \simeq -(\epsilon/x)^2$. Prove this for yourself.)

The relative error is often multiplied by 100 and called the *percentage error*.

The rather cumbersome phrase *absolute relative error* means simply the absolute value $|\epsilon/X|$ of the relative error.

EXAMPLE 2.7 *Case 1.* Let $X = 1$ and $\epsilon = 0.01$. Then the relative error is $0.01/1 = 0.001 \equiv 1\%$.

Case 2. Let $X = 1000$ and $\epsilon = 0.01$. Then the relative error is $0.01/1000 = 0.000\,01 \equiv 0.001\%$.

The *errors* in Cases 1 and 2 are the same but the value x in Case 2 is *relatively* much more accurate than that in Case 1.

EXAMPLE 2.8 If we use 3.142 as an approximation to $\pi = 3.141\,59 \ldots$ then

$$\text{the error } \epsilon = -0.000\,41 \ldots$$

$$\text{the absolute error } |\epsilon| = 0.000\,41 \ldots$$

$$\text{the relative error} = \frac{-0.000\,41 \ldots}{3.141\,59 \ldots}$$

and   the absolute relative error $= \dfrac{0.000\,41\ldots}{3.141\,59\ldots}$.

In practice we would not bother with the dots and would quite happily use the approximations

$$\epsilon \simeq -0.0004, \qquad |\epsilon| \simeq 0.0004$$

$$\text{relative error} \simeq -\frac{0.0004}{3.1} \simeq -0.0001$$

absolute relative error $\simeq 0.0001$.

EXERCISE 2.8

Calculate the relative and absolute relative errors in the approximations
(a) 2.718 to $e = 2.718\,28\ldots$ (b) 2.24 to $\sqrt{5} = 2.236\,06\ldots$.

2.9 ERROR BOUNDS AND RELATIVE ERROR BOUNDS

Practical numerical calculations are subject to errors and, if we require a certain accuracy in answers, it is usual to use at least one more decimal place throughout the calculation than is required in the answer. Then at the end of the calculation the answer can be rounded to the required accuracy. In doing so we will be asserting that the error in the answer is less than say 0.5×10^{-n} for nD; i.e., that 0.5×10^{-n} is the maximum possible absolute error. As this happens very often let us write down a definition for it.

The *maximum absolute error* is the maximum possible value of $|\epsilon|$.

Similarly the *maximum absolute relative error* is the maximum possible value of $|\epsilon/X|$, or in practice frequently $|\epsilon/x|$.

Such maxima are called *error bounds* and *relative error bounds* respectively and are the quantities we usually work with.

EXAMPLE 2.9   (a) If $x = 5.926$ (to 3D) then

$$5.9255 \leqslant x \leqslant 5.9265$$

and clearly the error in x cannot be greater than 0.5×10^{-3} in magnitude, i.e., $|\epsilon| \leqslant 0.5 \times 10^{-3}$. Thus the error bound is $\frac{1}{2} \times 10^{-3}$. The relative error bound is $\frac{1}{2} \times 10^{-3}/5.926$ approximately, which is 8.4×10^{-5} approximately.

(b) If $x = 0.026$ (to 3D) then

$$0.0255 \leqslant x \leqslant 0.0265$$

and the error bound is $\frac{1}{2} \times 10^{-3}$ as in (a). The relative error bound however is approximately $\frac{1}{2} \times 10^{-3} / 0.026$, which is 0.02 or 2%, approximately.

Parts (a) and (b) demonstrate that the number of correct decimal places determines the error bound, whereas the relative error bound depends on the number of correct significant figures. This is proved more precisely in Section 2.11.

EXERCISE 2.9

Find the error bounds and approximate relative error bounds for the numbers stated as

(a) -21.59 (to 2D)   (b) 0.512×10^{-3} (to 3S).

2.10   ERRORS IN ARITHMETIC PROCESSES

In this section we will have a look at what happens to the error bounds in the four basic arithmetic operations of addition, subtraction, multiplication and division. We will see that in elementary cases there are simple rules to guide us. In this course we shall not take this sort of analysis much further but the ideas are important in themselves and as prerequisites for certain analyses beyond the scope of this course.

2.10.1   ERRORS IN SUMS AND DIFFERENCES

Let X_1 and X_2 be defined exactly and let x_1, x_2 be approximations to X_1 and X_2 with corresponding errors ϵ_1, ϵ_2.

Then by the definition of error

$$\epsilon_1 = X_1 - x_1 \qquad \text{and} \qquad \epsilon_2 = X_2 - x_2.$$

If $X = X_1 + X_2$ and $x = x_1 + x_2$ then the error in this sum, ϵ, is given by

$$\epsilon = X - x$$

$$\therefore \quad \epsilon = (X_1 + X_2) - (x_1 + x_2)$$

$$\therefore \quad \epsilon = (X_1 - x_1) + (X_2 - x_2) = \epsilon_1 + \epsilon_2.$$

Since we rarely know the actual errors we must convert this to an expression involving error bounds. To do this we apply the triangle inequality ($|a + b| \leqslant |a| + |b|$ — see Appendix 1) to the right-hand side of this equation

$$|\epsilon_1 + \epsilon_2| \leqslant |\epsilon_1| + |\epsilon_2|$$

so that $|\epsilon| \leqslant |\epsilon_1| + |\epsilon_2|$.

As we have said, in a practical situation we are likely only to know the error bounds e_1 and e_2 for x_1 and x_2 and this means that the only information we have concernining ϵ_1 and ϵ_2 is

$$|\epsilon_1| \leqslant e_1 \quad \text{and} \quad |\epsilon_2| \leqslant e_2$$

$$\therefore \quad |\epsilon| \leqslant e_1 + e_2 .$$

This inequality can be expressed by saying that the absolute error in a sum of two numbers is at most the sum of the error bounds for the two numbers. In other words, $e_1 + e_2$ is an error bound for $x_1 + x_2$.

EXAMPLE 2.10   Consider the sum

$$3.433 + 5.728 = 9.161.$$

Taking $x_1 = 3.433$ and $x_2 = 5.728$, assumed correct to 3D, then $e_1 = 0.5 \times 10^{-3}$ and $e_2 = 0.5 \times 10^{-3}$ so that the error in the answer is at most 1×10^{-3}. (Why? Remember — you should be checking all that we say in these examples.)

So the answer is

$$9.161 \pm 0.001.$$

How can we quote it to a nD or nS? To be strict the answer is

9.16 (to 2D)

but to give more information we could write

$$9.161 \pm 0.001 \quad \text{(to 3D)}.$$

Of course if we did 2000 additions like this the error in the answer *could* be as big as 1. (Check this statement!) However we would hope that this would be unlikely, i.e., that some of the errors would be of opposite sign and hence cancel out to some extent.

Now consider

$$3.4 + 5.728 = 9.128.$$

In this case, assuming the data are correctly rounded, $e_1 = 0.5 \times 10^{-1}$ and $e_2 = 0.5 \times 10^{-3}$ so that the error in the answer is at most 0.0505. There is no point in quoting error bounds accurately so we would usually take the error bound in this case to be 0.5×10^{-1}. Anyway, we can see that in this second example the error bound in the answer is about the same size as the error bound in the first member of the left-hand side so that we should only give an answer to 1D, i.e., 9.1 (to 1D). Generally speaking we try to keep a consistent number of decimal places in calculations involving addition since we can see that the error in the result would be dominated by the largest error in the data. For example, if we have data given to 6D and need a value of π in the addition then it should be given to 6D too.

EXERCISE 2.10

Calculate the values of the following expressions and analyse the behaviour of the errors in each case. Assume that the numbers are correctly rounded.

(a) $3.127 + 5.938 + 4.271$ (b) $0.124 + 3.5791 + 8.613\,12$.

EXERCISE 2.11

Defining $X_1, X_2, x_1, x_2, \epsilon_1, \epsilon_2, e_1$ and e_2 as above prove that, if $X = X_1 - X_2$, $x = x_1 - x_2$ and $\epsilon = X - x$ then

$$\epsilon = \epsilon_1 - \epsilon_2 \quad \text{but} \quad |\epsilon| \leqslant e_1 + e_2.$$

(*Hint*: Try the method in the text.) Notice that numerical mathematics is not just 'sums' — it is 'pure mathematics' as well!

This result means that the absolute error in the result of a subtraction is less than the sum of the error bounds of the two numbers involved (exactly as in addition), whereas the *true* error is the difference of the errors.

We can combine the two results for addition and subtraction as follows.

The absolute error in a sum or difference of two numbers is less than (or equal to) the sum of the maximum errors in the two numbers. In other words $e_1 + e_2$ is an error bound for $x_1 \pm x_2$.

EXAMPLE 2.11   Consider the difference $\pi - e$, taking $\pi = 3.14$ (to 2D) and $e = 2.72$ (to 2D).   Then the *actual* error in $3.14 - 2.72 = 0.42$ is $\epsilon_1 - \epsilon_2$, i.e.,

$$(3.141\,59\ldots - 3.14) - (2.718\,28\ldots - 2.72) \simeq 0.0016 + 0.0017$$

$$= 0.0033.$$

On the other hand, if we are given simply $3.14 - 2.72$ without knowing anything else then $e_1 = 0.005 = e_2$ and the error bound for the result (0.42) is $e_1 + e_2 = 0.01$. In other words we have

$$(3.14 \pm 0.005) - (2.72 \pm 0.005) = 0.42 \pm 0.01.$$

EXERCISE 2.12

Calculate the values of the following expressions and analyse the behaviour of the errors in each case. Assume that the numbers are correctly rounded.

(a) $3.127 - 5.938 + 4.271$ (b) $0.124 - 3.5791 - 8.613\,12$.

2.10.2   ERRORS IN PRODUCTS AND QUOTIENTS

Using the definitions of the last section if we now let

$$X = X_1 X_2, \quad x = x_1 x_2,$$

and the error in the product be ϵ, then

$$\epsilon = X - x = X_1X_2 - x_1x_2.$$

The analysis in this case turns out to depend on relative errors so we first assume that the relative error bounds for x_1 and x_2 are (say) r_1 and r_2 respectively, i.e.

$$\frac{|\epsilon_1|}{|x_1|} \leqslant r_1 \quad \text{and} \quad \frac{|\epsilon_2|}{|x_2|} \leqslant r_2.$$

(We shall use the more practical $|\epsilon/x|$ rather than $|\epsilon/X|$.)

It can be shown (using the triangle inequality again) that the relative error in x_1x_2 satisfies

$$\frac{|\epsilon|}{|x_1x_2|} \leqslant r_1 + r_2 + r_1r_2.$$

Do it for yourself! The method is the same as before. You start off like this:

$$\frac{\epsilon}{x_1x_2} = \frac{X_1X_2 - x_1x_2}{x_1x_2}$$

$$= \frac{(x_1 + \epsilon_1)(x_2 + \epsilon_2) - x_1x_2}{x_1x_2}.$$

Can you finish the proof? Just follow the lines of the two previous proofs!

Usually r_1 and r_2 are small so that r_1r_2 is very small. (If $r_1 = 0.5 \times 10^{-3}$ and $r_2 = 0.5 \times 10^{-4}$, $r_1r_2 = 0.25 \times 10^{-7}$ which is *much* smaller than r_1 or r_2). In consequence the following rule is used.

The absolute relative error in a product of two numbers is taken to be less than the sum of the maximum absolute relative errors in the two numbers. In other words $r_1 + r_2$ is a relative error bound for x_1x_2.

EXAMPLE 2.12 Consider $3.528 \times 4.734 = 16.701\,552$. Now how many decimal places are reliable?

Well $e_1 = \frac{1}{2} \times 10^{-3}$ and $e_2 = \frac{1}{2} \times 10^{-3}$.

$$\therefore \qquad r_1 \simeq \frac{\frac{1}{2} \times 10^{-3}}{3.528} \simeq 0.000\,142$$

$$\text{and} \qquad r_2 \simeq \frac{\frac{1}{2} \times 10^{-3}}{4.734} \simeq 0.000\,106$$

$$\therefore \quad r_1 + r_2 \simeq 0.000\,248 \simeq 0.000\,25.$$

We can convert relative error bounds to error bounds by multiplying by the number itself therefore the maximum error $\simeq 0.000\,25 \times 16.7 \simeq 0.0042$. So the answer may only be correct to 2D, i.e.

$$3.528 \times 4.734 = 16.70 \ (\text{to 2D}) \ (\text{or 4S}).$$

Notice that the answer has no more *significant figures* than the data. Generally, the answer will have no more correct significant figures than the least accurate data. This is because relative errors are as closely connected to significant figures as errors are to decimal places. (We shall make this statement more precise in Section 2.11.)

EXERCISE 2.13

Calculate the values of the following expressions giving the accuracy in terms of significant figures. Assume that the numbers are correctly rounded.

(a) 4.27×3.1 (b) 24.6×0.0052 (c) 0.0024×0.0006.

The error analysis for division is similar to that for multiplication but the algebra is less pleasant. This time $X = X_1/X_2$ and $x = x_1/x_2$, otherwise all the symbols are as before. Then

$$\epsilon = X - x = \frac{X_1}{X_2} - \frac{x_1}{x_2} = \frac{x_1 + \epsilon_1}{x_2 + \epsilon_2} - \frac{x_1}{x_2}.$$

So the relative error is

$$\frac{\epsilon}{x_1/x_2} = \frac{\epsilon x_2}{x_1} = \frac{x_2(x_1 + \epsilon_1)}{x_1(x_2 + \epsilon_2)} - 1$$

$$= \frac{1 + \epsilon_1/x_1}{1 + \epsilon_2/x_2} - 1.$$

As usual -- check the algebra carefully!

We have found the relative error of the quotient in terms of the relative errors of x_1 and x_2, but, as before, we would prefer a result in terms of relative error bounds. To do this we have to make some kind of approximation. One way is to rewrite the last line above by doing the division by $1 + \epsilon_2/x_2$.

$$\frac{\epsilon}{x_1/x_2} = \left\{ 1 + \left[\left(\frac{\epsilon_1}{x_1} - \frac{\epsilon_2}{x_2} \right) \Big/ \left(1 + \frac{\epsilon_2}{x_2} \right) \right] \right\} - 1$$

$$= \left(\frac{\epsilon_1}{x_1} - \frac{\epsilon_2}{x_2} \right) \Big/ \left(1 + \frac{\epsilon_2}{x_2} \right).$$

Now if we assume that ϵ_2/x_2 is small compared with one (as it would be in the majority of cases) then

$$\frac{\epsilon}{x_1/x_2} \simeq \frac{\epsilon_1}{x_1} - \frac{\epsilon_2}{x_2}.$$

Now we can get our relative error bound result, since

$$\left|\frac{\epsilon}{x_1/x_2}\right| = \left|\frac{\epsilon_1}{x_1} - \frac{\epsilon_2}{x_2}\right|$$

$$\leqslant \left|\frac{\epsilon_1}{x_1}\right| + \left|\frac{\epsilon_2}{x_2}\right| \quad \text{(why?)}$$

$$\leqslant r_1 + r_2 \quad \text{(why?)}.$$

Thus the result is the same for division as it was for multiplication.

The results can be put together as

the absolute relative error in a product or quotient of two numbers is taken to be less than the sum of the relative error bounds in the two numbers, or $r_1 + r_2$ *is a relative error bound for* $x_1 x_2$ *and for* x_1/x_2.

EXAMPLE 2.13   Consider $\dfrac{3.528}{4.734} = 0.745\,247\,148\,3\ldots$.

Now, again, how many decimal places are reliable? As before,

$$e_1 = e_2 = \tfrac{1}{2} \times 10^{-3}$$

$$r_1 \simeq \frac{\tfrac{1}{2} \times 10^{-3}}{3.528} \simeq 0.000\,142$$

$$r_2 \simeq \frac{\tfrac{1}{2} \times 10^{-3}}{4.734} \simeq 0.000\,106$$

$$\therefore \quad r_1 + r_2 = 0.000\,248 \simeq 0.000\,25.$$

This means that the absolute relative error in the answer is less than 0.000 25, which in turn means that the error bound for the answer is approximately 0.000 25 × 0.75, · i.e., 0.000 19. We can now say with certainty that

$$\frac{3.528}{4.734} = 0.745 \text{ (to 3S)}$$

or, if we wish to give a little more information, that

$$\frac{3.528}{4.734} = 0.7452 \pm 0.0002 \text{ (to 4S)}$$

meaning that the answer lies between 0.7450 and 0.7454. (The answer can also be written in terms of the relative error bound as $0.7452 \times (1 \pm 0.000\,25)$.)

EXERCISE 2.14

Calculate the values of the following expressions giving the accuracy in each case in terms of the number of significant figures or as an error bound. Assume that the numbers are correctly rounded.

(a) $\dfrac{4.27}{3.1}$ (b) $\dfrac{24.6}{4.1}$ (c) $\dfrac{0.0024}{0.0006}$ (d) $\dfrac{5.312 \times 0.027}{11.97}$.

2.10.3 LOSS OF SIGNIFICANCE

So in addition and subtraction it is errors and decimal places that are important, and in multiplication and division it is relative errors and significant figures. Well, not quite! Whether we use errors or relative errors, decimal places or significant figures is very much a matter of context, principally whether all the numbers in a calculation are of the same order of magnitude or not. We have seen that although we calculate the relative error bound in a product or quotient we may still quote the answer in terms of an error bound. Similarly there are situations where, after addition or subtraction, it is the relative error bound that we want, not the error bound that we have calculated.

EXAMPLE 2.14   Consider the addition

$$49.85 + 0.15 = 50.00.$$

If $x_1 = 49.85$, $x_2 = 0.15$, $x = 50.00$, then

$$e_1 = e_2 = 0.005$$

so $|\epsilon| \leqslant 0.01$

gives the error bound for 50.00 and

$$|\epsilon/50| \leqslant 0.0002 \equiv 0.02\%$$

gives the relative error bound for 50.00. This latter compares with relative error bounds

$$r_1 \simeq \frac{0.005}{50} = 0.0001 \equiv 0.01\%$$

$$r_2 \simeq \frac{0.005}{0.15} = 1/30 \equiv 10/3\%$$

for x_1 and x_2 separately, i.e., the answer is *relatively* more accurate than the data.

On the other hand, consider the subtraction

$$50.00 - 49.85 = 0.15.$$

If $x_1 = 50.00$, $x_2 = 49.85$, $x = 0.15$, then

$$e_1 = e_2 = 0.005$$

so $\qquad |\epsilon| < 0.01$

gives the error bound for 0.15, much as before.

But the relative error bound for 0.15 is now

$$|\epsilon/0.15| \leqslant 0.1/0.15 \equiv 20/3\%$$

compared with relative error bounds

$$r_1 \simeq r_2 \simeq \frac{0.005}{50} \equiv 0.01\%$$

for x_1 and x_2. In other words, although the absolute error/decimal place accuracy is the same in both the addition and the subtraction, the absolute relative error is much worse for the subtraction than for the addition. This is because we have subtracted nearly equal quantities and consequently the number of significant figures in the answer (2) is much less than in the data (4).

EXERCISE 2.15

Reconsider Exercise 1.2 parts (d) (ii) and (iii). Suppose that $\sqrt{40.001}$ and $\sqrt{40.000}$ can be calculated correct to 8D. Analyse the resulting errors in

$$\sqrt{40.001} - \sqrt{40.000}$$

and

$$0.001/(\sqrt{40.001} + \sqrt{40.000})$$

assuming that the data are exact.

Which is the better way to do the calculation?

2.10.4   SUMMARY

This section has probably been very strange for you. As we said earlier these are important ideas for numerical mathematics; error analysis is one of its most important activities. However, this is an introductory course so that the consequences of this section will not be pursued to any great extent.

The results we have found are:

(a) If e_1 and e_2 are error bounds for x_1 and x_2 then $e_1 + e_2$ is an error bound for $x_1 \pm x_2$.

(b) If r_1 and r_2 are relative error bounds for x_1 and x_2 then $r_1 + r_2$ is an approximate relative error bound for $x_1 x_2$ and x_1/x_2.

The practical consequences are that we should aim to work with

(a) a consistent number of decimal places when addition and subtraction are the principal operations involved;

(b) a consistent number of significant figures when multiplication and division are the principal operations involved;

(c) at least one more figure or place than given in any data and should round answers at the end of the calculation, stating the accuracy obtained.

EXERCISE 2.16

We have analysed all the basic arithmetic operations separately, but they usually appear together of course. To finish off this part of Lesson 2 calculate the value of the following expression and give the accuracy of the result in some suitable form:

$$\frac{6.10}{1.01}(3.52 + 0.116).$$

2.11 COMPUTER STORAGE OF NUMBERS

In practice much of the problem solving done by numerical mathematicians involves the use of digital computers to do the long and laborious calculations. Numbers are usually stored in computers in what is called *floating-point form*. A number is in floating-point form when it is expressed as a number between 0.1 and 1 multiplied by an appropriate power of 10.

EXAMPLE 2.15 $50 = 0.5 \times 10^2$ in floating-point form

$0.000\,031 = 0.31 \times 10^{-4}$ in floating-point form

$-9.2701 = -0.927\,01 \times 10^1$ in floating-point form.

Here we can see that in floating-point form the important element is the number of significant figures quoted and usually the number of significant figures used by a machine is fixed. (In fact computers do their arithmetic by expressing the numbers in binary form (using a base 2) instead of the decimal form we usually use (with a base 10). But we need not worry about that.) Computers usually store a number (for future use) using at least 8 significant decimal digits, i.e., $513.728\,946\,3$ could be stored as $0.513\,728\,95 \times 10^3$ if 8 digits are stored.

EXAMPLE 2.16   Consider a (hypothetical) machine which stores 4 decimal significant figures. All numbers stored will be of the form

$(\pm 0.\times \times \times \times)10^n.$

In this crude machine

$$1/3 \text{ would be stored as } (0.3333)10^0$$
$$1 \text{ would be stored as } (0.1000)10^1$$
$$-0.002 \text{ would be stored as } (-0.2000)10^{-2}$$
$$\pi \text{ would be stored as } (0.3142)10^1.$$

In fact every number in the range $3.1415 \leqslant x \leqslant 3.1425$ would also be stored as $(0.3142)10^1$. (Check this statement!) Generally a number stored in the form $(\pm 0.\text{x x x x })10^n$ will have an error and the maximum absolute error will be $(0.000\,05)10^n$. This will depend on n, but the maximum absolute relative error (or relative error bound) is

$$\frac{(0.000\,05)10^n}{(0.\text{x x x x })10^n} \leqslant \frac{0.000\,05}{0.1000} = 0.0005$$

$$= 5 \times 10^{-4}$$

(check these calculations!) which is independent of n. In an m-digit decimal machine the relative error bound would be 5×10^{-m}. *At this stage you should just realise that storing numbers in a computer introduces errors that might not otherwise occur.* (Notice that we have here proved the connection between relative error and significant figures mentioned in Sections 2.9 and 2.10.)

2.12 COMPUTER ARITHMETIC

An obvious question now is: just how do computers do arithmetic with these floating-point numbers? The answer is that, usually, each arithmetic operation $(+, -, \times, \div)$ is performed using twice as many digits for the result as for the numbers in storage. This is because if we multiply two three-digit decimal numbers together we can get a six-digit decimal number. The result then obtained is rounded before being returned to storage or used for some other calculation.

EXAMPLE 2.17   To calculate

$$6.10(3.52 + 0.116)/1.01$$

using three-digit, decimal, floating-point arithmetic the sequence of the calculations would be

$$3.52 + 0.116 = 3.636\,00 \text{ to 6S}$$
$$\rightarrow 3.64 \text{ to 3S}$$
$$6.10 \times 3.64 = 22.2040 \text{ to 6S}$$
$$\rightarrow 22.2 \text{ to 3S}$$

$$22.2 \div 1.01 = 21.9802 \text{ to 6S}$$

$$\rightarrow 22.0 \text{ to 3S.}$$

The computer answer is thus 22.0 and in fact this is the answer correct to 3S (assuming the data to be exact, i.e., not having any errors). But the rounding errors *can* accumulate quite badly, especially when thousands of operations are involved (as is frequently the case). (Have you checked the calculations above?).

EXERCISE 2.17

Perform these two calculations using 2-digit, decimal, floating-point arithmetic, and decide which is the better answer. Can you draw a general conclusion about summing a long sequence of numbers using a computer?

(a) $((1.0 + 0.50) + 0.14) + 0.042$

(b) $1.0 + (0.5 + (0.14 + 0.042))$.

EXERCISE 2.18

Recalculate your answers to Exercise 1.2 parts (d) (ii) and (iii) using 5-digit, floating-point, decimal arithmetic. Comment on the answers compared with your original answers.

2.13 POSTSCRIPT

In this Lesson we have concentrated on errors inherent in numbers and errors introduced by the use of computers. The most important thing for you to remember is that the results of all but the simplest of calculations will have some errors in them, however difficult it might be to analyse those errors.

Wherever possible in this course we shall analyse the errors implicit in a given algorithm or method, and we shall stress ways in which we can check the accuracy of the answers we obtain.

ANSWERS TO EXERCISES

2.1   (a) If n is a natural number then n is an integer and n is a real number.

(b) If z is an integer then z is a real number.

2.2 $r = \frac{1}{2}$ is such that r is real but r is not an integer.

2.3 $x \in I$ in cases (b), (c), (d)

      $x \notin I$ in cases (a), (e), (f).

2.4   (a) 2.4665 and 2.4675   (b) 5.438 65 and 5.438 75

      (c) 0.002 177 5 and 0.002 178 5   (d) 1.45 and 1.55

      (e) 1.985 and 1.995   (f) 12.5 and 13.5.

2.5   (a) 1.225 and 1.235   (b) 4.278 45 and 4.278 55

      (c) 0.002 755 and 0.002 765   (d) 11.45 and 11.55

      (e) 1.985 and 1.995   (f) 129.5 and 130.5   (g) 125 and 135.

2.6

x	4S	3S	2S	3D	2D
(a) 5.987 24	5.987	5.99	6.0	5.987	5.99
(b) 11.55	11.55	11.6	12	—	11.55
(c) 0.0172	—	0.0172	0.017	0.017	0.02
(d) 7.8854	7.885	7.89	7.9	7.885	7.89

2.7   (a) $\epsilon = X - x = 2.718\,28\ldots - 2.718 = 0.000\,28\ldots$

        $|\epsilon| = 0.000\,28\ldots$

      (b) $\epsilon = -0.003\,93\ldots, \quad |\epsilon| = 0.003\,93\ldots.$

2.8   (a) $\epsilon/X = 0.000\,28\ldots/2.718\,28\ldots \simeq 0.0001$

        $|\epsilon/X| \simeq 0.0001$

      (b) $\epsilon/X \simeq -0.0018, \quad |\epsilon/X| \simeq 0.0018.$

2.9   (a) $e = 0.005, \quad r \simeq 0.005/21.59 \simeq 0.000\,23$

      (b) $e = 0.5 \times 10^{-6}, \quad r \simeq 10^{-3}.$

2.10   (a) Formal sum = 13.336, error bound = $3 \times 0.0005 = 0.0015$.

So the answer satisfies $13.3345 \leqslant X \leqslant 13.3375$.

$\therefore$ sum = 13.33 or 13.34 (to 2D).

(b) Formal sum = 12.316 22, error bound = 0.000 555 $\simeq$ 0.0006.

$\therefore$ sum = 12.316 or 12.317 (to 3D).

(Note how the 4D and 5D accuracy of two of the numbers is wasted by associating them with a 3D number.)

2.11 $\epsilon = \epsilon_1 - \epsilon_2$ is straightforward.

     $|\epsilon| \leqslant |\epsilon_1| + |-\epsilon_2|$ by the triangle inequality

       $= |\epsilon_1| + |\epsilon_2|$

       $\leqslant e_1 + e_2.$

2.12   (a) Formal answer = 1.46, error bound = $3 \times 0.0005 = 0.0015$.

$\therefore$ Answer $= 1.460 \pm 0.002$ (to 3D).

(b) Formal answer $= -12.068\,22$, error bound $= 0.000\,555 \simeq 0.0006$.

$\therefore$ Answer $= -12.068$ or -12.069 (to 3D).

2.13 (a) Formal answer $= 13.237$, rel. error bound $\simeq \dfrac{0.005}{4.27} + \dfrac{0.05}{3.1} \simeq 0.017$.

$\therefore$ Answer $= 13.237(1 \pm 0.017)$

$= 13.237 \pm 0.225$

$= 13$ (to 2S) or 0.13×10^2 (to 2S) preferably.

(There is a small possibility that the answer is larger than 13.5.)

(b) Formal answer $= 0.127\,92$.

Rel. error bound $\simeq \dfrac{0.05}{24.6} + \dfrac{0.000\,05}{0.0052} \simeq 0.012$.

$\therefore$ Answer $= 0.127\,92(1 \pm 0.012)$

$= 0.127\,92 \pm 0.0015$

$= 0.128 \pm 0.002$ (to 3S).

(c) Formal answer $= 0.144 \times 10^{-5}$.

Rel. error bound $\simeq \dfrac{0.5 \times 10^{-4}}{0.24 \times 10^{-2}} + \dfrac{0.5 \times 10^{-4}}{0.6 \times 10^{-3}} \simeq 0.1$.

$\therefore$ Answer $= 0.144 \times 10^{-5} (1 \pm 0.1)$

$= 0.144 \times 10^{-5} \pm 0.144 \times 10^{-6}$

$= 0.1 \times 10^{-5}$ or 0.2×10^{-5} (to 1S).

(It is bound to be difficult to quote an answer when the data have only 1 or 2 significant figures in them.)

2.14 (a) $1.3774\ldots \times (1 \pm 0.017) = 1.4$ (to 2S).

(b) $6.0 \times (1 \pm 0.014) = 6.0 \pm 0.1$ (to 1D).

(c) $4.0 \times (1 \pm 0.1) = 4$ (to 1S) (or 0.4×10^1 (to 1S) preferably).

(d) Formal answer $= 0.011\,981 \ldots$.

Rel. error bound $\simeq \dfrac{0.5 \times 10^{-3}}{5.312} + \dfrac{0.5 \times 10^{-3}}{0.027} + \dfrac{0.5 \times 10^{-2}}{11.97}$

$\simeq 0.019$.

$\therefore$ Answer $= 0.011\,981 \ldots \times (1 \pm 0.019)$.

$= 0.012$ (to 2S).

Notice again that (in a product and/or quotient) the answer has no more correct significant figures than the least accurate data.

2.15 Let $x_1 = \sqrt{40.001}$, $x_2 = \sqrt{40.000}$. Then $e_1 = e_2 = \frac{1}{2} \times 10^{-8}$.

The error bound for $x_1 \pm x_2$ is 1×10^{-8}.

Hence the relative error bound for $x_1 + x_2$ is $10^{-8}/(x_1 + x_2) \simeq 0.8 \times 10^{-9}$.

This is also the relative error bound for $0.001/(x_1 + x_2)$ because the constant 0.001 is to be treated as exact. Hence the answer for $x_1 - x_2$ is

$$0.7905\,642\,5 \times 10^{-4} \pm 10^{-8} = (0.7906 \pm 0.0001) \times 10^{-4} \text{ (to 4S)}$$

whereas the answer for $0.001/(x_1 + x_2)$ is

$$0.790\,564\,474 \times 10^{-4} (1 \pm 0.8 \times 10^{-9})$$

$$= 0.790\,564\,474 \times 10^{-4} \pm 0.64 \times 10^{-13}$$

$$= (0.790\,564\,474 \pm 10^{-9}) \times 10^{-4} \text{ (to 9S)}.$$

The second approach (avoiding subtraction of nearly equal quantities) is clearly more accurate.

2.16 Error bound for $3.52 + 0.116$ is 0.0055. Hence its rel. error bound of $0.0055/3.636$. Hence total rel. error bound of

$$\frac{0.005}{6.10} + \frac{0.005}{1.01} + \frac{0.0055}{3.636} \simeq 0.007.$$

Hence

$$\text{answer} = 21.96(1 \pm 0.007)$$

$$= 21.96 \pm 0.15$$

$$= 22 \text{ or } 0.22 \times 10^2 \text{ (to 2S) or } 22.0 \pm 0.2 \text{ (to 3S)}.$$

2.17 (a) $1.5 + 0.14 = 1.64 \rightarrow 1.6$

   $1.6 + 0.042 = 1.642 \rightarrow \underline{1.6}$

(b) $0.14 + 0.042 = 0.182 \rightarrow 0.18$

   $0.18 + 0.5 = 0.68$

   $0.68 + 1.0 = 1.68 \rightarrow \underline{1.7}.$

The second answer is better. (Exact answer $= 1.682$.)

In general sum numbers in increasing order of magnitude. (Consider what happens if you sum

$$1.0 + 0.5 + 0.14 + 0.042 + 0.042 + 0.042 + \ldots$$

starting at the left and using two-digit arithmetic!)

FURTHER READING

(a) Ribbans, Book 1, Chapter 1, gives a very simple introduction to the ideas we have discussed in this lesson. One word of caution — what we simply call the error Ribbans calls the absolute error.

(b) Pennington, Chapter 3, gives a much fuller account, similar in detail to our own. The notation is different, but most mathematics books differ from each other in this respect: it is one of the little annoyances of life worth overcoming as soon as possible.

(c) Noble, Vol. 1, Chapter 1, Sections 1.1—1.4, contains the same basic ideas though stated quite briefly, and also describes the statistical approach to error propagation in long calculations.

(d) Hosking, Joyce, and Turner, 'Steps 1—4' is a fairly full treatment of errors.

(e) Dixon, Chapter 2, gives an even fuller account. Section 2.6.4 uses a Taylor series, which you may not have met.

LESSON 2 — COMPREHENSION TEST

1. Arrange the following into equivalent pairs, e.g., $1 < x < 2$ is equivalent to $x \in (1, 2)$:

$$1 < x < 2 \qquad x \in [-1, 1]$$
$$x > 0 \qquad x \in (0, \infty)$$
$$x \leqslant 1 \qquad x \in [0, \infty)$$
$$-1 \leqslant x \leqslant 1 \qquad x \in (-\infty, 1]$$
$$x \geqslant 0 \qquad x \in (-\infty, 1)$$
$$x < 1 \qquad x \in (1, 2).$$

2. Which of the following numbers becomes 3.6 when rounded correctly to 2S:

$$3.45 \quad 3.55 \quad 3.65 \quad 3.549 \quad 3.649 \quad 3.551 \quad 3.651?$$

3. 1.1 is obtained by some process as an approximation to 1. Give its error and relative error.

4. 2.214 is obtained as correct to 3D. Give its error bound.

5. 0.015 is obtained as correct to 2S. Give its relative error bound.

6. Give the error bound for the sum $2.214 + 0.015$.

7. Give an approximate relative error bound for the product 2.2×0.015.

8. Compute $1.1 \times 2.4 + 2.1 \times 1.4$ using two-digit, decimal, floating-point arithmetic.

ANSWERS

1 $1 < x < 2, x \in (1, 2)$; $x > 0, x \in (0, \infty)$; $x \leqslant 1, x \in (-\infty, 1]$;
   $-1 \leqslant x \leqslant 1, x \in [-1, 1]$; $x \geqslant 0, x \in [0, \infty)$; $x < 1, x \in (-\infty, 1)$.

2 3.55, 3.65, 3.649, 3.551.

3 Error $= -0.1$, Relative error $= -0.1$.

4 0.5×10^{-3}.

5 1/30.

6 0.001.

7 ~ 0.06.

8 5.5.

PART B NUMERICAL SOLUTION OF NON-LINEAR ALGEBRAIC EQUATIONS

Lesson 3 Finding First Approximations to Solutions

OBJECTIVES

At the end of this lesson you should be able to

(a) use sketch and tabulation methods to find first approximations to the roots of equations;

(b) use standard tables to evaluate functions;

(c) use Synthetic Division to evaluate polynomials and their derivatives;

(d) draw a flow chart describing the Synthetic Division algorithm.

CONTENTS

3.1 INTRODUCTION

In Lessons 3 through 6 we shall be concerned with the problem of finding a solution, or all the solutions, of a single, non-linear algebraic equation of the form

$$f(x) = 0. \qquad\qquad\qquad\qquad\qquad\qquad\qquad\qquad\text{[3.1]}$$

We exclude linear equations in one unknown

$$ax + b = 0$$

because their solution is trivial. There are very few other kinds of equation that we can solve as easily. Among these are quadratic equations, e.g.

$$x^2 + 0.01x + 1 = 0$$

and certain trigonometric equations, e.g.

$$3\cos\theta + 4\sin\theta - 5 = 0.$$

But it takes very little alteration to these simple equations to make them virtually unsolvable by 'pure-mathematical methods'. For example, neither of

$$x^3 + 0.01x + 1 = 0$$
$$3\sqrt{\cos\theta} + 4\sqrt{\sin\theta} - 5 = 0$$

yield their roots without recourse to considerable manipulation, if then. (In fact both cubic and quartic polynomial equations have formula solutions, but these are very complicated. A famous result of pure mathematics is that this is not true for quintic and higher degree polynomial equations.)

There is a lot of good to be said for manipulative skill and finding roots exactly if possible; it may give considerable insight into the underlying problem just to rearrange an equation and to find information about roots. However, when it comes to actually finding the numerical values of roots, a well-chosen numerical method will usually do this quickly and as accurately as can be justified.

All the numerical methods for finding roots of non-linear algebraic equations are iterative. That is, given initial information about a solution, they refine the information, producing a sequence of successively more accurate values for the solution. As far as we are concerned these methods are of two types.

The first of these are called *bracketing methods*, and we shall consider them in Lesson 4. They require as initial information an interval containing a solution. They then produce a smaller interval still containing the solution. This is repeated until either the interval is sufficiently small or the value of $f(x)$ (see Equation [3-1]) at an end-point of the interval is sufficiently small.

The second class of methods we shall call *one-point methods*. In this case the method starts with a point near the solution and produces a sequence of points, each nearer to the solution than its predecessor. Each point is calculated from the position of its predecessor. We shall consider one-point methods in Lessons 5 and 6.

Before considering these however we shall spend some time discussing functions and derivatives, since we often find that students at this level have difficulty with them.

3.2 GRAPHS AND FUNCTIONS

We started this lesson with the general equation [3-1] involving the symbol $f(x)$. What does it mean? In this section we shall attempt to give you some confidence in reading and manipulating function notation.

3.2.1 GRAPHS

Before defining functions themselves we consider the more familiar idea of graphs.

EXAMPLE 3.1 The graphs of $y = x$ and $y = x^2$ are sketched in Fig. 3.1.

A graph consists of a set of points. Every point on a graph can be described by a pair of coordinates (x, y). x and y are connected by the equation of the graph so that each point on the graph has coordinates which satisfy *exactly* the equation of the graph. Given a particular value for x and the equation of the graph we can, in theory, find the value (or values) of y corresponding to this value of x.

It is useful to have an idea of what the shape of a graph is when we know its equation. Some common ones (which you should know) are shown in Fig. 3.2.

The examples shown in Fig. 3.2 have a very important property in common. For each *suitable* point on the x-axis ($x = -1$ will not do in graph (d) for example) there is a unique point on the graph and therefore a unique point on the y-axis. (Note that the reverse is not necessarily true. Some points on the y-axis can correspond to many points on the x-axis.)

If we take $y = x^2$ as a specific example, then, given any point $x = a$ on the x-axis (point A in Fig. 3.3(a)) there corresponds the single point $y = a^2$ on the y-axis (point C). These two coordinates define the unique point (a, a^2) on the graph (point B).

On the other hand a graph like that shown in Fig. 3.3(b) has an equation such that, for some x-values at least, there correspond more than one y-value and hence more than one point on the graph. For the x-value represented by point D there are three points E, F and G on the graph corresponding to three y-values. The three y-values are represented by points P, Q and R.

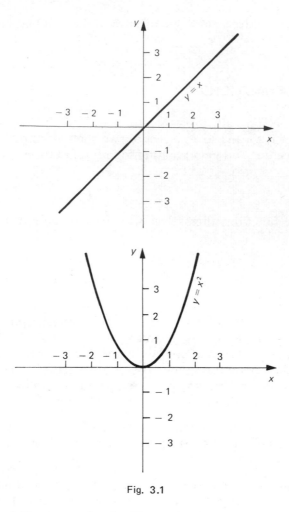

Fig. 3.1

Thus, in each of the examples in Fig. 3.2 there is a very special relation between (as it happens) the set of points forming the x-axis and the set of points forming the y-axis. This relation we call a *function*.

3.2.2   FUNCTIONS

We have seen that a function is defined when, given a suitable number x, we can calculate by some rule a unique corresponding number y:

The rule may be expressed by an equation, of which these are examples from Fig. 3.2

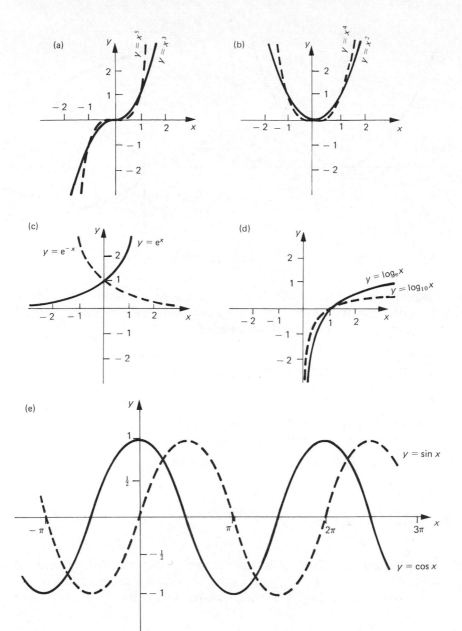

Fig. 3.2

$$y = x^3, \qquad y = \sin x.$$

When we wish to talk of a general function we give it a symbolic name, such as f, and the rule is written

$$y = f(x)$$

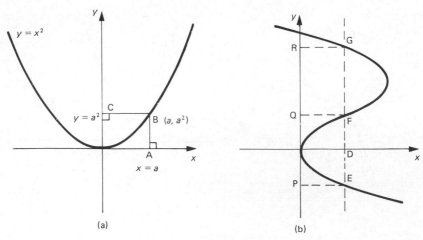

Fig. 3.3

meaning 'apply f to x to obtain y'. The brackets are important; fx looks like a multiplication, *which $f(x)$ is certainly not.* Here are some more functions; note how we sometimes have to restrict x.

$$f_1(x) = \log_e x, \quad x > 0$$
$$f_2(x) = \sqrt{x(2-x)}, \quad 0 \leqslant x \leqslant 2$$
$$f_3(x) = e^x/x, \quad x \neq 0.$$

The root in f_2 is the positive square root only; if we wanted both square roots we would place $\pm$ before the square root sign.

You should note that there is an essential difference between equations like

$$y = x^3$$

which define the number y, and equations like

$$f_1(x) = \log_e x, \quad x > 0$$

which define the function f_1. In the former y depends for its existence on the value of x. In the latter f_1 does not depend for its existence on x — it has an existence of its own. We could equally well write

$$f_1(\theta) = \log_e \theta, \quad \theta > 0$$

or use any other symbol instead of x.

One of the problems of understanding functions is that we confuse a function f with its value $f(x)$ for some given x. We talk for example of the function x^2 because we are too lazy to say 'the function that turns x into x^2' or 'the function f such that $f(x) = x^2$.' Sometimes we *can* make the distinction easily. For example, $\sin x$ is a number (e.g., $\sin \pi/2 = 1$), whereas sine is a function — given a number (an angle if you like) sine can operate on it to produce another number.

3.2.3 NOTATION

Frequently some difficulty is found with function notation. Here is an example (and some exercises) to help you to understand it.

EXAMPLE 3.2   If the value of a function f is given by

$$f(x) = x^3 + 4x^2 - 3x + 5, \quad x \in \mathbb{R},$$

then

$$f(2) = 2^3 + 4(2^2) - 3(2) + 5 = 23$$
$$f(-1) = (-1)^3 + 4(-1)^2 - 3(-1) + 5 = 11$$

and

$$f(b) = b^3 + 4b^2 - 3b + 5.$$

Notice that x is replaced by 2, − 1, b (or whatever) everywhere it appears in the definition of f(x).

EXERCISE 3.1

If $f(x) = x^4 + 3x^2 - 5x + 11$, $x \in \mathbb{R}$, write down (and where possible evaluate)

(a) $f(-1)$ (b) $f(2)$ (c) $f(-a)$ (d) $f(2y)$ (e) $f(x-1)$
(f) $f(2z+1)$ (g) $f(x^2)$ (h) $f(\sin x)$.

Apart from functions we shall also be involved with derivatives. The derivative of a function f is usually denoted by f'. If f is defined in terms of x (by defining $f(x)$) then $f'(x)$ is commonly written as $\dfrac{df}{dx}$. Similarly higher derivatives are indicated by $f'', f''', \ldots$ and their values by $f''(x)$, $f'''(x), \ldots$ or $\dfrac{d^2f}{dx^2}, \dfrac{d^3f}{dx^3}, \ldots$. Both of these notations have disadvantages.

In the first case we run out of dashes and it is common to indicate an nth derivative by $f^{(n)}(x)$. To avoid the problem we shall use this notation for *all* derivatives, so that f' will be written as $f^{(1)}$, f'' as $f^{(2)}$, and so on. *The superscript in brackets will indicate the number of times f has been differentiated.*

The reason for not using $\dfrac{df}{dx}, \dfrac{d^2f}{dx^2}, \ldots$ is that the notation is rather clumsy.

For example, to specify the value of the first derivative at $x = 2$ we would have to write

$$\left.\frac{df}{dx}\right|_{x=2} \quad \text{or possibly} \quad \frac{df}{dx}(2)$$

whereas in our notation this would simply be $f^{(1)}(2)$.

EXAMPLE 3.3   Let $f(x) = x^3 - 3x^2 + 7x - 11$, $x \in \mathbb{R}$. Then

$$\left.\begin{aligned}
f^{(1)}(x) &= 3x^2 - 6x + 7 \\
f^{(2)}(x) &= 6x - 6 \\
f^{(3)}(x) &= 6 \\
f^{(4)}(x) &= f^{(5)}(x) = \ldots = 0.
\end{aligned}\right\} \quad x \in \mathbb{R}$$

When you are writing your solutions we ask you to be careful with the notation $f^{(1)}, f^{(2)}, \ldots$. Please use the brackets.

Lastly we have to admit that we may also use $\dfrac{dy}{dx}, \dfrac{d^2y}{dx^2}, \ldots$ where these and y are to be read as both numbers and functions. This is an almost universal and traditional laziness in mathematics.

EXERCISE 3.2

(a) $f(x) = x^4 - 3x^2 + 5x + 11$, $x \in \mathbb{R}$. Find the functions $f^{(1)}, f^{(2)}, f^{(3)}, f^{(4)}, f^{(5)}, f^{(99)}$ and evaluate $f^{(1)}(0), f^{(2)}(-1)$ and $f^{(5)}(10^6)$.

(b) $g(x) = \sin x \cos x$, $x \in \mathbb{R}$. Find the functions $g^{(1)}, g^{(2)}, g^{(3)}, g^{(4)}, g^{(8)}$ and evaluate $g^{(i)}(0)$ for $i = 1, 2, 3$ and 4.

3.2.4   POLYNOMIALS

One of the most useful types of function in numerical mathematics is the *polynomial*. It is useful because it can be evaluated very readily by hand or on a computer and can be differentiated and integrated easily. So to start with here is a definition of the term polynomial.

A polynomial function, p_n, of degree n in the variable x is a function defined by an equation of the form

$$p_n(x) = a_0 + a_1 x + a_2 x^2 + \ldots + a_n x^n, \quad x \in \mathbb{R}.$$

[*Remember*: $x \in \mathbb{R}$ means that x may be any real number.]

Note that a polynomial has a *finite* number of terms. The *degree* of the polynomial is the highest *power* of x, i.e., n. This is why we have the subscript n in $p_n(x)$. It indicates the degree of the polynomial.

$a_0, a_1, a_2, \ldots, a_n$ are called the coefficients of the powers of x, i.e., of $x^0, x^1, x^2, \ldots, x^n$.

The values of x which make the value of $p_n(x)$ equal to zero are called the zeros of $p_n(x)$ or are called the roots of the equation

$$p_n(x) = 0.$$

EXAMPLE 3.4 $p_2(x) = x^2 - 3x + 2$ is a polynomial of degree 2 in x and its zeros are $\{1, 2\}$.

EXERCISE 3.3

What are the degrees and the zeros of the following polynomials? Sketch their graphs.

(a) $x^2 - 6x - 7$ (b) $x^2 + 4x + 4$

(c) $(x + 1)(x + 2)(x - 1)$ (d) $x^3 - 6x^2 + 11x - 6$

*(e) $x^3 + x^2 - 2$ *(f) $x^4 - 1$.

Because all the terms in a polynomial are of the same kind, $a_i x^i$, we can use the *summation notation* to abbreviate the general form

$$p_n(x) = a_0 + a_1 x + a_2 x^2 + \ldots + a_n x^n$$

$$= \sum_{i=0}^{n} a_i x^i.$$

Σ is the Greek capital letter *sigma*; it is understood that i takes all integer values between 0 and n inclusive. When you see the summation sign you should mentally picture the complete sum. Here are some more examples.

$$\sum_{r=1}^{10} r = 1 + 2 + 3 + \ldots + 10$$

$$\sum_{i=0}^{m} \frac{x^i}{i!} = 1 + x + \frac{x^2}{2!} + \frac{x^3}{3!} + \ldots + \frac{x^m}{m!}$$

$$\sum_{k=1}^{N} (\mu - x_k)^2 = (\mu - x_1)^2 + (\mu - x_2)^2 + \ldots + (\mu - x_N)^2.$$

We shall often need to refer to a set of integers in *arithmetic progression*, i.e., in the general form

$$j, j + k, j + 2k, j + 3k, \ldots, j + nk = m.$$

We shall use the notation $i = j(k)m$ to mean that i takes the values j to m in steps of k. For instance, $i = 1(1)10$ means $i = 1, 2, 3, \ldots, 10$ in turn; $i = n(-1)0$ means $i = n, n - 1, n - 2, \ldots, 2, 1, 0$ in turn. In summation notation $\sum\limits_{k=1}^{N}$, for example, implies $k = 1(1)N$.

We are now ready to return to the main business of this lesson.

Before we can start on the iterative methods themselves we need two skills. Firstly we need to be able to find initial approximations to roots, either as a bracketing interval or as a nearby point. Secondly, we need to be able to evaluate $f(x)$ in the equation $f(x) = 0$ as efficiently as possible. This lesson deals with these two skills.

3.3 LOCATING ROOTS OF EQUATIONS

In some cases it is important for the initial approximation to a solution of an equation to be close to the true value; in others it is not. It is obvious that in any case a good starting value for an iterative method will reduce the number of iterations required to find an accurate solution. On the other hand, it is pointless to do more work finding a good first approximation than would be needed by an iterative method to improve a poor first approximation.

A compromise is called for. We find the best initial approximation we can while using only the simplest of techniques. These include sketching graphs, tabulating the function values $f(x)$ for various x, and using linear interpolation.

Consider first the use of graph sketching. To solve $f(x) = 0$ graphically we can sketch the graph of $y = f(x)$ and note where it crosses the x-axis ($y = 0$). If f is complicated this will not be very convenient. A simple alternative exists if f can be split into two parts f_1 and f_2 so that

$$f(x) = f_1(x) - f_2(x).$$

We can then sketch the graphs of f_1 and f_2 and note their points of intersection (where $f_1(x) = f_2(x)$ and hence $f(x) = 0$).

EXAMPLE 3.5   To find first approximations to the roots of

$$e^x - 3 = 0.$$

Fig. 3.4 shows a sketch of $y = e^x - 3$.

We observe that $x = 1.1$ is a rough value for the only root. (The true solution is $x = 1.0986$ to 4D.) Note that this sketch is made without evaluating $e^x - 3$ at all accurately. If you evaluate $e^x - 3$ at $x = 0, 1$ and 2 then there is no need to sketch the graph — as we shall see shortly.

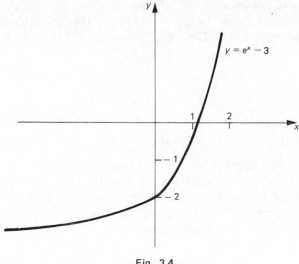

Fig. 3.4

EXAMPLE 3.6 To find first approximations to the roots of

$$0.5 - E + 0.2 \sin E = 0.$$

We cannot easily sketch $y = 0.5 - E + 0.2 \sin E$ without evaluating the right-hand side. But we can split the equation thus

$$\sin E = 5E - 2.5$$

and sketch $y = \sin E$ and $y = 5E - 2.5$ (see Fig. 3.5).

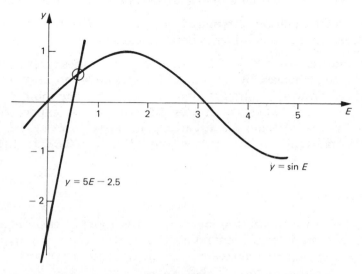

Fig. 3.5

The only root is near $x = 0.65$. If we require an initial interval (for a bracketing method) then $[\frac{1}{2}, 1]$ would do nicely. (It must be certain to contain the solution.) The true solution in this case is $x = 0.6155$ to 4D.

EXERCISE 3.4

Use sketches to estimate the locations of the roots of these equations.

(a) $x - e^{-x} = 0$ (b) $1 - x - \sin x = 0$ (c) $x^4 - x - 1 = 0$.

It is perhaps tempting to think that, in order to produce the sketches, one should first tabulate the functions. If you do this then there is no need to draw the graph! You simply observe where the function $f(x)$ changes sign — if you can find it! In reality it may be a matter of hard work and good luck in finding an approximate root. In a text book things are easier!

EXAMPLE 3.7   To locate the roots of $8x^4 - 8x^2 + 1 = 0$. Let $f(x) = 8x^4 - 8x^2 + 1$. Here are some evaluations, in the order that seems sensible. (Note that f is an 'even' function. That is, $f(-x) = f(x)$.)

x	0	± 1	± 2	$\pm \frac{1}{2}$
$f(x)$	1	1	97	$-\frac{1}{2}$

So f changes sign in each of the intervals $-1 < x < -\frac{1}{2}$, $-\frac{1}{2} < x < 0$, $0 < x < \frac{1}{2}$ and $\frac{1}{2} < x < 1$. Since a quartic polynomial equation can have at most four real roots we need look no further. (Try sketching the graph of f.)

If a single approximation is required we can use a technique called *linear interpolation*, which we consider in detail in Lesson 14.

For the interval $0 < x < \frac{1}{2}$, for example, $f(x)$ varies from $f(0) = 1$ to $f(\frac{1}{2}) = -\frac{1}{2}$, i.e., changes by $1\frac{1}{2}$ units. Using simple proportion (which is equivalent to drawing a straight line between the end points $(0, 1)$ and $(\frac{1}{2}, -\frac{1}{2})$ and seeing where it crosses the x-axis) we expect $f(x)$ to be zero when x is $1/1\frac{1}{2}$ times the distance from $x = 0$ to $x = \frac{1}{2}$, i.e., at $x = \frac{1}{3}$. Fig. 3.6 illustrates this. In this case the true solution is 0.38 to 2D.

EXERCISE 3.5

Use tabulation to estimate the location of the roots of these equations. In each case give an interval containing the root and an approximate root within that interval. (To evaluate e^x see the next section.)

(a) $x^5 - x - 1 = 0$ (b) $e^x - 3x = 0$.

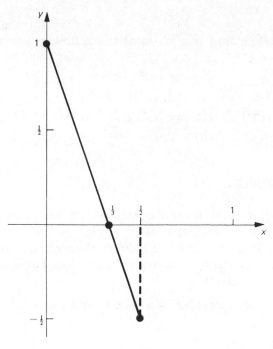

Fig. 3.6

3.4 USING STANDARD TABLES

Solving $f(x) = 0$ obviously involves evaluating $f(x)$ for various x. If f involves transcendental functions — trigonometric, exponential, logarithmic, etc. — then this is a non-trivial operation. If the equation is solved using a computer program then values of the standard functions are provided through mathematical software. For example, to evaluate $y = e^x - 3x$ in FORTRAN or BASIC we simply write

$$Y \ = \ EXP(X) - 3*X$$

If the working is done by hand then there are two possibilities. You can either use a modern, scientific, electronic calculator with the standard functions 'hard-wired' in, or a set of mathematical tables. Simple versions of these tables are known by all schoolchildren. You are recommended to look at and to learn to use a more advanced set of tables, such as *Chambers' Shorter 6-figure Mathematical Tables.* They are at the very least of historical interest, though their practical value is admittedly waning. It is also instructive to consider the problem of building such a set of tables.

We do not propose to describe here the technique of using tables, but here are some look-up problems for you to practise on.

*EXERCISE 3.6

Use standard tables to find the values of the following, correct to the given precision.

(a) $\sin 1$ (6S)

(b) $\sin 1.051$ (6S)

(c) $\sin 1.0513$ (6S)

(d) $\log_e 1.2345$ (6D)

(e) $\log_{10} 1.2345$ (6D)

(f) $\cos 57°22'10''$ (6D)

(g) $e^{-3.219}$ (6D)

(h) $\sqrt{0.474}$ (8S).

3.5 SYNTHETIC DIVISION

In the preceding section we concentrated on transcendental functions because they cannot be evaluated easily using just the algebraic operations $+$, $-$, $\times$, $\div$, $\uparrow$ (raising to a power). Despite their relative simplicity (which accounts for their popularity) polynomials are also worth considering from the point of view of evaluation.

To see why this is so, consider how many multiplications are required to evaluate

$$1 + 2x + 3x^2 + 4x^3.$$

Answer: 5. And how many to evaluate

$$1 + x(2 + x(3 + 4x)) ?$$

Answer: 3. The two polynomials have identical values of course. The second (nested) version is clearly more efficient for evaluation. Let's check the general case. How many multiplications in each of these:

$$a_0 + a_1 x + a_2 x^2 + \ldots + a_n x^n \qquad [3\text{-}2]$$

$$a_0 + x(a_1 + x(a_2 + \ldots + x(a_{n-1} + a_n x) \ldots)) ? \qquad [3\text{-}3]$$

Answer: $2n - 1$ and n respectively. There is a clear advantage in using [3-3] rather than [3-2] in, say, a computer program requiring many hundreds of evaluations of polynomials, particularly if the degree is at all high.

EXERCISE 3.7

Arrange these polynomials into their nested forms.

(a) $3 + x + 2x^2 + 2x^3$

(b) $2 - x + 4x^2 - 3x^3 + x^5$.

There is an obvious disadvantage to the form [3-3]. For a high-degree polynomial it will be extremely tedious to write it out. Moreover as the coefficients a_i may be given as data, i.e., a string of numbers, it seems pointless to have to translate them into the nested form. We avoid this by evaluating the polynomial by a sequence of simple, identical operations, collectively known as the method of Synthetic Division. There are many ways of deriving this method. Here is one based on the Remainder Theorem.

THEOREM

Let p_n be a polynomial of degree n and suppose $p_n(x)$ is divided formally by $x - \alpha$, then

$$p_n(\alpha) = R$$

where R is the remainder of the division.

PROOF

Let q_{n-1} be the quotient polynomial, of degree at most $n-1$, i.e.

$$p_n(x) = (x - \alpha)q_{n-1}(x) + R.$$ [3-4]

Then clearly $p_n(\alpha) = R$.

So, to evaluate $p_n(\alpha)$, divide $p_n(x)$ by $(x - \alpha)$ and take the remainder. To work out the details let the given polynomial p_n be

$$p_n(x) = \sum_{i=0}^{n} a_i x^i$$

and let the quotient polynomial q_{n-1} be

$$q_{n-1}(x) = \sum_{i=0}^{n-1} b_i x^i.$$

Substituting into [3-4] we obtain

$$\sum_{i=0}^{n} a_i x^i = (x - \alpha) \sum_{i=0}^{n-1} b_i x^i + R$$

$$= \sum_{i=0}^{n-1} b_i x^{i+1} - \sum_{i=0}^{n-1} \alpha b_i x^i + R.$$

(If you are not sure of this write it out with $n = 4$, say.)

If the left- and right-hand sides are to be the same for all x we can equate coefficients of $x^n, x^{n-1}, \ldots, x^2, x, 1$, giving

$$x^n: \quad a_n = b_{n-1}$$

$$x^{n-1}: \quad a_{n-1} = b_{n-2} - \alpha b_{n-1}$$

$$x^{n-2}: \quad a_{n-2} = b_{n-3} - \alpha b_{n-2}$$

. . . . . .

$$x: \qquad a_1 = b_0 - \alpha b_1$$
$$1: \qquad a_0 = R - \alpha b_0.$$

All of these except the first and last have exactly the same form

$$a_i = b_{i-1} - \alpha b_i \quad i = \overline{n-1}(-1)1. \qquad\qquad [3\text{-}5]$$

(Recall that $i = \overline{n-1}(-1)1$ means $i = n-1, n-2, n-3, \ldots, 3, 2, 1$.)

If we define b_n and b_{-1} by

$$b_n = 0 \qquad b_{-1} = R$$

then *all* the relations above are of the form [3-5] and we can alter the range of the index i to $i = n(-1)0$ (i.e., $i = n, n-1, n-2, \ldots, 1, 0$). Finally, it is the b_i values we want, so we turn [3-5] round to give

$$\boxed{\; b_{i-1} = a_i + \alpha b_i \quad i = n(-1)0. \;} \qquad\qquad [3\text{-}6]$$

This is the Synthetic Division algorithm. Given $a_0, a_1, \ldots, a_n$ and α we calculate $b_{n-1}, b_{n-2}, \ldots, b_0, b_{-1}$ *in that order*. Then b_{-1} is $p_n(\alpha)$, the value we want.

Note that this *is* the same as nesting the polynomial because

$$
\begin{aligned}
p_n(\alpha) &= b_{-1} \\
&= a_0 + \alpha b_0 \\
&= a_0 + \alpha(a_1 + \alpha b_1) \\
&= a_0 + \alpha(a_1 + \alpha(a_2 + \alpha b_2)) \\
&= \cdots \\
&= a_0 + \alpha(a_1 + \alpha(a_2 + \alpha(\ldots + \alpha(a_{n-1} + \alpha a_n)\ldots))).
\end{aligned}
$$

If executed by hand the Synthetic Division algorithm is tabulated as follows, the calculation working from left to right.

You will observe that the division by $x - \alpha$ is not done explicitly in the Synthetic Division algorithm.

EXAMPLE 3.8   Given $p(x) = 4x^3 + 3x^2 + 2x + 1$, evaluate

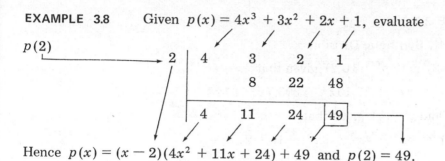

Hence $p(x) = (x - 2)(4x^2 + 11x + 24) + 49$ and $p(2) = 49$.

EXERCISE 3.8

Evaluate by Synthetic Division

(a) $p(-2)$ given $p(x) = 2x^3 + 2x^2 + x + 3$

(b) $p(3)$ given $p(x) = x^5 - 3x^3 + 4x^2 - x + 2$.

(Don't forget the zero coefficient of x^4!)

We can extend the Synthetic Division algorithm to the evaluation of the derivatives of polynomials. Since

$$p_n(x) = (x - \alpha)q_{n-1}(x) + R$$

$$\frac{d}{dx}p_n(x) = q_{n-1}(x) + (x - \alpha)\frac{d}{dx}q_{n-1}(x)$$

$$\therefore \quad p_n^{(1)}(\alpha) = q_{n-1}(\alpha).$$

Hence to evaluate $p_n^{(1)}(\alpha)$, we evaluate $q_{n-1}(\alpha)$, by a second stage of Synthetic Division.

EXAMPLE 3.9   To evaluate $p(2)$ *and* $p^{(1)}(2)$ given that $p(x) = 4x^3 + 3x^2 + 2x + 1$.

$$
\begin{array}{r|rrrr}
2 & 4 & 3 & 2 & 1 \\
  &   & 8 & 22 & 48 \\
\hline
  & 4 & 11 & 24 & \boxed{49} = p(2) \\
  &   & 8 & 38 & \\
\hline
  & 4 & 19 & \boxed{62} & = p^{(1)}(2)
\end{array}
$$

Note that the remainder at the first stage is *not* a part of the second stage. (The coefficients of q_{n-1} are $b_{n-1}, b_{n-2}, \ldots, b_0$ and do not include b_{-1}.)

EXERCISE 3.9

Evaluate by Synthetic Division

(a) $p(10.2)$ and $p^{(1)}(10.2)$ given that

$$p(x) = 3x^3 + 30.14x^2 - 296.7x - 3126$$

(b) $p(\frac{1}{2})$ and $p^{(1)}(\frac{1}{2})$ given that

$$p(x) = 4x^4 - 4x^3 + 9x^2 - 8x + 2.$$

EXERCISE 3.10

Draw a flow chart to describe the Synthetic Division algorithm [3-6], as though for a computer subroutine, i.e.,

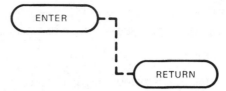

At the entry point assume that $a_n, \ldots, a_0$ and α are available. At the exit (return) point $b_{n-1}, \ldots, b_{-1}$ should have been calculated.

*EXERCISE 3.11

Extend the algorithm [3-6] to include the evaluation of $p_n^{(1)}(\alpha)$, and adapt your flow chart from Exercise 3.10 accordingly. (*Hint*: start by letting

$$q_{n-1}(x) = (x - \alpha)r_{n-2}(x) + S$$

where $r_{n-2}(x) = \sum\limits_{i=0}^{n-2} c_i x^i$; then follow similar lines to the development of the basic algorithm.)

*EXERCISE 3.12

What do you get if you continue the Synthetic Division process to a third stage? A fourth stage? A kth stage?

ANSWERS TO EXERCISES

3.1 (a) 20   (b) 29

(c) $a^4 + 3a^2 + 5a + 11$ (d) $16y^4 + 12y^2 - 10y + 11$

(e) $x^4 - 4x^3 + 9x^2 - 15x + 20$ (f) $16z^4 + 32z^3 + 36z^2 + 10z + 10$

(g) $x^8 + 3x^4 - 5x^2 + 11$ (h) $\sin^4 x + 3 \sin^2 x - 5 \sin x + 11.$

3.2 (a) $f^{(1)}(x) = 4x^3 - 6x + 5$

$f^{(2)}(x) = 12x^2 - 6$

$f^{(3)}(x) = 24x$

$f^{(4)}(x) = 24$

$f^{(5)}(x) = 0$

$f^{(99)}(x) = 0$

$f^{(1)}(0) = 5, \quad f^{(2)}(-1) = 6, \quad f^{(5)}(10^6) = 0$

(b) $g(x) = \sin x \cos x = \frac{1}{2} \sin 2x$

$g^{(1)}(x) = \cos 2x$

$g^{(2)}(x) = -2 \sin 2x$

$g^{(3)}(x) = -4 \cos 2x$

$g^{(4)}(x) = 8 \sin 2x$

$g^{(8)}(x) = 128 \sin 2x$

$g^{(1)}(0) = 1, \quad g^{(2)}(0) = 0, \quad g^{(3)}(0) = -4, \quad g^{(4)}(0) = 0.$

3.3

	Degree	Zeros
(a)	2	$7, -1$
(b)	2	$-2, -2$
(c)	3	$-1, -2, 1$
(d)	3	$1, 2, 3$
(e)	3	$1, -1 \pm i$
(f)	4	$\pm 1, \pm i.$

3.4 Here are the solutions correct to 3S. You should be reasonably near them.

(a) 0.567   (b) 0.511   (c) $-0.724, 1.22.$

3.5 As above, and your intervals should contain the roots.

(a) 1.17   (b) 0.619, 1.51

3.6 (a) 0.841 471   (b) 0.867 920
(c) 0.868 069   (d) 0.210 666
(e) 0.091 491   (f) 0.539 220
(g) 0.039 995   (h) 0.688 476 58.

3.7 (a) $3 + x(1 + x(2 + 2x))$

(b) $2 + x(-1 + x(4 + x(-3 + x(0 + x))))$.

Note how we have made this example fit the standard format. A golden rule in computation is to eliminate as much variation as possible from a method and to make it as formal as possible.

3.8 (a)

-2	2	2	1	3
		-4	4	-10
	2	-2	5	-7

Hence $2x^3 + 2x^2 + x + 3 = (x + 2)(2x^2 - 2x + 5) - 7$ and $p(-2) = -7$.

(b)

3	1	0	-3	4	-1	2
		3	9	18	66	195
	1	3	6	22	65	197	$= p(3)$

3.9 (a) $p(10.2) = 167.0$, $p^{(1)}(10.2) = 1255$, each to 4S.

(b) $p(\frac{1}{2}) = p^{(1)}(\frac{1}{2}) = 0$. (What does this tell you about the graph of $p(x)$ at $x = \frac{1}{2}$?)

3.10

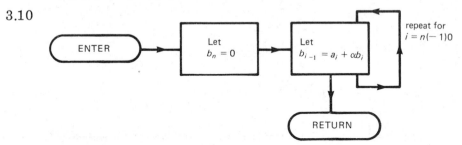

3.11 $c_{i-1} = b_i + \alpha c_i$, $i = \overline{n-1}(-1)0$ [3-7]

where $c_{n-1} = 0$ and $c_{-1} = S = p_n^{(1)}(\alpha)$.

Putting [3-6] and [3-7] together presents a slight problem. Should we find all the b_i values first and then all the c_i values (the obvious approach)? Or can we make the whole thing more efficient? If you choose the former then the extension of the flow chart above is simply a matter of letting $c_{n-1} = 0$ at the start and adding an extra loop below the one for calculating the b_i values. For extra efficiency the calculation could proceed as follows.

$$b_{n-1} = a_n + \alpha b_n$$

$$b_{n-2} = a_{n-1} + \alpha b_{n-1}$$

$$c_{n-2} = b_{n-1} + \alpha c_{n-1}$$

$$b_{n-3} = a_{n-2} + \alpha b_{n-2}$$

$$c_{n-3} = b_{n-2} + \alpha c_{n-2}$$

etc. until

$$b_0 = a_1 + \alpha b_1$$

$$c_0 = b_1 + \alpha c_1$$

$$b_{-1} = a_0 + \alpha b_0$$

$$c_{-1} = b_0 + \alpha c_0.$$

Since $b_n = 0$ we can separate off the first equation and take the rest in pairs.

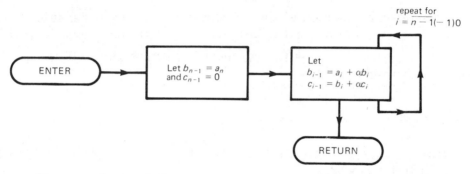

repeat for
$i = n-1(-1)0$

ENTER

Let $b_{n-1} = a_n$
and $c_{n-1} = 0$

Let
$b_{i-1} = a_i + \alpha b_i$
$c_{i-1} = b_i + \alpha c_i$

RETURN

If you understand about subroutines (procedures) in computing, you could now write a program which evaluates a polynomial and its derivative by inputting its coefficients and a value for α, calling the above subroutine, and outputting the results (b_{-1} and c_{-1}).

3.12   The third stage gives $2p_n^{(2)}(\alpha)$, the fourth $6p_n^{(3)}(\alpha)$, and the kth gives $(k-1)! \, p_n^{(k-1)}(\alpha)$.

FURTHER READING

For a very full treatment of Synthetic Division see Stark, Section 3.14. This treatment *follows* his description of methods of solution. Hence in one of his examples he assumes you have met the Newton–Raphson method. Also he chooses to label his coefficients so that b_0 is the remainder rather than b_{-1}.

Stark also discusses finding initial approximations in Sections 3.12 and 3.13, but is more realistic than us — he assumes that sketching is out of the question (equations are too complicated usually) and that tabulation will be done by computer (it takes too long by hand in general).

For a very brief and simple version of the finding of initial approximations see Ribbans, Book 1, Sections 6.0 and 6.1.

For slightly more detail see Hosking, Joyce and Turner, 'Step 6'.

For a brief treatment of Synthetic Division see Dixon, Chapter 3.

LESSON 3 – COMPREHENSION TEST

1. Let $f(x) = \sqrt{x}$, $x \geqslant 0$. Write down $f(2)$, $f(x+1)$, $f(c^2)$, $f^{(1)}(2)$, $f^{(2)}(2)$.

2. Write down the general form of a polynomial of degree n.

3. What values does j take if $j = 0(2)8$?

4. Write down a condition which will ensure that $f(x) = 0$ has at least one solution in the interval $[a, b]$, assuming that f is continuous in $[a, b]$.

5. Let $p_3(x) = x^3 - 2x + 10$. Evaluate $p_3(-3)$ and $p_3^{(1)}(-3)$ by Synthetic Division.

ANSWERS

1   $f(2) = \sqrt{2}$; $f(x+1) = \sqrt{x+1}$, $x \geqslant -1$; $f(c^2) = c$; $f^{(1)}(2) = 1/2\sqrt{2}$; $f^{(2)}(2) = -1/8\sqrt{2}$.

2   $p_n(x) = a_0 + a_1 x + a_2 x^2 + \ldots + a_n x^n \equiv \sum\limits_{i=0}^{n} a_i x^i$.

3   $j = 0, 2, 4, 6, 8$.

4   $f(a)$ and $f(b)$ have opposite signs, i.e., $f(a)f(b) < 0$.

5

$$
\begin{array}{r|rrrr}
-3 & 1 & 0 & -2 & 10 \\
   &   & -3 & 9 & -21 \\
\hline
   & 1 & -3 & 7 & \boxed{-11} \;= p_3(-3) \\
   &   & -3 & 18 & \\
\hline
   & 1 & -6 & \boxed{25} \;= p_3^{(1)}(-3). &
\end{array}
$$

Lesson 4 **Bracketing Methods**

OBJECTIVES

At the end of this lesson you should

(a) understand the interval-reduction strategy of bracketing methods;

(b) be able to apply the bisection method;

(c) be able to apply the false-position method;

(d) appreciate the various convergence criteria used;

(e) be able to draw flow charts describing the bisection and false-position methods.

CONTENTS

4.1   INTRODUCTION

In Lesson 3 we described methods for locating an interval (a, b) containing roots of an algebraic equation $f(x) = 0$. In essence these methods reduce to finding a and b such that $f(a)$ and $f(b)$ have opposite signs. Then, if f is continuous in (a, b) (if its graph has no breaks therein), there must be at least one point x in (a, b) such that $f(x) = 0$.

In Lesson 3 we only considered examples where the interval contained just one root — see Fig. 4.1 for the two possibilities.

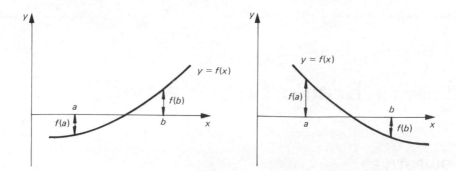

Fig. 4.1

It should be clear however that $f(a)$ and $f(b)$ having opposite signs is not a sufficient condition for a *single* root in (a, b). Fig. 4.2 shows two examples where $f(a)f(b) < 0$ but where there is more than one root in (a, b).

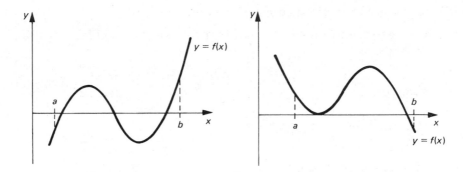

Fig. 4.2

The methods described in this lesson assume that the initial interval contains just one root. By their nature they will always find a root, but you will never know from the results whether it is the only root in the initial interval, or just one of many. Sometimes this is important, sometimes not.

In fact we describe two methods in this lesson. The *bisection* method and the *false position* method. Both are iterative processes that progressively shrink the initial interval in such a way that a root is always contained in (bracketed by) the latest interval.

4.2   REDUCING THE INTERVAL

If we can be certain that we have a single root of $f(x) = 0$ in (a, b), what can we do to reduce the length of the interval containing the root? First of all we can choose any point c *inside* (a, b) and evaluate $f(c)$. To make things more precise let us suppose that $f(a) > 0$ and $f(b) < 0$. Then there are just two possibilities (as shown in Fig. 4.3).

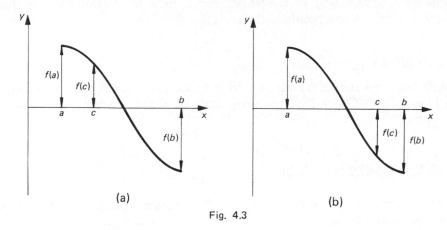

(a)   (b)

Fig. 4.3

These possibilities are:

(a) $f(c) > 0$, when the root will be in (c, b);

(b) $f(c) < 0$, when the root will be in (a, c).

In either case we have found a smaller interval containing a root.

Anticipating the general case (when the signs of $f(a)$ and $f(b)$ are only known to be opposite), it would be better to describe the two possibilities as

(a) $f(c)f(b) < 0$, when the root will be in (c, b);

(b) $f(c)f(a) < 0$, when the root will be in (a, c);

so that we clearly retain the opposite-sign property. In fact we need not test both $f(c)f(b)$ and $f(c)f(a)$. If $f(c)f(a) > 0$ then, unless $f(c) = 0$ (when c is exactly the solution required), $f(c)f(b) < 0$ is automatically true. That is, if (b) is false then (a) must be true. (We shall ignore the possibility that $f(c)$ may be zero. This fluke is unlikely to occur in practice, but can easily be allowed for if you wish.)

To construct an iterative method we simply allow (c, b) (in case (a)) or (a, c) (in case (b)) to be the new bracketing interval (a, b), and then repeat the whole process as often as required. For example, to obtain the root with an error of at most e, the length $b - a$ is successively reduced until it is less than e when any point in the final interval (a, b) will be within e of the true solution.

Summing up:

(a) if $f(a)f(c) < 0$ the root is in (a, c), so let $b = c$;

(b) otherwise the root is in (c, b), so let $a = c$;

(a) and (b) to be repeated until $b - a < e$ is true.

Fig. 4.4 shows a flow chart for the process we have devised. To be suitable for automatic computation it requires two improvements. Firstly we need to specify precisely how c is to be chosen. Secondly, it is possible that, for some ways of choosing c, the length $b - a$ may never reduce to less than e. We shall clarify these points in the sections that follow.

EXERCISE 4.1

Draw diagrams similar to Fig. 4.3 to indicate the possibilities when $f(a) < 0$ and $f(b) > 0$. Check that the flow chart of Fig. 4.4 works just as well for this case.

4.3 THE BISECTION METHOD

A fairly obvious way of choosing c in the bracketing approach is to make it the midpoint of (a, b), i.e., $c = \frac{1}{2}(a + b)$. As a result of this choice the interval (a, b) is bisected at each stage of the iterative process, which is then called the bisection method.

It is clear that, in this case, the interval length $b - a$ *can* be made as small as we like (and the solution found as accurately as we like) simply by doing sufficient bisections. Hence the test 'Is $b - a < e$?' is suitable for stopping the bisection process. One of the pleasant properties of this method is that the number of iterations required to obtain a specified accuracy in the solution is predetermined. Suppose that we wish to find the root correct to kD, i.e., with an error of at most $\frac{1}{2} \times 10^{-k}$. To achieve this all we need do is to stop the bisecting when $b - a < 10^{-k}$ and give the answer as $\frac{1}{2}(a + b)$. If $l = b - a$ is the length of the *initial* interval, the lengths of the succeeding intervals are $\dfrac{l}{2}, \dfrac{l}{2^2}, \dfrac{l}{2^3}, \dots, \dfrac{l}{2^p}$ after $1, 2, 3, \dots, p$ bisections, respectively.

Thus kD accuracy will be attained after p bisections when $\dfrac{l}{2^p} < 10^{-k}$ for the smallest value of p — or, if you like, as soon as p attains a value such that $2^p > l \times 10^k$. The solution given will be the $(p + 1)$th midpoint.

EXAMPLE 4.1   The roots of $f(x) = x^2 - x - 2 = 0$ are $x = -1, 2$ in fact.

Suppose that we simply observe that

$$f(1) = 1 - 1 - 2 = -2, \quad f(4) = 16 - 4 - 2 = 10$$

so that there is at least one root of $f(x) = 0$ in the interval $(1, 4)$. Note that $f^{(1)}(x) = 2x - 1$, which is positive throughout $(1, 4)$. Therefore there is only one root in $(1, 4)$.

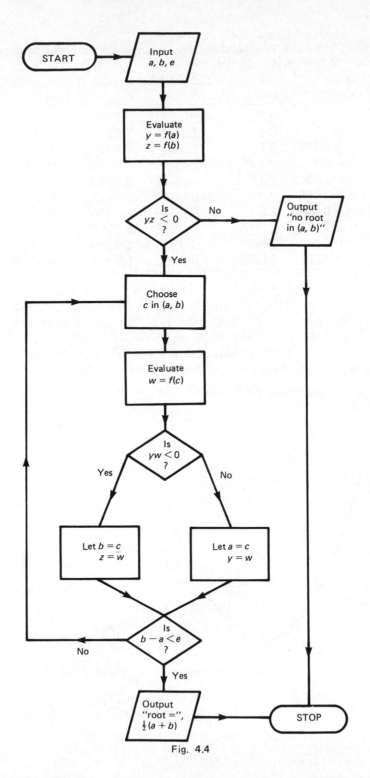

Fig. 4.4

Let us take $a = 1$, $b = 4$, and apply the bisection method to find the root of $x^2 - x - 2 = 0$ in $(1, 4)$.

a	$f(a)$	b	$f(b)$	c	$f(c)$	$b - a$
1	-2	4	10	2.5	1.75	3
1	-2	2.5	1.75	1.75	-0.6875	1.5
1.75	-0.6875	2.5	1.75	2.125	0.3906	0.75
1.75	-0.6875	2.125	0.3906	1.9375	-0.1836	0.375
1.9375	-0.1836	2.125	0.3906	2.0313	0.0949	0.1875
1.9375	-0.1836	2.0313	0.0949	1.9844	-0.0466	0.0938
1.9844	-0.0466	2.0313	0.0949	2.0079	0.0238	0.0469
1.9844	-0.0466	2.0079	0.0238	1.9962	-0.0114	0.0234
1.9962	-0.0144	2.0079	0.0238	2.0020	0.0060	0.0117
1.9962	-0.0114	2.0020	0.0060	1.9991	-0.0027	0.0058

If we require the root correct to 2D the last entry in the $b - a$ column must be less than 0.01. So the root is $1.9991 = 2.00$ to 2D. We also notice that the value of $f(1.9991)$ is -0.0027 which is suitably small. We will return to this point later. Note also that we would expect to make p bisections, where p is the smallest integer satisfying $2^p > l \times 10^k = 3 \times 10^2$. This requires $p = 9$, which is confirmed by the calculation.

Fig. 4.5 gives a diagrammatic interpretation of the bisection method for this example.

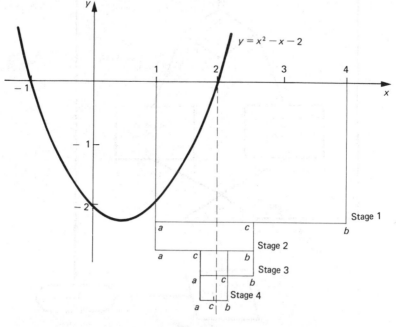

Fig. 4.5

EXERCISE 4.2

Draw a flow diagram for the bisection method (adapt Fig. 4.4 and include in it a facility for printing $a, f(a), b, f(b), c$ and $f(c)$ after each bisection and, at the end of the calculation, the number of bisections performed to attain the required accuracy).

EXERCISE 4.3

Use the bisection method to find the root of $\sqrt{x} = 4x - 2$ that lies between $x = \frac{1}{2}$ and $x = 1$. Work with 4D and give your answer correct to 2D.

4.4 THE FALSE-POSITION METHOD

How else might we choose c in a scientific way? The simplicity of the choice of c as the midpoint of (a, b) is a very attractive property of the bisection method, as is the guaranteed convergence arising from the fact that the bracketing interval can be made as small as we like. The disadvantage of the bisection method is that it is, in general, extremely slow.

Notice that, although we calculate $f(a)$, $f(b)$ and then a succession of $f(c)$ values, we only ever use their signs. Intuitively, we could increase the convergence rate by using the values as well. This leads to the idea of choosing c as the point where the straight line joining the points $(a, f(a))$, $(b, f(b))$ cuts the x-axis, as in Fig. 4.6.

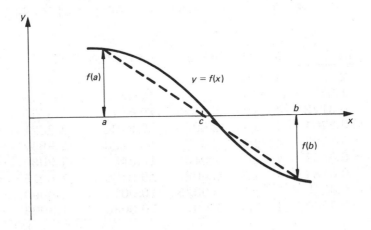

Fig. 4.6

The equation of the straight line through $(a, f(a))$, $(b, f(b))$ is

$$y = p_1(x) = f(a) + \frac{x - a}{b - a}(f(b) - f(a)).$$

(Check that $p_1(a) = f(a)$ and $p_1(b) = f(b)$.) If $p_1(x) = 0$ at $x = c$ (i.e., $p_1(c) = 0$) then

$$0 = f(a) + \frac{c-a}{b-a}(f(b) - f(a))$$

so that

$$c = \frac{af(b) - bf(a)}{f(b) - f(a)}.$$

[4-1]

(Don't forget to work this out for yourself!)

This result can also be written as

$$c = a - \frac{(b-a)f(a)}{f(b) - f(a)}$$

[4-2]

or

$$c = b - \frac{(b-a)f(b)}{f(b) - f(a)}.$$

[4-3]

EXERCISE 4.4

Do we have to consider the two cases (a) $f(a) > 0$, $f(b) < 0$ and (b) $f(a) < 0$, $f(b) > 0$ in applying the false-position rule? Why?

EXAMPLE 4.2

Consider the problem of Example 4.1 again, i.e., to find the root of $x^2 - x - 2 = 0$ in $(1, 4)$, this time using the false-position method.

a	$f(a)$	b	$f(b)$	$b-a$	$f(b) - f(a)$	c	$f(c)$
1	-2	4	10	3	12	1.5	-1.25
1.5	-1.25	4	10	2.5	11.25	1.7778	-0.6172
1.7778	-0.6172	4	10	2.2222	10.6172	1.9070	-0.2704
1.9070	-0.2704	4	10	2.0930	10.2704	1.9621	-0.1123
1.9621	-0.1123	4	10	2.0379	10.1123	1.9847	-0.0457
1.9847	-0.0457	4	10	2.0153	10.0457	1.9939	-0.0183
1.9939	-0.0183	4	10	2.0061	10.0183	1.9975	-0.0075
1.9975	-0.0075	4	10	2.0025	10.0075	1.9990	-0.0030
1.9990	-0.0030	4	10	2.0010	10.0030	1.9996	-0.0012

We observe that the successive values of c have converged to the solution $x = 2.00$ to 2D. Note however that the interval lengths $b - a$ have *not* reduced to 10^{-3} as in the bisection method. This means that we must reconsider our convergence criterion — 'Is $b - a < e$?' will not do for the false-position method.

Fig. 4.7 gives a diagrammatic interpretation of the first few iterations of the false-position method for Example 4.2.

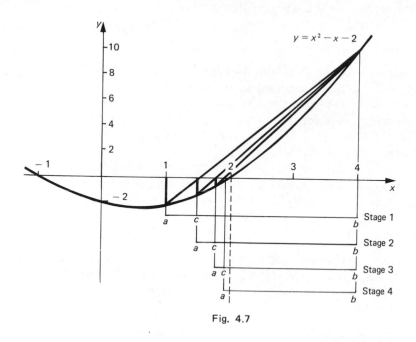

Fig. 4.7

4.5 CONVERGENCE CRITERIA

The obvious alternative criterion for the false-position method is to stop the process when the difference between successive values of c is 'sufficiently small'. In the flow chart of Fig. 4.4 we would introduce a new variable (c^*, say) which would always be the new value of c, i.e.,

$$c^* = \frac{af(b) - bf(a)}{f(b) - f(a)}$$

(see Equation [4-1]). Then we would test $|c^* - c|$ for its smallness, where c is the 'old' value of c. If $|c^* - c|$ is too large we would set $c = c^*$ and continue with the interval reduction stage. To complete this strategy we would require an initial value of c; $c = a$ or $c = b$ would be suitable.

EXERCISE 4.5

Draw a flow chart for the false-position method, using the strategy outlined above. Include output of $a, f(a), b, f(b), c$ and $f(c)$, as indicated for the bisection method in Exercise 4.2.

There remains the question of how small to make $|c^* - c|$. It is not sufficient in general to make it less than $\frac{1}{2} \times 10^{-k}$ to obtain kD accuracy. Consider for example the successive values of c obtained in Example 4.2. The last five, with their successive differences are

1.9847		1.9939		1.9975		1.9990		1.9996.
	0.0092		0.0036		0.0015		0.0006	

Stopping when successive values are less than 0.005 apart would give 1.9975 as the last value, which does happen to be correct to 2D. But if the convergence rate were slower this might not be the case. When we compute by hand we naturally allow for this. (In Example 4.2 we did two more iterations 'to make sure'.) Although it is often possible to be more scientific the simplest way to automate this convergence criterion is to ask for $|c^* - c|$ to be, say, ten times smaller than the acceptable error in the solution, and to inspect the complete sequence of c values produced by the computer program to ensure that convergence really has occurred to the required precision.

EXERCISE 4.6

Apply the false-position method to the problem of Exercise 4.3, viz. to solve $\sqrt{x} = 4x - 2$ for the root in $(\frac{1}{2}, 1)$, working with 4D and obtaining the root correct to 2D. Compare the number of iterations required with that for the bisection method.

Note that our convergence criteria are designed to ensure that the root obtained is near the true solution. If the equation is $f(x) = 0$, and if c is the approximate root obtained, then they do not guarantee that $f(c)$ is 'small'. As an extreme example consider the equation

$$x^{99} - 1 = 0$$

and suppose that 1.1 is found as an approximation to the solution $x = 1$. Then

$$f(1.1) = 1.1^{99} - 1 \simeq 1.25 \times 10^4.$$

If it is important for $f(x)$ to be 'small' at the approximate solution then a double criterion must be applied. For the false-position method we might have

(a) if $|c^* - c| < e_1$ and $|f(c^*)| < e_2$ then stop;

(b) otherwise continue with interval reduction.

In particular it is often sensible for e_1 to equal e_2.

*EXERCISE 4.7

Adapt your false-position flow chart to add the requirement that $|f(c^*)| < e_2$ before the process is stopped.

Note that the criterion $|f(c^*)| < e_2$ by itself is equally dangerous. It is possible for $|f(c^*)|$ to be small while $|c^* - c|$ remains unacceptably large. You might like to construct an example for which this is true.

One last point should be made with respect to automatic use of these methods on a digital computer. If extreme accuracy is required so that $f(a), f(b)$ and $f(c)$ are very small, it is possible, due to rounding errors, for some of these function values to be small negative when they should be small positive (and vice versa). This would clearly upset the whole interval-reduction process.

ANSWERS TO EXERCISES

4.1

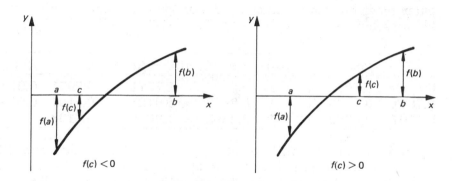

4.2 In Fig. 4.4 replace $\boxed{\text{Choose } c}$ by $\boxed{\text{Set } c = \dfrac{a+b}{2}}$.

Add $\diagup\text{Output } a, f(a), b, f(b), c, f(c)\diagup$ after the evaluation of $f(c)$. To obtain the number of bisections

add $\boxed{\text{Let } p = 0}$ after the evaluation of $f(a)$ and $f(b)$,

add $\boxed{\text{Add 1 to } p}$ after the evaluation of $f(c)$, and

add $\diagup\text{Output } p\diagup$ after the output of $\frac{1}{2}(a + b)$.

4.3 $f(x) = \sqrt{x} - 4x + 2$, $f(\frac{1}{2}) \simeq 0.7$, $f(1) = -1$; therefore there is a root in $(\frac{1}{2}, 1)$.

a	$f(a)$	b	$f(b)$	c	$f(c)$	$b-a$
0.5	0.7071	1	-1	0.75	-0.1340	0.5
0.5	0.7071	0.75	-0.1340	0.625	0.2906	0.25
0.625	0.2906	0.75	-0.1340	0.6875	0.0792	0.125
0.6875	0.0792	0.75	-0.1340	0.7188	-0.0274	0.0625
0.6875	0.0792	0.7188	-0.0274	0.7032	0.0256	0.0312
0.7032	0.0256	0.7188	-0.0274	0.7110	-0.0008	0.0156
0.7032	0.0256	0.7110	-0.0008	0.7071	0.0125	0.0078

$b-a$ is now less than 0.01. Therefore the required root is $0.7071 = 0.71$ to 2D. $f(0.7071)$ is also small.

4.4 No. Linear interpolation does not depend on the signs of the data values.

4.5 If you want to test your flow chart to see if it is correct, write a computer program from it and apply the program to some equations with known solutions.

4.6

a	$f(a)$	b	$f(b)$	$b-a$	$f(b)-f(a)$	c	$f(c)$
0.5	0.7071	1	-1	0.5	-1.7071	0.7071	0.0125
0.7071	0.0125	1	-1	0.2929	-1.0125	0.7107	0.0002
0.7107	0.0002	1	-1	0.2893	-1.0002	0.7108	

By good fortune the first c value gives the solution correct to 2D. Note, however, that, as described in the lesson, we continue until two successive c values agree to 3D in order to be confident of the 2D value, viz. 0.71.

FURTHER READING

For a short, clear account see Stark, Sections 3.8—3.11 which includes some computer programs. For a more complete account and an extension of the interpolation approach see Phillips and Taylor, Sections 7.1—7.3.

Another brief account is given by Hosking, Joyce and Turner, 'Steps 7 and 8', and a slightly fuller description of the false-position method by Dixon, Section 7.5.

LESSON 4 — COMPREHENSION TEST

1. True or False?

(a) The bisection method is safe but slow.

(b) The length of the bracketing interval always tends to zero with the false-position method.

(c) The false-position method is generally faster than the bisection method.

2. Let $x^2 - 6x + 5 = 0$ and let $[0, 4]$ be the initial interval for a bracketing method of solution. What is the next interval using:

(a) the bisection method

(b) the false-position method?

ANSWERS

1 (a) True;   (b) False;   (c) True.

2 (a) $[0, 2]$;   (b) $[0, 2\frac{1}{2}]$.

LESSON 4 – SUPPLEMENTARY EXERCISES

4.8. Find bracketing intervals for all the roots of the following equations and use both bisection and false-position methods to find all the roots correct to 3D:

(a) $4x - 4 \sin x - 1 = 0$

(b) $x^2 - 20 \sin x = 0$

(c) $2^x - 2x^2 - 1 = 0$

(d) $xe^x - 2 = 0$.

* 4.9. When the false-position method is applied to a convex or concave function (as in Fig. 4.7) one end point remains fixed throughout the process. In such cases convergence can be improved by the following rule: if an end-point, say $(a, f(a))$ remains unchanged in two consecutive iterations, replace it by $(a, \frac{1}{2}f(a))$.

(a) Repeat Example 4.2 using this modified false-position method.

(b) Amend your flow charts of Exercises 4.5 and 4.7 to include the modification.

Lesson 5 One-Point Iterative Methods

OBJECTIVES

At the end of this lesson you should

(a) be able to rearrange equations $F(x) = 0$ into the form $x = f(x)$;

(b) be able to analyse a given one-point iterative method $x_{n+1} = f(x_n)$ for its convergence properties;

(c) be able to draw a flow chart for a general one-point iterative method;

(d) be able to derive and apply Newton's method.

CONTENTS

5.1 ONE-POINT AND MULTI-POINT ITERATIVE METHODS

Before we go any further in the discussion of methods for solving non-linear algebraic equations we should make the meaning of some of the terms used quite clear.

Generally by an *iterative method* we mean a process in which we find a first approximation to the solution of a problem and then, using the properties of the problem (usually in the form of equations), calculate a sequence of items (usually numbers) which we hope will prove to be better and better estimates of the solution required. Of course we should show, before we start, that the method to be employed will produce a sequence of numbers that *converges* to the required solution. Otherwise we may waste a great deal of time. Showing that the numerical method is going to work may require some pure mathematics — which is why this subject is called numerical mathematics.

We have seen in the bracketing methods an approach which starts with two points and from them derives two other points which bracket the solution more closely. This process forms one iteration and as two points are used to start the iteration and two points are the result of the iteration we call this process a *two-point iterative method*. If n points are involved at both beginning and end of the iteration we call such a process an *n-point iterative method*. One-point iterative methods are common and easy to follow and are the subject of this lesson.

5.2 THE GENERAL ONE-POINT ITERATIVE METHOD

In a one-point iterative method each iteration starts from a single number and produces another single number. If we let the initial approximation be x_0 and succeeding approximations be $x_1, x_2, \ldots$, then the general step uses x_n to obtain x_{n+1}.

Suppose that the formula for doing this is

$$x_{n+1} = f(x_n) \quad n = 0, 1, 2, \ldots \tag{5-1}$$

where f is some well-behaved function.

Let x_0 be given, then

$$x_1 = f(x_0) \qquad x_2 = f(x_1) \qquad x_3 = f(x_2)$$

and so on. The numbers $x_0, x_1, x_2, x_3, \ldots$, form a *sequence*. In Exercise 1.1 you constructed the sequence

$$1, \tfrac{1}{2}, \tfrac{1}{4}, \tfrac{1}{8}, \ldots$$

from the formula $x_{n+1} = \tfrac{1}{2}x_n$ with $x_0 = 1$. It is pretty clear that the elements of this sequence become infinitesimally close to zero as we move along the sequence. We say that the sequence *converges* to the *limit* zero and we write $x_n \to 0$ as $n \to \infty$. (Read $\to$ as 'tends to'.)

It is impossible for us to treat convergence properly; it is a considerable part of the study of pure mathematics. We shall rely on your having an intuitive notion of the convergence of sequences.

Suppose now that the sequence generated by Equation [5-1] converges to the number X, i.e., $x_n \to X$ as $n \to \infty$. This means that, again intuitively, if n is large enough the numbers x_{n+1} and x_n in [5-1] are virtually indistinguishable from X so that $X = f(X)$. In other words X is a solution of the equation $x = f(x)$. (Strictly speaking we are assuming that $f(x_n) \to f(X)$ as $x_n \to X$. Hence the need for f to be a 'well-behaved' function — in fact f needs to be continuous in some interval containing X.)

As an example consider the equation $e^x = x^2$. It has a root near $x = -0.7$. Rearrange the equation into $x = -\sqrt{e^x} = -e^{x/2}$, implying the iterative process $x_{n+1} = -e^{x_n/2}$. With $x_0 = -0.7$ we obtain the sequence

$$x_1 = -e^{-0.7/2} = -0.705$$

$$x_2 = -e^{-0.705/2} = -0.703$$

$$x_3 = -e^{-0.703/2} = -0.7036$$

$$x_4 = -e^{-0.7036/2} = -0.7034$$

$$x_5 = -0.703\,49$$

$$x_6 = -0.703\,46.$$

It seems clear that this sequence is converging and that the root is -0.7035 correct to 4D.

This suggests an approach to solving a general equation $F(x) = 0$. If we can rearrange it into the form $x = f(x)$ in such a way that the roots of $F(x) = 0$ (or at least those we want) are also roots of $x = f(x)$, then it is possible that the iterative process $x_{n+1} = f(x_n)$ will find a root or roots of $F(x) = 0$.

Notice that there will be many rearrangements of $F(x) = 0$ into $x = f(x)$. As a trivial example of this, $x = x + cF(x)$ is such a rearrangement for any non-zero constant c. More interestingly, reconsider the equation $e^x = x^2$, then we have these obvious rearrangements

$$x = \pm e^{1/2x}$$

$$x = 2\ln x$$

$$x = e^x/x$$

but there are many, many more that are less obvious. For example,

$$x = x - \frac{e^x - x^2}{e^x - 2x}$$

$$x = \frac{\lambda x}{1 + \lambda} + \frac{e^x}{x(1 + \lambda)} \quad (\lambda \neq -1)$$

both of which are in some sense equivalent to $e^x = x^2$.

EXERCISE 5.1

Show that $\pm\sqrt{a}$ are roots of the following equations

(a) $x = \frac{1}{2}x^2 + x - \frac{1}{2}a$ (b) $x = 3x^2 + x - 3a$ (c) $x = a/x$

(d) $x = \dfrac{1}{2}\left(x + \dfrac{a}{x}\right)$ (e) $x = \dfrac{1}{3}\left(\dfrac{x^3}{a} + 2x\right)$.

(These are therefore all 'equivalent' to $x^2 - a = 0$.) Hence write down corresponding iterative processes of the form $x_{n+1} = f(x_n)$, for finding $\pm\sqrt{a}$.

EXERCISE 5.2

Write down as many rearrangements as you can of $\ln x = x^2 - x - 1$ in the form $x = f(x)$.

EXAMPLE 5.1   Let us try to find the root $x = 2$ of the equation $x^2 - x - 2 = 0$ by rewriting it in the form $x = 1 + (2/x)$ and applying the corresponding iterative process $x_{n+1} = 1 + (2/x_n)$. We shall start at $x_0 = 3$ and work with 4D.

n	0	1	2	3	4	5	6
x_n	3	1.6667	2.2000	1.9091	2.0476	1.9767	2.0118
$x_n - x_{n-1}$		-1.3333	0.5333	-0.2909	0.1385	-0.0709	0.0349

n	7	8	9	10	11	12	13
x_n	1.9942	2.0029	1.9986	2.0007	1.9996	2.0002	1.9999
$x_n - x_{n-1}$	-0.0175	0.0087	-0.0043	0.0021	-0.0011	0.0006	-0.0003

The differences between successive approximations are clearly decreasing steadily, in fact by a factor of 2 at each step. We have stopped the process at the point where $|x_n - x_{n-1}| < 0.0005$. As we pointed out in Lesson 4 this does not guarantee that x_n is correct to 3D; that depends on the *rate* of convergence. In this particular example we observe that the x_n oscillate and that the solution therefore lies between x_n and x_{n-1} at each stage. Hence the solution is 2.000 to 3D (since x_{12} and x_{13} agree to 3D and the solution lies between them).

In general we advise that $|x_n - x_{n-1}|$ is made ten times smaller than the error that is acceptable in the solution, and the sequence of approximations should be studied to assess whether the required accuracy has been obtained.

EXERCISE 5.3

Calculate the sequence of values generated by the iterative process

$$x_{n+1} = \frac{x_n^2 + 2}{2x_n - 1} \quad \text{with } x_0 = 0.6$$

stopping when the sequence has converged to within 3D accuracy. What equation have you solved approximately and what is the exact value of the root found?

The next exercise leads into the next section. You should therefore complete it before reading on. It can be done by hand in a few minutes!

EXERCISE 5.4

Calculate the first three iterates x_1, x_2, x_3 generated by the process $x_{n+1} = x_n^2 - 2$, with $x_0 = 3$. Is this process going to find a solution of $x^2 - x - 2 = 0$?

Before we analyse the convergence or otherwise of the sequence generated by $x_{n+1} = f(x_n)$ we shall consolidate the method itself with a flow chart.

EXERCISE 5.5

Draw a flow chart for the general one-point method $x_{n+1} = f(x_n)$, with input data x_0 and e (the convergence tolerance on $|x_{n+1} - x_n|$), and each approximation x_n as output. Note that arrays are not required. Interpret the iterative process as $x_{\text{new}} = f(x_{\text{old}})$ and the convergence test as $|x_{\text{new}} - x_{\text{old}}| < e$ so that only two adjacent iterates are required at any time.

5.3   ERROR ANALYSIS FOR $x_{n+1} = f(x_n)$

We have seen that in some problems (Exercise 5.3) the sequence $\{x_0, x_1, x_2, \ldots, x_n, \ldots\}$ seems to tend to a limit but that in others (Exercise 5.4) this is not so. Why should this be the case and how can we tell if a sequence generated by an iterative process is going to converge or not?

In fact we have to look at the influence of two elements of a process:

(a) the form of the process, $x_{n+1} = f(x_n)$; and

(b) the choice of the first element x_0.

For convergence it turns out that both of these have to be chosen carefully.

To see how this comes about we will do some mathematical analysis of the general case. This has the great advantage that the conclusions can be applied to any particular problem.

Let $X = f(X)$, i.e., let X be a root of $x = f(x)$.

Let $X = x_n + \xi_n$, $n = 0, 1, 2, \ldots$ (true value = approximate value + error), and hence also $X = x_{n+1} + \xi_{n+1}$. Now if f is a continuous function in the interval with endpoints at X and x_n, and differentiable inside that interval, then we can apply the Mean Value Theorem to it. This is explained in some detail in Appendix 2 for interested readers. Applying the theorem as stated there with $b = x_n$, $a = X$ and $\xi = \eta_n$, we obtain (see Fig. 5.1 for $n = 0$)

$$f(x_n) = f(X) + (x_n - X) f^{(1)}(\eta_n)$$

where η_n is a number between x_n and X but is otherwise unknown. Substituting this, with $x_n - X = -\xi_n$ and $x_{n+1} = X - \xi_{n+1}$, into $x_{n+1} = f(x_n)$ we find

$$X - \xi_{n+1} = f(X) - \xi_n f^{(1)}(\eta_n).$$

Since $X = f(X)$ (by definition of X) this becomes

$$\xi_{n+1} = \xi_n f^{(1)}(\eta_n), \quad n = 0, 1, 2, \ldots. \qquad [5\text{-}2]$$

Hence if $|f^{(1)}(\eta_n)| < 1$ the error in x_n will be reduced in x_{n+1}, and if $|f^{(1)}(\eta_n)| > 1$ the error will be increased. If $f^{(1)}(\eta_n)$ is positive the errors will be the same sign (x_n and x_{n+1} will be the same side of X), if $f^{(1)}(\eta_n)$ is negative the errors will have opposite signs (x_n and x_{n+1} will lie on opposite sides of X).

But we require the errors to decrease at *each* stage, and, not only that, but to decrease to zero.

It is clear that to reduce the error at each stage we require $|f^{(1)}(\eta_i)| < 1$ for all $i = 0, 1, 2, \ldots$. Suppose therefore that $|f^{(1)}(x)| < 1$ throughout some interval $a < x < b$ which contains X, i.e., $a < X < b$. If we choose x_0 in (a, b) then $\eta_0 \in (a, b)$ too since it lies between x_0 and X. Hence $|f^{(1)}(\eta_0)| < 1$ so that $|\xi_1| < |\xi_0|$ from Equation [5-2] with $n = 0$. Therefore x_1 is nearer X than x_0, i.e., $x_1 \in (a, b)$ too. This argument can be repeated *ad infinitum*. At each stage x_n is nearer X than x_{n-1}, hence $x_n \in (a, b)$ and hence $\eta_n \in (a, b)$ so that $|f^{(1)}(\eta_n)| < 1$.

Fig. 5.1 illustrates this progression.

Suppose in particular that $|f^{(1)}(x)| \leqslant M < 1$ where M is a constant, then from Equation [5-2]

$$|\xi_{n+1}| \leqslant M |\xi_n|, \quad n = 0, 1, 2, \ldots.$$

Applying this successively we obtain

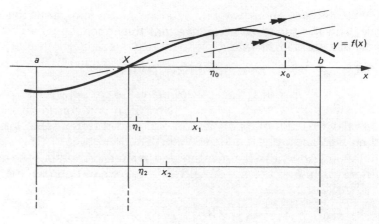

Fig. 5.1

$$|\xi_1| \leqslant M|\xi_0|$$

$$|\xi_2| \leqslant M|\xi_1| \leqslant M^2|\xi_0|$$

etc., until

$$|\xi_n| \leqslant M^n|\xi_0|.$$

But $M^n \to 0$ as $n \to \infty$ because $0 \leqslant M < 1$; therefore $|\xi_n| \to 0$ as $n \to \infty$, i.e., $x_n \to X$ as $n \to \infty$. We have proved this theorem.

THEOREM

If   (a) $|f^{(1)}(x)| < 1$ for all $x \in (a, b)$;

 (b) there exists $X \in (a, b)$ such that $X = f(X)$;

 (c) $x_0 \in (a, b)$;

then the sequence $x_1, x_2, \ldots$ generated by $x_{n+1} = f(x_n)$ converges to a solution of $x = f(x)$ in (a, b).

COROLLARY

If $|f^{(1)}(x)| < 1$ for all $x \in \mathbb{R}$ then the sequence generated by $x_{n+1} = f(x_n)$ converges to a solution of $x = f(x)$ for any $x_0 \in \mathbb{R}$.

EXAMPLE 5.2   Consider the recurrence relation $x_{n+1} = 1 + (2/x_n)$ used in Example 5.1.

Here $f(x) = 1 + (2/x)$; therefore

$$f^{(1)}(x) = -\frac{2}{x^2}.$$

Now usually we do not know the solution X — the whole point of the investigation is to find it. It is therefore not possible in general to check that the conditions of the theorem hold. In this particular example we know the roots $X = -1$ and 2 and $f^{(1)}$ is simple enough for us to investigate where it is less than one in magnitude.

In fact

$$\left| -\frac{2}{x^2} \right| < 1$$

$$\Rightarrow \quad x^2 > 2$$

i.e. $\quad x > \sqrt{2}$

or $\quad x < -\sqrt{2}.$

Since $|f^{(1)}(x)| < 1$ for $x \in (\sqrt{2}, \infty)$ we would expect convergence to the root $X = 2$ (which we happen to know in advance), from any initial point $x_0 \in (\sqrt{2}, \infty)$ — in fact we used $x_0 = 3$. Note also that $f^{(1)}(x)$ is negative throughout $(\sqrt{2}, \infty)$ indicating the error oscillation that we observed in Example 5.1, and that $f^{(1)}(2) = -\frac{1}{2}$ indicating that the errors should approximately halve at each iteration, which we also observed in Example 5.1. (We have used Equation [5-2] here, with $f^{(1)}(\eta_n) \simeq -\frac{1}{2}$.)

EXERCISE 5.6

Why would $x_{n+1} = 1 + (2/x_n)$ not, under any circumstances, produce a sequence converging to the root $X = -1$ of $x^2 - x - 2 = 0$? Calculate x_1 to x_6 from $x_0 = -1\frac{1}{2}$. Explain your results.

EXERCISE 5.7

Analyse why the iterative process used in Exercise 5.3 was successful in finding a root of $x^2 - x - 2 = 0$, whereas that used in Exercise 5.4 failed.

5.4 CONVERGENCE ANALYSIS IN PRACTICE

If the equation to be solved is very complicated we may be fortunate even to find a rearrangement $x = f(x)$ and an approximate solution x_0. In such cases the test for convergence will often be to 'try it and see'.

Even if it is reasonable to find an expression for $f^{(1)}(x)$ it may be unreasonable to find the intervals within which its modulus is less than unity. We cannot evaluate $f^{(1)}(X)$ because we do not know X. In practice, therefore, we are left with evaluating $f^{(1)}(x_0)$. We cannot guarantee a convergent sequence with $|f^{(1)}(x_0)| < 1$ alone — but it would be worth trying a process with this property.

We have not mentioned the rate of convergence yet. Equation [5-2] indicates that the speed at which the errors decrease depends on the size of the $f^{(1)}(\eta_i)$ — the smaller they are the faster the convergence.

Although $|f^{(1)}(x)| < 1$ for $x \in (a, b)$, with X and $x_0 \in (a, b)$, is the theoretical condition for convergence, $|f^{(1)}(x)| < 0.5$ is required for a realistic convergence rate, and $|f^{(1)}(x)| < 0.2$ would be the condition if your lazy authors had to do the calculation by hand!

To summarise: in practices evaluate $f^{(1)}(x_0)$. If $|f^{(1)}(x_0)| < \frac{1}{2}$, then the calculation is worth trying.

EXAMPLE 5.3 $x = -e^x$ has a solution near $x = -\frac{1}{2}$. Using the equation as it stands, $f(x) = -e^x$, $f^{(1)}(x) = -e^x$ and $f^{(1)}(-\frac{1}{2}) = -e^{-1/2} \simeq -0.75$. The sequence generated by $x_{n+1} = -e^{x_n}$, $x_0 = -\frac{1}{2}$, would probably converge slowly, and in oscillating fashion, to the nearby root.

EXAMPLE 5.4   In Lesson 4 we investigated the equation $\sqrt{x} = 4x - 2$, and found a solution in the interval $(\frac{1}{2}, 1)$. Consider the rearrangement $x = \frac{1}{4}(\sqrt{x} + 2)$. Here

$$f(x) = \tfrac{1}{4}(\sqrt{x} + 2)$$
$$f^{(1)}(x) = \tfrac{1}{8}x^{-1/2}.$$

So $\frac{1}{8} < |f^{(1)}(x)| < \sqrt{2}/8$ if $x \in (\frac{1}{2}, 1)$, which implies fairly quick convergence using $x_{n+1} = \frac{1}{4}(\sqrt{x_n} + 2)$.

Now consider the rearrangement $x = (4x - 2)^2$. Here

$$f(x) = (4x - 2)^2$$
$$f^{(1)}(x) = 8(4x - 2).$$

Over the interval $(\frac{1}{2}, 1)$, $f^{(1)}(x)$ varies linearly from 0 to 16. Despite the zero value at $x = \frac{1}{2}$ the process $x_{n+1} = (4x_n - 2)^2$ is most unlikely to give a convergent sequence.

EXERCISE 5.8

The equation $x^3 - 3x + 1 = 0$ has approximate roots $-1.9, 0.3, 1.5$. Using the rearrangement $x = (x^3 + 1)/3$ calculate 4 iterations for each root. Investigate the theoretical convergence in each case and compare your analysis with the evidence of the actual calculations. Work with 3D.

Find first approximations to the two roots of

$$\ln x = x^2 - x - 1$$

by sketching the graphs of $y = \ln x + 1$ and $y = x^2 - x$. Find two re-arrangements of the equation (one for each root) that will produce convergent sequences of approximations. (See your answer to Exercise 5.2.) Don't give up too easily; try all your rearrangements at both roots before looking up the answers.

5.5 GRAPHICAL INTERPRETATIONS OF CONVERGENCE AND DIVERGENCE

Now that we know in what cases the process $x_{n+1} = f(x_n)$ converges or diverges $(|f^{(1)}(x)| < 1$ or $|f^{(1)}(x)| > 1)$, it is possible to interpret these cases graphically.

Let us consider the case in which $-1 < f^{(1)}(x) < 0$.

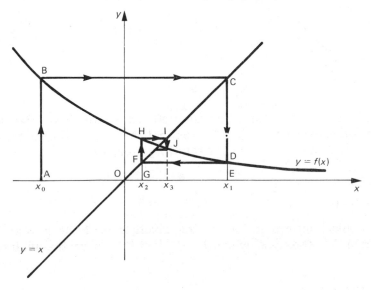

Fig. 5.2

In Fig. 5.2 we have drawn a graph $y = f(x)$ for which $-1 < f^{(1)}(x) < 0$, and the graph of $y = x$, and chosen a value for x_0 (see the point A). Hence $AB = f(x_0)$. Now $x_1 = f(x_0)$, so if we draw BC parallel to the x-axis to meet $y = x$ at C then $OE = EC = AB$, so that $x_1 = OE$. Similarly $DE = f(x_1) = FG$, and $x_2 = OG$, etc. It is clear that applying the recurrence relation is equivalent to following the path $A \rightarrow B \rightarrow C \rightarrow D \rightarrow F \rightarrow H \rightarrow I \rightarrow J \rightarrow$ etc. Equally clearly this spiral gives x coordinates that are nearer and nearer to the solution of $x = f(x)$ where the two graphs meet.

EXERCISE 5.10

Give a graphical interpretation of the convergence of the iterative process $x_{n+1} = f(x_n)$ for the case when $0 < f^{(1)}(x) < 1$ near the required root. The graphs should look as in Fig. 5.3.

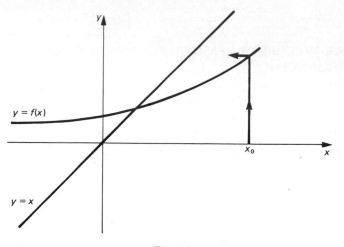

Fig. 5.3

If $-\infty < f^{(1)}(x) < -1$ near the required root we have the graphical interpretation shown in Fig. 5.4. We see that the path of the approximations spirals outwards and away from the solution, indicating divergence.

EXERCISE 5.11

Give a graphical interpretation of the divergence of the iterative process $x_{n+1} = f(x_n)$ for the case when $1 < f^{(1)}(x) < \infty$ near the required root.

5.6 NEWTON'S METHOD

We have seen that some one-point iterative methods generate convergent sequences for some roots. It would be satisfying to have a rearrangement which always worked for all roots, and with a fast rate of convergence. Newton's method (also known as the Newton–Raphson method) does just this. (Well almost. It can be caught out.)

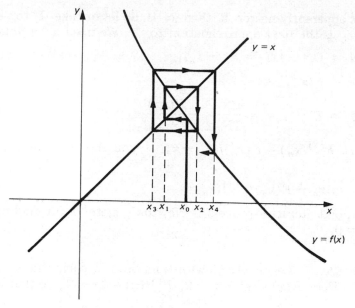

Fig. 5.4

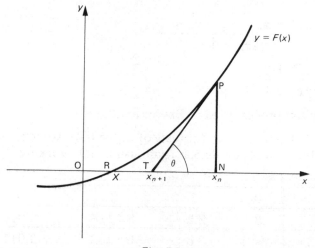

Fig. 5.5

The easiest way to derive Newton's method is to do it geometrically. We don't look at the rearrangement equation $x = f(x)$ but at the equation $F(x) = 0$, which is the one we want to solve anyway.

Referring to Fig. 5.5, let x_n be the latest approximation to the root X of $F(x) = 0$ and P be the point $(x_n, F(x_n))$ on the graph of $y = F(x)$. PT is the tangent to $y = F(x)$ at P.

Now T is apparently nearer R than N is. So let us take T to be $(x_{n+1}, 0)$ where x_{n+1} is the next approximation to X. We need a formula for x_{n+1}.

$$\text{TN} = \text{ON} - \text{OT} = x_n - x_{n+1}$$

$$\text{PN} = F(x_n)$$

$$\tan \theta = F^{(1)}(x_n) = \frac{\text{PN}}{\text{TN}} = \frac{F(x_n)}{x_n - x_{n+1}}.$$

Looking at $F^{(1)}(x_n) = F(x_n)/(x_n - x_{n+1})$ you should be able to rearrange it into

$$x_{n+1} = x_n - F(x_n)/F^{(1)}(x_n) \qquad \text{[5-3]}$$

This one-point iterative process is Newton's method for finding any root of $F(x) = 0$.

EXAMPLE 5.5   Let us use Newton's method to solve the equation $x^2 - x - 2 = 0$. Here $F(x) = x^2 - x - 2$, $F^{(1)}(x) = 2x - 1$, so that

$$x_{n+1} = x_n - \frac{x_n^2 - x_n - 2}{2x_n - 1}$$

$$= \frac{2x_n^2 - x_n - x_n^2 + x_n + 2}{2x_n - 1}$$

i.e. $x_{n+1} = \dfrac{x_n^2 + 2}{2x_n - 1}.$

This is exactly the process used in Exercise 5.3.

We will work it from $x_0 = 3$ (instead of $x_0 = 0.6$) to compare the results directly with those of Example 5.1, again using 4D working.

n	0	1	2	3	4
x_n	3	2.2	2.0118	2.0000	2.0000
$x_n - x_{n-1}$		-0.8	-0.1882	-0.0118	0

We observe a startlingly fast convergence, occurring after 3 iterations compared with 12 for the same conditions in Example 5.1. There must be something to this Newton chap!

Well, why is the method so much better? We can analyse it using our previous theory. Note that it is based on a rearrangement of $F(x) = 0$ into $x = x - F(x)/F^{(1)}(x)$ so that

$$f(x) = x - F(x)/F^{(1)}(x).$$

Hence

$$f^{(1)}(x) = 1 - \{F^{(1)}(x)F^{(1)}(x) - F(x)F^{(2)}(x)\}/\{F^{(1)}(x)\}^2$$
$$= F(x)F^{(2)}(x)/\{F^{(1)}(x)\}^2 \qquad\qquad [5\text{-}4]$$

after some manipulation — which you should check, as usual. Now at a root X of $F(x) = 0$ we have $F(X) = 0$ and hence $f^{(1)}(X) = 0$ — *provided* $F^{(1)}(X) \neq 0$. We can conclude that $f^{(1)}(x)$ is pretty small near the root and this explains the speedy convergence.

Note that, not only do we need $F^{(1)}(X) \neq 0$ for fast convergence, but that if $F^{(1)}(x_n)$ is very small for any x_n in the sequence then the method may at worst fail and at best converge very slowly (look at equation [5-3]).

EXERCISE 5.12

Use Newton's method to find the roots of $x^3 - 3x + 1 = 0$ near -1.9, $0.3, 1.5$, by calculating three iterations for each root and working to 7D for each root. Compare your results with those of Exercise 5.8.

EXERCISE 5.13

Use Newton's method to find the smaller root of $\ln x = x^2 - x - 1$ correct to 4D. (See Exercise 5.9.)

5.7   APPLICATIONS OF NEWTON'S METHOD

Before the advent of inexpensive electronic calculators the speed of Newton's method was much used to calculate square roots, cube roots, etc., and reciprocals without division. To find $\sqrt{a}$, for example, we solve $x^2 - a = 0$ by the process

$$x_{n+1} = x_n - \frac{x_n^2 - a}{2x_n} = \tfrac{1}{2}(x_n + a/x_n).$$

EXERCISE 5.14

Evaluate $\sqrt{2}$ to 4D using Newton's method and starting with $x_0 = 2$.

EXERCISE 5.15

Show that Newton's method applied to $(1/x) - a = 0$ gives the process $x_{n+1} = x_n(2 - ax_n)$ for finding the reciprocal of a without division. Evaluate $1/191$ correct to 4S starting with $x_0 = 0.005$.

It would be a mistake to think that Newton's method is the answer to all our problems with non-linear algebraic equations. For one thing, the equation to be solved may be so complicated that using the expression $F(x_n)/F^{(1)}(x_n)$ cannot be contemplated. For another, if $F^{(1)}(x)$ is very small near the root sought, then Newton's method can slow down considerably, and even not converge at all. Lastly if x_0 is not a *good* first approximation, Newton's method may converge slowly or to the wrong root.

In the next lesson we consider some alternative methods which aim to speed up the basic one-point method. These can often be as fast as Newton's method at its best, but with less calculation per iteration.

ANSWERS TO EXERCISES

5.1 (a) $x = \frac{1}{2}x^2 + x - \frac{1}{2}a \Rightarrow x^2 = a$, as required.

The process is $x_{n+1} = \frac{1}{2}x_n^2 + x_n - \frac{1}{2}a$.

Similarly the others, except that (e) has the extra root $x = 0$.

5.2 For example, $x = e^{x^2 - x - 1}$

$$x = x^2 - 1 - \ln x$$
$$x = (1 + x + \ln x)/x$$
$$x = \sqrt{1 + x + \ln x}.$$

5.3 The sequence is

$\qquad$ 0.6 11.8 6.25 3.5707 2.4017 2.0424 2.0006 2.0000 2.0000.

We have found the root $x = 2$ of $x^2 - x - 2 = 0$.

5.4 $x_1 = 7$, $x_2 = 47$, $x_3 = 2207$. This is clearly diverging. So $x_{n+1} = f(x_n)$ doesn't always work!

5.6 Because $|f^{(1)}(x)| > 1$ for x near -1 $\qquad$ (for $x \in (-\sqrt{2}, \sqrt{2})$ to be precise).

$$x_0 = -\tfrac{3}{2}, \; x_1 = -\tfrac{1}{3}, \; x_2 = -5, \; x_3 = \tfrac{3}{5},$$
$$x_4 = \tfrac{13}{3}, \quad x_5 = \tfrac{19}{13}, \quad x_6 = \tfrac{45}{19}.$$

The sequence oscillates in divergent fashion until $x_4 > \sqrt{2}$. From then on convergence to $x = 2$ occurs.

5.7 For $x_{n+1} = \dfrac{x_n^2 + 2}{2x_n - 1}$

$$f(x) = \frac{x^2 + 2}{2x - 1}$$

$$f^{(1)}(x) = \frac{(2x - 1)2x - (x^2 + 2)2}{(2x - 1)^2}$$

$$= \frac{2(x^2 - x - 2)}{(2x - 1)^2}$$

$$= 0 \quad \text{at } x = 2, \quad \text{hence 'small' near } x = 2.$$

Hence there is rapid convergence.

For $x_{n+1} = x_n^2 - 2$

$$f(x) = x^2 - 2$$

$$f^{(1)}(x) = 2x = 4 \quad \text{at } x = 2, \text{ hence } |f^{(1)}(x)| > 1 \text{ near } x = 2.$$

Hence there is divergence.

5.8 (a) $x_0 = -1.9 \Rightarrow -1.953, -2.150, -2.979, -8.152$.

Apparently this is diverging.

Now $f(x) = \frac{1}{3}(x^3 + 1)$, $f^{(1)}(x) = x^2$ and $f^{(1)}(-1.9) = 3.61 > 1$. This indicates divergence.

(b) $x_0 = 0.3 \Rightarrow 0.342, 0.347, 0.347, 0.347$.

Apparently this is converging fast. This time $f^{(1)}(0.3) = 0.09$, i.e., quite small, as expected.

(c) $x_0 = 1.5 \Rightarrow 1.458, 1.366, 1,183, 0.885$.

Obviously this is not converging to the root near 1.5, but probably is to the root near 0.347. Here $f^{(1)}(1.5) = 2.25$, explaining the initial divergence. In fact, however close we start to the left of the root near 1.5 we converge to the root near 0.347. If we start on the right (try $x_0 = 1.6$) we diverge to $+\infty$.

5.9 Roots are 0.4 and 1.8 approximately.

$x = e^{x^2 - x - 1}$ gives convergence to the root near 0.4.

$x = (1 + x + \ln x)/x$ gives convergence to the root near 1.8. You may find others that work.

5.10

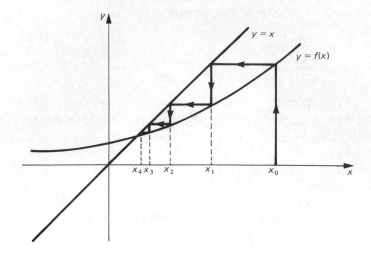

5.11

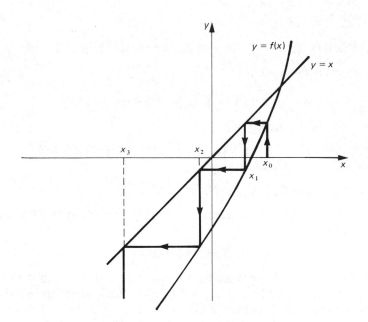

5.12 $x_0 = -1.9 \Rightarrow -1.879\,693\,5$
$-1.879\,385\,3$
$-1.879\,385\,2.$

$x_0 = 0.3 \Rightarrow 0.346\,520\,1$
$0.347\,296\,1$
$0.347\,296\,4.$

$$x_0 = 1.5 \Rightarrow \qquad 1.533\,333\,3$$
$$1.532\,090\,6$$
$$1.532\,088\,9.$$

5.13   0.2984.

FURTHER READING

For an extensive but essentially simple description of one-point iteration, biased towards computational problems, see Stark, Chapter 3, up to Section 3.4 inclusive.

For a treatment containing a comprehensive account of convergence properties see Dixon, Sections 7.1—7.3.

LESSON 5 – COMPREHENSION TEST

1. If a sequence generated by $x_{n+1} = x_n^2 - 1/x_n$ converges with limit X, what equation is X the solution of?

2. State the condition on f such that $x_{n+1} = f(x_n)$ gives a sequence converging to a solution X of $x = f(x)$.

3. Which of these recurrence relations defines a sequence that converges to the root $X = 1$ of $x^2 - 6x + 5 = 0$:

(a) $x_{n+1} = 6 - 5/x_n$,   (b) $x_{n+1} = (x_n^2 + 5)/6$,   (c) $x_{n+1} = \sqrt{(6x_n - 5)}$?

4. (a) State Newton's method for the general equation $F(x) = 0$.

(b) Is Newton's method generally very fast or very slow?

(c) Under what condition on F may Newton's method become slower than normal, or even fail?

ANSWERS

1   $x = x^2 - 1/x$, or $x^3 - x^2 - 1 = 0$.

2   $|f^{(1)}(x)| < 1$ for all x in an interval containing X and x_0.

3   (a) $f^{(1)}(1) = 5$ — diverges;
   (b) $f^{(1)}(1) = \frac{1}{3}$ — converges;
   (c) $f^{(1)}(1) = 3$ — diverges.

4   (a) $x_{n+1} = x_n - F(x_n)/F^{(1)}(x_n)$.

   (b) Very fast.

   (c) $F^{(1)}(X) = 0$, or $F^{(1)}(x) = 0$ near X, or even $F^{(1)}(x)$ very small near X.

LESSON 5 – SUPPLEMENTARY EXERCISES

5.16. Find rearrangements of the equations in Exercise 4.8 in the form $x = f(x)$ such that $x_{n+1} = f(x_n)$ produces a sequence converging to the root sought. Do this for each root of each equation. Hence reobtain all the roots correct to 3D.

5.17. Show that $x_{n+1} = a/x_n$ can never produce a sequence converging to $\pm\sqrt{a}$ unless $x_0 = \pm\sqrt{a}$ exactly.

5.18. Apply Newton's method to find all of the roots of each equation in Exercise 4.8 correct to 6D.

5.19. Use the formula for Newton's method to derive a process for finding $\sqrt[n]{a}$, $a > 0$, for any positive integer n. Hence find $\sqrt[5]{2}$ correct to 6D.

5.20. This exercise demonstrates how Newton's method may not always converge quickly or predictably. Show that Newton's method applied to $x^3 - 3x = 0$ reduces to

$$x_{n+1} = 2x_n^3/3(x_n^2 - 1).$$

Calculate the first few iterations for each of these values of x_0: 0.81, 0.78, 0.775, 0.7747, 0.77, $\sqrt{0.6}$.

*5.21. Rewrite Equation [4-3] with $c = x_{n+1}$, $b = x_n$ and $a = x_{n-1}$, and consider the result as defining a two-point iterative method (not necessarily bracketing, i.e., the solution not necessarily being between x_{n-1} and x_n). This is the *secant* method. Use it to obtain one of the roots found in Exercise 5.18 and compare both the formula and the effectiveness of the method with Newton's method.

*5.22. Devise a scheme for finding any number of the roots of $x \tan x = 1$. Find the first ten positive roots correct to 4S.

Lesson 6 **Acceleration and Relaxation**

OBJECTIVES

At the end of this lesson you should

(a) appreciate the need for speeding up simple one-point iterative processes;

(b) be able to derive and use the Aitken acceleration formula;

(c) understand what relaxation means;

(d) be able to apply relaxation to an iterative process;

(e) be able to derive and apply the modified Newton's method.

CONTENTS

6.1   INTRODUCTION

We mentioned in Lesson 5 that an iterative process $x_{n+1} = f(x_n)$ derived from a simple rearrangement of $F(x) = 0$ may generate a sequence x_1, $x_2, \ldots$ that converges too slowly to be useful (assuming we can find a convergent process at all!) The obvious alternative is to try Newton's method, but in some cases this is equally impractical; for example, because

(a) $F^{(1)}(x)$ may be small or zero near the required root,

or because

(b) $F^{(1)}(x)$ may be very complicated (or even impossible) to derive and/or evaluate.

If we can't use Newton's method and $x_{n+1} = f(x_n)$ is very slow, what can we do about it? The methods discussed in this lesson aim to take $x_{n+1} = f(x_n)$ (which will be assumed to work, albeit slowly) and to modify it in such a way as to speed up convergence of the sequence it generates.

6.2   THE ORDER OF A PROCESS

We proved in Lesson 5 that $x_{n+1} = f(x_n)$ generates a convergent sequence if $|f^{(1)}(x)| < 1$ near the required root X of $x = f(x)$, and if x_0 is also near enough to X. We also observed that if the value of $|f^{(1)}(x)|$ is near unity the convergence is very slow, whereas if it is near zero (as for Newton's method — provided $F^{(1)}(x)$ is not also small) then the convergence is significantly faster. In Lesson 12 you will find a careful analysis of the background to these observations. We shall show that

(a) if $f^{(1)}(X) \neq 0$ then

$$\xi_{n+1} = \xi_n f^{(1)}(\eta_n) \qquad\qquad [6\text{-}1]$$

where η_n lies between x_n and X as in Equation [5-2];

(b) if $f^{(1)}(X) = 0$ but $f^{(2)}(X) \neq 0$ then

$$\xi_{n+1} = -\tfrac{1}{2}\xi_n^2 f^{(2)}(\eta_n) \qquad\qquad [6\text{-}2]$$

where η_n lies between x_n and X again.

In case (a) $x_{n+1} = f(x_n)$ is called a *first-order process*, or is said to exhibit *linear convergence*. ξ_{n+1} is, approximately, a constant fraction of ξ_n. In case (b) $x_{n+1} = f(x_n)$ is called a *second-order process*, or is said to exhibit *quadratic convergence*. ξ_{n+1} is now roughly proportional to ξ_n^2, so that if, for example, ξ_n is of the order of 10^{-3}, then ξ_{n+1} is of the order of 10^{-6}, ξ_{n+2} of the order of 10^{-12}, and so on. In fact a second-order process will approximately double the number of correct decimal places at each iteration. No wonder that Newton's method is so popular!

Yes — Newton's method is second-order. We proved this in effect in Lesson 5 by showing that if $f(x) = x - F(x)/F^{(1)}(x)$, $F(X) = 0$ and $F^{(1)}(X) \neq 0$, then $f^{(1)}(X) = 0$. If $F^{(1)}(X) = 0$ it can be shown that Newton's method 'slows down' to first-order. Strictly speaking, $f^{(1)}(X) = 0$ is not a sufficient condition for a second-order process. We should also show that $f^{(2)}(X) \neq 0$. This is true for Newton's method; you may like to prove it for yourself. If $f^{(2)}(X) = 0$ then we are into the realms of third- and higher-order processes, which are even faster. In practice we are usually content to observe that, if $f^{(1)}(X) = 0$, then $x_{n+1} = f(x_n)$ is 'at least second-order'.

EXAMPLE 6.1   Consider again applying $x_{n+1} = 1 + 2/x_n$ to finding the root $X = 2$ of $x^2 - x - 2 = 0$ (as in Example 5.1).

n	0	1	2	3	4	5	6
x_n	3	1.6667	2.2000	1.9091	2.0476	1.9767	2.0118
ξ_n	-1	0.3333	-0.2	0.0909	-0.0476	0.0233	-0.0118
ξ_n/ξ_{n-1}		-0.33	-0.6	-0.45	-0.52	-0.49	-0.51

We observe that the ratio ξ_n/ξ_{n-1} is settling down to about -0.5. Since $f(x) = 1 + 2/x$ and hence $f^{(1)}(2) = -2/2^2 = -\frac{1}{2}$, that is exactly what we would expect from Equation [6-1].

Now consider applying Newton's method to the same problem, as in Example 5.3, but with more figures so that we can study the error behaviour.

n	0	1	2	3	4
x_n	3	2.2	2.0118	2.000 045 7	2.000 000 001
ξ_n	-1	-0.2	-0.0118	$-0.000 045 7$	-10^{-9}
ξ_n/ξ_{n-1}^2		0.2	0.29	0.33	0.5

The ratio 0.5 is not accurate because we have only one significant figure in ξ_4. The other ratios indicate the approximate proportionality of ξ_n and ξ_{n-1}^2, and we observe the corresponding doubling (roughly) of correct decimal places.

We can now rephrase the point of this lesson: if Newton's method is impractical, but we can find a first-order process $x_{n+1} = f(x_n)$ which converges, but too slowly, can we modify it to make it at least second-order? The answer is yes, and we present two ways of doing so.

6.3 AITKEN ACCELERATION

We have just seen evidence that, at least in some cases, it is possible to assume that the $f^{(1)}(\eta_n)$ term in Equation [6-1] is approximately constant, so that $f^{(1)}(\eta_{n+1}) = f^{(1)}(\eta_n) = k$, say, for all n. Then we can write

$$\xi_{n+1} = k\xi_n \qquad \xi_{n+2} = k\xi_{n+1}$$

and eliminating k gives

$$\frac{\xi_{n+2}}{\xi_{n+1}} = \frac{\xi_{n+1}}{\xi_n},$$

i.e., $\xi_{n+2}\xi_n = \xi_{n+1}^2$.

Substituting the definitions of the ξ_i we have

$$(X - x_{n+2})(X - x_n) = (X - x_{n+1})^2.$$

Rearranging to find X gives

$$X = \frac{x_{n+2}x_n - x_{n+1}^2}{x_{n+2} - 2x_{n+1} + x_n}$$

$$= x_{n+2} - \frac{(x_{n+2} - x_{n+1})^2}{x_{n+2} - 2x_{n+1} + x_n}. \qquad [6\text{-}3]$$

(Do check the algebra!) We seem to have found the solution exactly, but we assumed $f^{(1)}(\eta_{n+1}) = f^{(1)}(\eta_n)$. If they are approximately equal we might expect

$$x_{n+2}^* = x_{n+2} - \frac{(x_{n+2} - x_{n+1})^2}{x_{n+2} - 2x_{n+1} + x_n} \qquad [6\text{-}4]$$

to be a lot closer to X than x_{n+2}. This is Aitken's acceleration technique. (It is often written in the form

$$x_{n+2}^* = x_{n+2} - \frac{(\nabla x_{n+2})^2}{\nabla^2 x_{n+2}} \qquad [6\text{-}5]$$

using the backward-difference notation (Lesson 16) and is often called Aitken's 'delta-squared' process.)

To complete this section we ought to show that Aitken's method is essentially second-order. This is quite difficult, so we shall not put you off by attempting it here. Suffice to say that, at least intuitively, we have eliminated the $f^{(1)}(X)$ term of the error $X - x_{n+2}$ in calculating x_{n+2}^*.

6.4   APPLYING AITKEN ACCELERATION

Here is one way of applying Equation [6-4].

(a) Given x_0 calculate $x_1 = f(x_0)$ and $x_2 = f(x_1)$.

(b) Are x_1 and x_2 sufficiently close? If yes — stop.

(c) Otherwise use Aitken's method to find x_2^* from x_0, x_1 and x_2.

(d) Let $x_0 = x_2^*$ and return to (a).

In diagrammatic form the process looks like this:

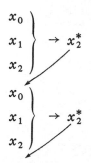

continuing until x_1 and x_2 are sufficiently close. Here, from Equation [6-4],

$$x_2^* = x_2 - \frac{(x_2 - x_1)^2}{x_2 - 2x_1 + x_0} = x_2 - \frac{(x_2 - x_1)^2}{(x_2 - x_1) - (x_1 - x_0)}.$$

Each stage of the calculation is therefore laid out thus:

$$\left\{ \begin{array}{l} x_0 \\ x_1 \\ x_2 \end{array} \right. \quad \begin{array}{c} x_1 - x_0 \\ \\ x_2 - x_1 \end{array} \quad (x_2 - x_1) - (x_1 - x_0)$$

$$x_2^* = \ldots .$$

EXERCISE 6.1

Draw a flow chart for the process described by (a)—(d) above. Input data should be x_0 and the convergence tolerance e (chosen so that $|x_1 - x_2| < e$ is the stopping criterion). The output should include all the calculated values, i.e., x_1, x_2 and x_2^* at each iteration.

Fig. 6.1 shows the sort of pattern that we expect this method to show.

EXAMPLE 6.2   We have already shown in Section 6.2 that when $x_{n+1} = 1 + 2/x_n$ is applied to finding the root $X = 2$ of $x^2 - x - 2 = 0$, then the

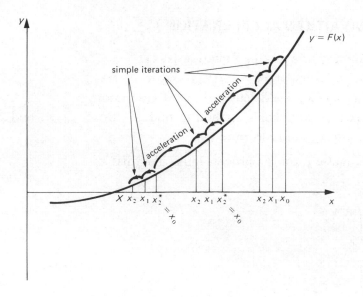

Fig. 6.1

errors decrease by a factor of approximately $-\frac{1}{2}$ at each iteration. Starting with $x_0 = 3$, $\xi_0 = -1$, we needed (Example 5.1) 13 iterations to obtain a 3D result.

Now observe the rate of convergence when the above version of Aitken's method is applied, with 5D accuracy as the aim.

$$\begin{cases} x_0 = 3 \\ x_1 = 1.666\,667 \\ x_2 = 2.200\,000 \end{cases} \quad \begin{matrix} -1.333\,333 \\ \\ 0.533\,333 \end{matrix} \qquad 1.866\,666$$

$$x_2^* = 2.200\,000 \quad - \quad \frac{(0.533\,33)^2}{1.866\,666} = 2.047\,619$$

$$\begin{cases} x_0 = 2.047\,619 \\ x_1 = 1.976\,744 \\ x_2 = 2.011\,765 \end{cases} \quad \begin{matrix} -0.070\,875 \\ \\ 0.035\,021 \end{matrix} \qquad 0.105\,896$$

$$x_2^* = 2.011\,765 \quad - \quad \frac{(0.035\,021)^2}{0.105\,896} = 2.000\,183$$

$$\begin{cases} x_0 = 2.000\,183 \\ x_1 = 1.999\,909 \\ x_2 = 2.000\,046 \end{cases} \quad \begin{matrix} -0.000\,274 \\ \\ 0.000\,137 \end{matrix} \qquad 0.000\,411$$

$$x_2^* = 2.000\,046 \ - \ \frac{(0.000\,137)^2}{0.000\,411} = 2.000\,000_3 .$$

We observe that we have now obtained 3D (2.000) in 4 iterations and 2 accelerations, and 6D (2.000 000) in 6 iterations and 3 accelerations.

Notice that a further acceleration would not be very accurate, because we would have at most 1S in the differences $x_2 - x_1$ and $x_1 - x_0$. Hence the errors in

$$\frac{(x_2 - x_1)^2}{(x_2 - x_1) - (x_1 - x_0)}$$

would dominate its true value. If further accuracy is required then more decimal places must be used in the working.

EXERCISE 6.2

Working with 6D use the iteration $x_{n+1} = -e^{x_n}$ to find x_1 to x_5, starting with $x_0 = -0.5$. (Sketch the graphs of $y = e^x$ and $y = -x$ to see that $x = -\frac{1}{2}$ is a reasonable approximate solution of $x + e^x = 0$.) Then rework the iteration with Aitken acceleration, starting with $x_0 = -0.5$ again, and obtaining the solution of $x + e^x = 0$ correct to 5D. Study the rates of convergence of the two processes.

6.5 RELAXATION

We now present an alternative method of achieving quadratic convergence. Reconsider $x_{n+1} = f(x_n)$ by rewriting it as

$$x_{n+1} = x_n + (f(x_n) - x_n). \qquad\qquad [6\text{-}6]$$

We can interpret this as saying that $f(x_n) - x_n$ is a *correction* to x_n, producing x_{n+1}. Now one-point iterative methods are slow when the correction at each stage is too small or too large. So an obvious idea is to modify the correction by a constant factor

$$x_{n+1} = x_n + c(f(x_n) - x_n). \qquad\qquad [6\text{-}7]$$

If $c > 1$ we sall this *over-relaxation* because we are making a larger than normal correction. If $c < 1$ we call it *under-relaxation* because the correction has been reduced.

Now the question arises as to whether c can be chosen so that the process [6-7] is second-order. Consider Equation [6-7] as being in the form $x_{n+1} = g(x_n)$ where

$$g(x) = x + c(f(x) - x).$$

For the second-order property we require $g^{(1)}(X) = 0$, i.e.

$$1 + c(f^{(1)}(X) - 1) = 0,$$

i.e. $c = 1/(1 - f^{(1)}(X))$.   [6-8]

This results in the second-order process

$$x_{n+1} = x_n + \frac{f(x_n) - x_n}{1 - f^{(1)}(X)}.$$ [6-9]

6.6 APPLYING RELAXATION

The problem with Equation [6-9] is that X is the value we are trying to find. The obvious alternative is to replace $f^{(1)}(X)$ by another constant, hopefully close to it. Since at the beginning of the iteration we only have x_0 we usually use $f^{(1)}(x_0)$. Then, if it turns out that x_0 is not very close to X we may modify our approximation as we proceed. With this replacement our process will no longer be strictly second-order.

EXAMPLE 6.3   To find a root of $x = \cos x$. We first sketch $y = x$ and $y = \cos x$ as in Fig. 6.2 and note that there is a root (the only root?) at about $x = 0.7$.

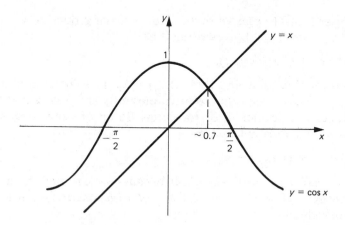

Fig. 6.2

As a comparison we first try the obvious one-point process $x_{n+1} = \cos x_n$. (Will it converge? Why?)

n	0	1	2	3	4	5
x_n	0.7	0.7648	0.7215	0.7508	0.7311	0.7444
$x_n - x_{n-1}$		0.0648	-0.0433	0.0293	-0.0197	0.0133

The successive iterates are oscillating about the solution, which means that the process is overcorrecting. We see however that the differences $x_n - x_{n-1}$ are decreasing in magnitude, but rather slowly, and typically of a first-order process. In fact, since $f(x) = \cos x$, $f^{(1)}(x) = -\sin x$ so that $f^{(1)}(0.7) \simeq -0.6$, which is the factor by which we would expect the errors to decrease at each stage.

Now let's try relaxation. Using $f^{(1)}(0.7) \simeq -0.6$ instead of $f^{(1)}(X)$ in Equation [6-9] we have

$$x_{n+1} = x_n + \frac{\cos x_n - x_n}{1 + 0.6}$$

$$= x_n + 0.625(\cos x_n - x_n)$$

i.e. $x_{n+1} = 0.375 x_n + 0.625 \cos x_n$.

Here are the results. Anticipating the increased convergence rate we have used 7D working.

n	0	1	2	3	4	5
x_n	0.7	0.740 526 4	0.739 018 3	0.739 088 2	0.739 085 0	0.739 085 1
$x_n - x_{n-1}$		0.041	-0.0015	0.000 070	$-0.000 003 2$	0.000 000 1

The original process had not obtained 2D accuracy in 5 iterations. Here we have 6D in 5 iterations. Even though the process is not strictly second-order (because we replaced $f^{(1)}(X)$ by $f^{(1)}(x_0)$) it is undoubtedly very quick when compared with the first-order original.

EXAMPLE 6.4   This example illustrates a mistake which can easily be made when relaxation is applied. When $F(x) = 0$ is rearranged into the form $x = f(x)$ it is obvious that any arithmetic operations performed must be performed with an accuracy at least as great as that required in the solutions, and for security to a much greater accuracy. Otherwise we shall solve an equation different to the original. If the relaxed process [6-7] is rearranged (as in Example 6.3) into the form

$$x_{n+1} = (1-c)x_n + cf(x_n)$$

then care must similarly be taken to ensure that the numerical values of

$1 - c$ and c do sum precisely to one, otherwise we shall not be solving the original problem.

As an example consider solving $\sqrt{x} = 4x - 2$ by relaxing the process $x_{n+1} = \frac{1}{4}(\sqrt{x_n} + 2)$. A natural development might be the following:

$$f(x) = \tfrac{1}{4}(\sqrt{x} + 2) \qquad f^{(1)}(x) = \tfrac{1}{8} x^{-1/2}.$$

The root is at approximately $x_0 = 0.7$. Hence

$$f^{(1)}(x_0) = \tfrac{1}{8}(0.7)^{-1/2} \simeq 0.149.$$

Therefore let $c = 1/(1 - 0.149)$, giving the relaxed process

$$x_{n+1} = x_n + \frac{\tfrac{1}{4}(\sqrt{x_n} + 2) - x_n}{1 - 0.149}$$

$$= -\frac{0.149}{0.851} x_n + \frac{\tfrac{1}{4}(\sqrt{x_n} + 2)}{0.851}$$

$$= -0.175 x_n + 0.294(\sqrt{x_n} + 2).$$

In this case we are solving

$$1.175 x = 0.294(\sqrt{x_n} + 2),$$

i.e. $\qquad x = 0.2502 \ldots (\sqrt{x} + 2),$

not the original equation. The safe rearrangement is

$$x_{n+1} = -0.175 x_n + 1.175\{\tfrac{1}{4}(\sqrt{x_n} + 2)\}$$

with any rounding done as part of the calculation process.

EXERCISE 6.3

Solve the equation $x + e^x = 0$ by relaxing the iterative process $x_{n+1} = -e^{x_n}$. Work with 6D, giving your answer correct to 5D, starting with $x_0 = -0.5$. Compare the number of evaluations of e^x with the number required by Aitken's method in Exercise 6.2.

In the above examples it was reasonable to estimate $f^{(1)}(X)$ by $f^{(1)}(x_0)$. In practice this may not be the case. If you have an interactive computing system, or a programmable calculator, it is perfectly reasonable to choose the relaxation parameter c by experiment. Since a computer/calculator takes most of the work out of applying the iterative process an ultra-fast convergence rate is not so important.

At the other end of the spectrum of problems the solution X may be known as an expression which enables $f^{(1)}(X)$ to be found and used in Equation [6-9]. Here is an example of that type.

EXERCISE 6.4

Starting from the first-order process $x_{n+1} = a/x_n$ for finding the square root of a (i.e., solving $x^2 = a$), use relaxation to derive the second-order process $x_{n+1} = \frac{1}{2}(x_n + a/x_n)$.

6.7 THE MODIFIED NEWTON'S METHOD

An alternative to approximating the constant $f^{(1)}(X)$ by the constant $f^{(1)}(x_0)$ is to replace it by $f^{(1)}(x_n)$ so that it becomes progressively nearer $f^{(1)}(X)$. It turns out that this is equivalent to Newton's method, and this suggests that we might study relaxation from the point of view of the original equation $F(x) = 0$.

Consider the 'rearrangement'

$$x = x - F(x)$$

and hence the iteration

$$x_{n+1} = x_n - F(x_n).$$

$F(x_n)$ is now the correction (instead of $f(x_n) - x_n$) and the relaxed version would be

$$x_{n+1} = x_n - cF(x_n).$$

To make this second-order we require the derivative of $x - cF(x)$ to be zero at the root, i.e.

$$1 - cF^{(1)}(X) = 0$$

i.e. $c = 1/F^{(1)}(X)$

giving

$$x_{n+1} = x_n - F(x_n)/F^{(1)}(X). \qquad [6\text{-}10]$$

We now see that the approximation $F^{(1)}(x_n)$ for $F^{(1)}(X)$ gives Newton's method, which is still second-order despite that approximation. We also see that the equivalent of our 'practical' relaxation would replace $F^{(1)}(X)$ by $F^{(1)}(x_0)$, giving

$$x_{n+1} = x_n - F(x_n)/F^{(1)}(x_0). \qquad [6\text{-}11]$$

This is often called the *modified Newton's method*. It is not strictly second-order, but can be quite fast if x_0 is close enough to X and $F^{(1)}(x)$ does not vary wildly near $x = X$. It has the great advantage that we do not have to recalculate $F^{(1)}(x)$ at each stage.

Equation [6-11] can be related, graphically, to Newton's method. In Newton's method the tangent to $y = F(x)$ is used in every iteration to find the next iterate. In our form of the relaxed iterative process the tangent at $x = x_0$ is used for the first iteration. For succeeding iterations lines are drawn through the points $(x_n, F(x_n))$ parallel to this first tangent at $(x_0, F(x_0))$ to find the points $(x_{n+1}, 0)$. Fig. 6.3 shows the geometrical relation between the points $(x_n, F(x_n))$ on the curve $y = F(x)$. The arrows indicate the parallel lines and the way in which the iterates are found geometrically.

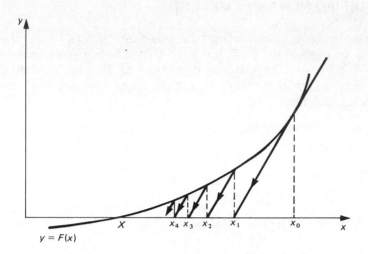

Fig. 6.3

EXAMPLE 6.5   Consider $\sqrt{x} = 4x - 2$ again. Let $F(x) = \sqrt{x} - 4x + 2$ and take $x_0 = 0.7$ as in Example 6.4. Then

$$F^{(1)}(0.7) = \tfrac{1}{2}(0.7)^{-1/2} - 4 \simeq -3.4$$

giving

$$x_{n+1} = x_n - \frac{\sqrt{x_n} - 4x_n + 2}{-3.4}.$$

Once again we must take care that any rearrangement of this does not alter the equation being solved. In fact it is better to leave it as it is; the size of the numerator at each stage tells us how well the process is converging (it should tend to zero quite quickly). Here are the results.

n	0	1	2	3
x_n	0.7	0.710 782	0.710 768	0.710 768
$F(x_n)$ (to 2S)	0.037	-0.50×10^{-4}	0.10×10^{-6}	

The full Newton's method could hardly converge much faster!

EXERCISE 6.5

According to Scheid (Schaum Series, *Numerical Analysis*) Leonardo of Pisa obtained 1.368 808 107 as a root of

$$x^3 + 2x^2 + 10x - 20 = 0$$

in the year 1225. Obtain Leonardo's result (yes — all 9D, if your calculator allows it) using

(a) $x_{n+1} = 20/(x_n^2 + 2x_n + 10)$;

(b) a relaxed version of (a) (perhaps the modified Newton's method);

(c) Newton's method;

starting with $x_0 = 1$ in each case.

Compare the effort required for the three methods.

ANSWERS TO EXERCISES

6.2 Using $x_{n+1} = -e^{x_n}$, $x_0 = -0.5$ gives

n	0	1	2	3	4	5
x_n	−0.5	−0.606 531	−0.545 239	−0.579 703	−0.560 065	−0.571 172
$x_n - x_{n-1}$		−0.106 531	0.061 292	−0.034 464	0.019 638	−0.011 107

We observe a slow convergence rate, consistent with a first-order process. Using Aitken's method gives

$$\begin{cases} x_0 = -0.5 \\ x_1 = -0.606\,531 \\ x_2 = -0.545\,239 \end{cases} \quad \begin{matrix} -0.106\,531 \\ \\ 0.061\,292 \end{matrix} \quad 0.167\,823$$

$$x_2^* = -0.545\,239 \quad - \quad \frac{(0.061\,292)^2}{0.167\,823} = -0.567\,624$$

$$\begin{cases} x_0 = -0.567\,624 \\ x_1 = -0.566\,871 \\ x_2 = -0.567\,300 \end{cases} \quad \begin{matrix} 0.000\,753 \\ \\ -0.000\,429 \end{matrix} \quad -0.001\,182$$

$$x_2^* = -0.567\,300 \quad - \quad \frac{(-0.000\,429)^2}{-0.001\,182} = -0.567\,144$$

$$\begin{cases} x_0 = -0.567\,144 \\ x_1 = -0.567\,143 \\ x_2 = -0.567\,143. \end{cases}$$

Hence the solution is certainly $-0.567\,14$ to 5D, with very little more work than that which obtained 1D in the original process.

6.3 $f(x) = -e^x$, $x_0 = -0.5$. Therefore $f^{(1)}(x_0) = -e^{-0.5} \simeq -0.6$.

Hence $c = 1/1.6 = 0.625$ and the relaxed process is

$$x_{n+1} = x_n + 0.625(-e^{x_n} - x_n)$$
$$= 0.375 x_n - 0.625 e^{x_n},$$

n	0	1	2	3	4
x_n	-0.5	$-0.566\,582$	$-0.567\,132$	$-0.567\,143$	$-0.567\,143$
$x_n - x_{n-1}$		$-0.066\,582$	$-0.000\,550$	$-0.000\,011$	$-0.000\,000$

giving $0.567\,14$ to 5D for certain. Newton's method would probably be marginally faster, i.e., require slightly less iterations, but each iteration would take about twice as long to calculate. The Aitken approach (Exercise 6.1) was a little longer and a little more complicated, requiring 6 evaluations of e^x as against 4 above.

6.4 The relaxed process is

$$x_{n+1} = x_n + c\left(\frac{a}{x_n} - x_n\right)$$

where $c = 1/(1 - f^{(1)}(X))$ and $f(x) = a/x$.

Now $f^{(1)}(x) = -a/x^2$ and $X = \sqrt{a}$, so $f^{(1)}(X) = -1$ and $c = \frac{1}{2}$.
Hence

$$x_{n+1} = x_n + \frac{1}{2}\left(\frac{a}{x_n} - x_n\right) = \frac{1}{2}\left(x_n + \frac{a}{x_n}\right).$$

6.5 (a) requires about 24 iterations.

(b) with $F(x) = x^3 + 2x^2 + 10x - 20$, $F^{(1)}(1) = 17$. Therefore use

$$x_{n+1} = x_n - \frac{x_n^3 + 2x_n^2 + 10x_n - 20}{17}.$$

This requires about 14 iterations. If we change to $F^{(1)}(1.4) \simeq 21.5$ when it is clear that the root is near to 1.4 (after about 2 or 3 iterations), i.e., use

$$x_{n+1} = x_n - \frac{x_n^3 + 2x_n^2 + 10x_n - 20}{21.5},$$

then we require only 6 iterations!

Alternatively with $f(x) = 20/(x^2 + 2x + 10)$

$$f^{(1)}(1) = -0.47$$

giving

$$x_{n+1} = x_n + \frac{20/(x_n^2 + 2x_n + 10) - x_n}{1 - (-0.47)}$$

$$= 0.320x_n + 0.680\{20/(x_n^2 + 2x_n + 10)\}.$$

(Don't forget that the two coefficients must sum exactly to unity.) This requires only 5 iterations! In this case $f^{(1)}(1.4) = -0.44$ and using this instead of -0.47 has little effect.

(c) Newton's method

$$x_{n+1} = x_n - \frac{x_n^3 + 2x_n^2 + 10x_n - 20}{3x_n^2 + 4x_n + 10}$$

is much more complicated but requires only 4 iterations.

FURTHER READING

Most books on numerical mathematics include a section on Aitken's acceleration method. See for example Ribbans, Book 2, Sections 5.0 and 5.1 which include a discussion of first and second-order processes and Newton's method. For a more mathematical treatment of Aitken's method, Newton's method and quadratic convergence see Dixon, Sections 7.6 and 7.7. Like most books this assumes a knowledge of Taylor series. We shall not cover this until Lesson 11, but it is still possible to benefit from the reference. Similarly Hosking, Joyce and Turner, 'Steps 9 and 10'.

A good reference is Phillips and Taylor, Section 7.5 where they consider relaxation and Aitken acceleration together. Their relaxation is described slightly differently; starting from

$$x = f(x)$$

they introduce λx thus

$$x + \lambda x = \lambda x + f(x)$$

and rearrange into

$$x = \frac{\lambda}{1 + \lambda}x + \frac{1}{1 + \lambda}f(x).$$

It is straightforward to show that their λ and our c are related by $c = 1/(1 + \lambda)$. The advantage of their choice is that the condition for second-order convergence is simply $\lambda = -f^{(1)}(X)$.

LESSON 6 – COMPREHENSION TEST

1. State the order of these iterative processes.

(a) $x_{n+1} = 6 - 5/x_n$ for solving $x^2 - 6x + 5 = 0$ for the root $x = 1$.

(b) Newton's method.

(c) Any relaxed one-point method.

(d) The Modified Newton's method.

2. What is the assumption underlying Aitken's acceleration technique?

3. A one-point method produced the following iterates. Which would accelerate the convergence: under- or over-relaxation:

    1.54   1.30   1.40   1.35   1.38   1.366   1.370   1.368   1.3691?

ANSWERS

1   (a) First-order since $\dfrac{d}{dx}(6 - 5/x) \neq 0$ at $x = 1$.

    (b) Second-order.

    (c) $x_{n+1} = x_n + (f(x_n) - x_n)/(1 - f^{(1)}(X))$ is second-order. In practice $f^{(1)}(X)$ is approximated and the process becomes first-order in theory, but is often still as fast as second-order. (Another example of the difference between theory, where zero is absolute, and practice, where it is not. We require $g^{(1)}(X) = 0$ for $x_{n+1} = g(x_n)$ to be second-order but $g^{(1)}(X) \simeq 0$ will imply just as fast a process.)

    (d) $x_{n+1} = x_n - F(x_n)/F^{(1)}(x_0)$ is first-order in theory, but the same remarks as in (c) apply.

2   The assumption is that, in $x_{n+1} = f(x_n)$, $f^{(1)}(x)$ is approximately constant near the root required.

3   Under-relaxation since the process is overshooting the root at each stage.

LESSON 6 – SUPPLEMENTARY EXERCISES

6.6. Reconsidering each process $x_{n+1} = f(x_n)$ used in Exercise 5.16, re-obtain all the roots of each equation

(a) by using Aitken's acceleration with the process;

(b) after first constructing a relaxed process;

(c) using the Modified Newton's method;

this time to 6D, and compare the work required with that for Newton's method in Exercise 5.18.

6.7. Show that applying Aitken acceleration to $x_{n+1} = a/x_n$ is equivalent to applying Newton's method to solving $x^2 = a$. (In this case acceleration makes a non-convergent method — see Exercise 5.17 — into a second-order method.)

*6.8. Show that Newton's method is first-order only when applied to the problem of finding a root X of $F(x) = 0$ such that $F^{(1)}(X) = 0$.

*6.9. Show that the modified Newton's method (Equation [6-11]) is a first-order process and converges to a root X of $F(x) = 0$ provided that

$$0 < \frac{F^{(1)}(x)}{F^{(1)}(x_0)} < 2$$

for all x in some interval containing x_0 and X.

*6.10. Use relaxation to improve your answer to Exercise 5.21, i.e. to provide a better computational scheme.

*6.11. Use any suitable iterative method to solve this classical problem:

A goat is tethered to the midpoint of one side of a square field of side $2s$ by a rope of length r. Find r (as a multiple of s) such that the goat can graze a region of area equal to half the area of the field.

(*Hint*: after representing the problem as a nonlinear equation involving r and s, convert it into an equation for $x = r/s$.)

PART C NUMERICAL SOLUTION OF SETS OF LINEAR ALGEBRAIC EQUATIONS

Lesson 7 **Matrices and Matrix Algebra**

OBJECTIVES

At the end of this lesson you should

(a)   understand what a matrix is;

(b)   be able to perform matrix addition and subtraction;

(c)   be able to multiply a matrix by a scalar;

(d)   be able to multiply a matrix by another matrix;

(e)   be able to express a set of linear equations in matrix form;

(f)   understand the ideas of matrix transpose and matrix inverse.

CONTENTS

7.1   INTRODUCTION

This lesson is largely theoretical, introducing the idea of a matrix (plural: matrices) and some matrix algebra. The reason for doing this is that in the next three lessons various methods of solving sets of n linear algebraic equations in n unknowns will be discussed and these are easy to express in terms of matrices. First of all, when we want to refer to a general set of such equations, writing it down in the form

$$a_{11}x_1 + a_{12}x_2 + \ldots + a_{1n}x_n = b_1$$
$$a_{21}x_1 + a_{22}x_2 + \ldots + a_{2n}x_n = b_2$$

$$\ldots \ldots$$

$$a_{n1}x_1 + a_{n2}x_2 + \ldots + a_{nn}x_n = b_n \qquad \text{[7-1]}$$

is both time and space consuming. Matrix notation avoids this. Secondly, when we use arithmetic operations to change the above set of n equations into another related set of n equations, describing how we do it can sometimes be made much clearer in matrix notation. Thirdly, in Lesson 9 we will come across problems which deal with the question of whether we have a 'good' solution or not. In this situation it is very much easier to analyse the problem using matrix notation than it is in other ways. Lastly in Lesson 10 we will come across iterative methods for solving sets of equations. In that case matrix notations can be used to replace what may appear as a jumble of letters and equations by just *one* equation.

Well of course nothing is for nothing. Such advantages can only be bought at a price. But the price is not high and is part of the currency of mathematicians *and* many others whose speciality lies in other fields. A smattering of Italian is useful when on holiday in Rome!

7.2 MATRICES

What follows may appear artificial, but bear with us — there is a pay-off in later lessons!

A matrix, so the mathematicians say, is a rectangular array of numbers. For example,

$$A = \begin{bmatrix} 3 & 2 & 1 \\ 4 & -6 & 7 \end{bmatrix}$$

is a matrix. This particular matrix has 2 *rows* and 3 *columns*. The brackets are important and tell us that we are dealing with matrices.

If

$$B = \begin{bmatrix} 2 & 4 \\ -3 & 1 \\ 6 & -5 \end{bmatrix}, \quad B \text{ has 3 } rows \text{ and 2 } columns.$$

Matrices may have any number of rows or columns.

What about the particular numbers that occur in our matrices **A** and **B** above? These numbers are called the entries in (or elements of) **A** and **B**. We could regard matrix **A**, say, as a rectangular piece of peg-board with 6 holes bored in it (see Fig. 7.1).

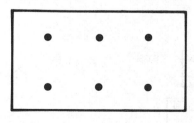

Fig. 7.1

The numbers could be written on pieces of paper stuck to matchsticks which could then be stuck into the appropriate holes. The holes stay the same but the numbers can be changed. Clearly **A** and **B** are matrices of different sorts as they have different shapes. It is not the particular numbers (which are the entries) but the *numbers of rows and columns* which form the major difference between **A** and **B**. In general a matrix with m rows and n columns is called an m by n (or $m \times n$) matrix.

There are various sorts of matrices that are sufficiently important to have particular names. A matrix with just one column, for example, is called a *column matrix* or a *column vector*. A matrix with just one row is called a *row matrix* or a *row vector*. For example,

$$\begin{bmatrix} 1 \\ 2 \\ -5 \end{bmatrix}$$ is a 3-element column vector, and

$$\begin{bmatrix} 2 & 0 & 6 & -3 \end{bmatrix}$$ is a 4-element row vector.

An $m \times n$ matrix with all its entries equal to zero is called a *zero matrix* or *null matrix* and is denoted by **0**. We shall meet other particular matrix types later.

7.3 MATRIX ADDITION AND SUBTRACTION

When a mathematician invents a new kind of mathematical object, such as a matrix, he usually has in mind some operations that he wishes to perform on them. In the next few sections we shall discuss adding and subtracting matrices, multiplying matrices by numbers, and multiplying matrices by matrices.

The natural rule for adding two matrices is that we add corresponding entries. For this to be a sensible rule the two matrices must have the same number of rows and the same number of columns.

If this is the case we say that the two matrices are *conformable for addition and subtraction*.

EXAMPLE 7.1 If

$$A = \begin{bmatrix} 3 & 2 & 1 \\ 4 & -6 & 7 \end{bmatrix} \quad \text{and}$$

$$B = \begin{bmatrix} 2 & -1 & 3 \\ -6 & 2 & 1 \end{bmatrix},$$

$$A + B = \begin{bmatrix} 3+2 & 2+(-1) & 1+3 \\ 4+(-6) & -6+2 & 7+1 \end{bmatrix}$$

$$= \begin{bmatrix} 5 & 1 & 4 \\ -2 & -4 & 8 \end{bmatrix}$$

and

$$A - B = \begin{bmatrix} 3-2 & 2-(-1) & 1-3 \\ 4-(-6) & -6-2 & 7-1 \end{bmatrix}$$

$$= \begin{bmatrix} 1 & 3 & -2 \\ 10 & -8 & 6 \end{bmatrix}.$$

EXERCISE 7.1

Calculate $A + B$ and $A - B$ in the following cases.

(a) $A = \begin{bmatrix} 4 & 3 \\ 2 & 1 \\ 6 & -5 \end{bmatrix}, \quad B = \begin{bmatrix} 2 & -1 \\ 3 & 4 \\ -7 & 2 \end{bmatrix}$

(b) $A = \begin{bmatrix} 1 & 2 & 3 \\ 4 & 5 & 6 \\ 7 & 8 & 9 \end{bmatrix}, \quad B = \begin{bmatrix} -3 & -4 & -5 \\ -6 & -7 & -8 \\ -9 & -1 & -2 \end{bmatrix}$

(c) $A = \begin{bmatrix} 3 \\ -2 \\ 1 \\ 5 \end{bmatrix}, \quad B = \begin{bmatrix} 4 \\ 7 \\ -5 \\ 2 \end{bmatrix}.$

You should notice that the matrices resulting from the addition or subtraction of matrices (which are conformable for these operations of course) have the same numbers of rows and columns as the matrices from which they are derived.

Note also that if A is an $m \times n$ matrix then $A - A = 0$, is the $m \times n$ zero matrix. Hence the zero matrix in matrix algebra behaves similarly to the number zero in ordinary algebra.

Two other important properties of matrix addition (and subtraction) are the *associative law*

$$A + (B + C) = (A + B) + C$$

which allows us to add together more than two matrices, and the *commutative law*

$$A + B = B + A.$$

These are simple consequences of the definition: 'add corresponding elements'.

7.4 SCALAR MULTIPLICATION OF MATRICES

If k is an ordinary number (scalar) and A a matrix, we mean by kA the matrix whose entries are k times those of A.

EXAMPLE 7.2   If

$$A = \begin{bmatrix} 1 & 2 & -4 \\ 6 & -3 & 5 \end{bmatrix} \quad \text{then}$$

$$3A = \begin{bmatrix} 3 & 6 & -12 \\ 18 & -9 & 15 \end{bmatrix} \quad \text{and} \quad (-2)A = \begin{bmatrix} -2 & -4 & 8 \\ -12 & 6 & -10 \end{bmatrix}.$$

We normally write $(-2)A$ as $-2A$. Similarly $-A$ is really $(-1)A$, so that subtraction is a special case of addition, as in ordinary algebra. That is,

$$A - B = A + (-B).$$

From now on we will assume that — unless stated to the contrary — the given matrices are conformable for addition and subtraction — mainly because it is boring to be saying it continually.

We can now combine addition and subtraction of matrices with scalar multiplication of matrices.

EXAMPLE 7.3   If

$$A = \begin{bmatrix} 3 & -4 \\ 2 & 1 \\ -5 & -2 \end{bmatrix} \quad \text{and} \quad B = \begin{bmatrix} 2 & -1 \\ 3 & 2 \\ 7 & 1 \end{bmatrix}$$

$$2A + 3B = \begin{bmatrix} 6+6 & -8-3 \\ 4+9 & 2+6 \\ -10+21 & -4+3 \end{bmatrix} = \begin{bmatrix} 12 & -11 \\ 13 & 8 \\ 11 & -1 \end{bmatrix}.$$

EXERCISE 7.2

If $A = \begin{bmatrix} 1 & -3 \\ 4 & 2 \end{bmatrix}$ and $B = \begin{bmatrix} 7 & 2 \\ -3 & 5 \end{bmatrix}$

find

(a) $4A - 3B$ (b) $2A + 3B$ (c) $-3A + 2B$.

7.5   MATRIX MULTIPLICATION – INTRODUCTION

If addition, subtraction and scalar multiplication of matrices is possible under the right conditions, can we multiply matrices together? The answer is yes — if they are *conformable for multiplication* — which is not necessarily the same as being conformable for addition and subtraction! We shall derive the conformability condition for multiplication in the next section. For the moment we consider an intermediate problem, that of multiplying a column vector by a matrix.

We start by defining the product of a row vector and a column vector (*in that order*) by

$$\begin{bmatrix} a & b & c \end{bmatrix} \begin{bmatrix} x \\ y \\ z \end{bmatrix} = ax + by + cz \qquad\qquad [7\text{-}2]$$

for the case when each has three entries. This is called an inner product. Note that the result looks like the kind of expression we have in linear equations. The definition for general rows and columns is the obvious extension

$$[a_1 \quad a_2 \quad \ldots \quad a_n] \begin{bmatrix} b_1 \\ b_2 \\ \cdot \\ \cdot \\ \cdot \\ b_n \end{bmatrix} = a_1 b_1 + a_2 b_2 + \ldots + a_n b_n \qquad [7\text{-}3]$$

$$= \sum_{i=1}^{n} a_i b_i.$$

Note that the row and the column must have the same number of entries. Here is an example.

$$[2 \quad 0 \quad -1 \quad 5] \begin{bmatrix} 4 \\ 3 \\ -6 \\ 2 \end{bmatrix} = (2)(4) + (0)(3) + (-1)(-6) + (5)(2)$$

$$= 24.$$

Now consider a set of four linear equations in x, y and z such as

$$a_1 x + b_1 y + c_1 z = u_1$$
$$a_2 x + b_2 y + c_2 z = u_2$$
$$a_3 x + b_3 y + c_3 z = u_3$$
$$a_4 x + b_4 y + c_4 z = u_4.$$

We can write them individually as

$$[a_1 \quad b_1 \quad c_1] \begin{bmatrix} x \\ y \\ z \end{bmatrix} = u_1$$

$$[a_2 \quad b_2 \quad c_2] \begin{bmatrix} x \\ y \\ z \end{bmatrix} = u_2$$

$$[a_3 \quad b_3 \quad c_3] \begin{bmatrix} x \\ y \\ z \end{bmatrix} = u_3$$

$$[a_4 \quad b_4 \quad c_4] \begin{bmatrix} x \\ y \\ z \end{bmatrix} = u_4.$$

It would be useful if we could combine these four equations into one matrix equation. This seems reasonable as all the left-hand sides have a 'common factor' in $\begin{bmatrix} x \\ y \\ z \end{bmatrix}$. Let's do so and then study the consequences.

$$\begin{bmatrix} a_1 & b_1 & c_1 \\ a_2 & b_2 & c_2 \\ a_3 & b_3 & c_3 \\ a_4 & b_4 & c_4 \end{bmatrix} \begin{bmatrix} x \\ y \\ z \end{bmatrix} = \begin{bmatrix} u_1 \\ u_2 \\ u_3 \\ u_4 \end{bmatrix}.$$

Note that the four row vectors each with 3 entries have formed a 4×3 matrix, and the four right-hand side numbers have formed a 4×1 matrix, i.e., a column vector. Now this equation only means what we want it to mean if we can read it row by row, i.e., the first row of the matrix times $\begin{bmatrix} x \\ y \\ z \end{bmatrix}$ equals u_1, the second row of the matrix times $\begin{bmatrix} x \\ y \\ z \end{bmatrix}$ equals u_2 and so on. Notice too that the matrix must have three columns to match the three entries in $\begin{bmatrix} x \\ y \\ z \end{bmatrix}$ for the inner products to be defined, and the number of rows in the matrix must equal the number of entries in the right-hand side vector. In general we *define* the product of an $m \times n$ matrix and an $n \times 1$ column vector to be the $m \times 1$ column vector whose *i*th entry $(i = 1(1)m)$ is obtained as the inner product of the *i*th row of the matrix with the given $n \times 1$ column vector.

Consider therefore the product $\mathbf{AX}$ where $\mathbf{A}$ is $m \times n$ and $\mathbf{X}$ is $n \times 1$ and let $\mathbf{A}_1, \mathbf{A}_2, \ldots, \mathbf{A}_m$ be the m rows (i.e., row vectors or $1 \times n$ matrices) of $\mathbf{A}$. Then the above definition says

$$\mathbf{AX} = \begin{bmatrix} \mathbf{A}_1 \\ \mathbf{A}_2 \\ \cdot \\ \cdot \\ \cdot \\ \mathbf{A}_m \end{bmatrix} \mathbf{X} = \begin{bmatrix} \mathbf{A}_1\mathbf{X} \\ \mathbf{A}_2\mathbf{X} \\ \cdot \\ \cdot \\ \cdot \\ \mathbf{A}_m\mathbf{X} \end{bmatrix} \qquad [7\text{-}4]$$

where each $\mathbf{A}_i\mathbf{X}$ is an inner product and hence just a real number.

EXAMPLE 7.4

(a) $\begin{bmatrix} 1 & 2 & -4 \\ 6 & -3 & 5 \end{bmatrix} \begin{bmatrix} 1 \\ -1 \\ 2 \end{bmatrix} = \begin{bmatrix} (1)(1) + (2)(-1) + (-4)(2) \\ (6)(1) + (-3)(-1) + (5)(2) \end{bmatrix}$

$$= \begin{bmatrix} -9 \\ 19 \end{bmatrix}.$$

Note the structure of these matrices. A 2×3 matrix times a 3×1 vector gives a 2×1 vector.

(b) The pair of linear equations

$$3x + 2y = 5$$
$$-x + y = -2$$

can be written as one matrix equation thus

$$\begin{bmatrix} 3 & 2 \\ -1 & 1 \end{bmatrix} \begin{bmatrix} x \\ y \end{bmatrix} = \begin{bmatrix} 5 \\ -2 \end{bmatrix}.$$

(Multiply out the left-hand side as a check.)

EXERCISE 7.3

Form these products.

(a) $\begin{bmatrix} 1 & 2 & 3 \end{bmatrix} \begin{bmatrix} -1 \\ 2 \\ -1 \end{bmatrix}$ (b) $\begin{bmatrix} 1 & 2 & 3 \\ 2 & 0 & -3 \end{bmatrix} \begin{bmatrix} -1 \\ 2 \\ -1 \end{bmatrix}$

(c) $\begin{bmatrix} 1 & 2 & 3 \\ 4 & 5 & 6 \\ 7 & 8 & 9 \end{bmatrix} \begin{bmatrix} 3 \\ -2 \\ 1 \end{bmatrix}$ (d) $\begin{bmatrix} 5 & 2 \\ 1 & 0 \\ -1 & 1 \end{bmatrix} \begin{bmatrix} 6 \\ -2 \end{bmatrix}$

(e) $\begin{bmatrix} 1 & 0 & 1 & 1 \\ 0 & 0 & 1 & 0 \\ 1 & 0 & 0 & 1 \\ 0 & 1 & 0 & 0 \end{bmatrix} \begin{bmatrix} 2 \\ -1 \\ 3 \\ -4 \end{bmatrix}.$

EXERCISE 7.4

Write these sets of equations in matrix form.

(a) $2x + 10y = 1$

(b) $5x - y = 0$
$\quad -3x + y = 4$

(c) $x_1 + 2x_2 + 3x_3 = 2$

$\quad\;\; 4x_1 + 5x_2 + 6x_3 = 8$

$\quad\;\; 7x_1 + 8x_2 + 9x_3 = 14$

(d) $5a + \;\; b = \;\; 1$

$\quad\;\; 2a - \;\; b = \;\; 2$

$\quad\quad\; a + 3b = \;\; 0$

$\quad -a - 3b = -6.$

If you compare (c) with part (c) of Exercise 7.3 you should be able to give solutions to the equations.

We observe that in each case in Exercise 7.4 the system of linear equations has a matrix representation of the form

$$AX = B$$

where **A** is a given matrix, **B** is a given column vector and **X** is a column vector containing the unknown quantities. If **A** is $m \times n$ then we have m equations in n unknowns, so that **B** will have m elements and **X** will have n elements.

Sometimes vectors are written using lower case letters, so that we may refer to the system of linear equations as $Ax = b$.

As far as this course is concerned it is the use of matrices in connection with linear equations that is of most interest.

7.6   MATRIX MULTIPLICATION – GENERAL RULES

We have defined the product of a row vector and a column vector, and of a matrix and a column vector. One more step gives us the general product of two matrices.

Consider matrices **A** and **B** where **A** is $m \times n$ and **B** is $n \times p$ and let **B** have columns (column vectors) $B_1, B_2, \ldots, B_p$ each with n elements. Then each of the p products AB_j is defined as in the previous section. We *define* the product **AB** to be the $m \times p$ matrix

$$AB = \begin{bmatrix} AB_1 & AB_2 & \ldots & AB_p \end{bmatrix}. \qquad\qquad [7\text{-}5]$$

Each product AB_j is a column with m elements.

EXAMPLE 7.5   Consider

$$\begin{bmatrix} 2 & 1 \\ -1 & 3 \\ 0 & 4 \end{bmatrix} \begin{bmatrix} 1 & -2 \\ 5 & 2 \end{bmatrix}$$

Now

$$\begin{bmatrix} 2 & 1 \\ -1 & 3 \\ 0 & 4 \end{bmatrix} \begin{bmatrix} 1 \\ 5 \end{bmatrix} = \begin{bmatrix} 7 \\ 14 \\ 20 \end{bmatrix}$$

and

$$\begin{bmatrix} 2 & 1 \\ -1 & 3 \\ 0 & 4 \end{bmatrix} \begin{bmatrix} -2 \\ 2 \end{bmatrix} = \begin{bmatrix} -2 \\ 8 \\ 8 \end{bmatrix}.$$

Hence

$$\begin{bmatrix} 2 & 1 \\ -1 & 3 \\ 0 & 4 \end{bmatrix} \begin{bmatrix} 1 & -2 \\ 5 & 2 \end{bmatrix} = \begin{bmatrix} 7 & -2 \\ 14 & 8 \\ 20 & 8 \end{bmatrix}.$$

We can think of this product in terms of inner products. If $\mathbf{A}$ had rows $\mathbf{A}_1, \mathbf{A}_2, \ldots, \mathbf{A}_m$ then each column $\mathbf{AB}_j$ has the form $\begin{bmatrix} \mathbf{A}_1\mathbf{B}_j \\ \mathbf{A}_2\mathbf{B}_j \\ \cdot \\ \cdot \\ \cdot \\ \mathbf{A}_m\mathbf{B}_j \end{bmatrix}$ with each

entry an inner product and hence a single number. Therefore $\mathbf{AB}$ could be defined by

$$\begin{bmatrix} \mathbf{A}_1\mathbf{B}_1 & \mathbf{A}_1\mathbf{B}_2 & \mathbf{A}_1\mathbf{B}_3 & \ldots & \mathbf{A}_1\mathbf{B}_p \\ \mathbf{A}_2\mathbf{B}_1 & \mathbf{A}_2\mathbf{B}_2 & \mathbf{A}_2\mathbf{B}_3 & \ldots & \mathbf{A}_2\mathbf{B}_p \\ \cdot & \cdot & \cdot & & \cdot \\ \cdot & \cdot & \cdot & & \cdot \\ \cdot & \cdot & \cdot & & \cdot \\ \mathbf{A}_m\mathbf{B}_1 & \mathbf{A}_m\mathbf{B}_2 & \mathbf{A}_m\mathbf{B}_3 & \ldots & \mathbf{A}_m\mathbf{B}_p \end{bmatrix}. \qquad [7\text{-}6]$$

Notice that each entry is just a number and therefore that $\mathbf{AB}$ is $m \times p$ as it should be.

Well that's all very difficult — another example should help to muddy the waters completely!

EXAMPLE 7.6

$$\begin{bmatrix} 2 & 3 \\ \hline -1 & 4 \\ \hline 6 & -7 \end{bmatrix} \begin{bmatrix} 1 & \vdots & 2 & \vdots & 3 \\ 4 & \vdots & 5 & \vdots & 6 \end{bmatrix} = \begin{bmatrix} \mathbf{A}_1 \\ \mathbf{A}_2 \\ \mathbf{A}_3 \end{bmatrix} \begin{bmatrix} \mathbf{B}_1 & \mathbf{B}_2 & \mathbf{B}_3 \end{bmatrix}$$

$$= \begin{bmatrix} A_1B_1 & A_1B_2 & A_1B_3 \\ A_2B_1 & A_2B_2 & A_2B_3 \\ A_3B_1 & A_3B_2 & A_3B_3 \end{bmatrix} = \begin{bmatrix} 14 & 19 & 24 \\ 15 & 18 & 21 \\ -22 & -23 & -24 \end{bmatrix}.$$

You should be able to see by now that the conformability condition for multiplication is as follows. **AB** *exists if the number of columns in* **A** *equals the number of rows in* **B**. (In Example 7.6 **A** had 2 columns and **B** had 2 rows.) This is because the elements of **AB** have the inner product form A_iB_j, where A_i is a row of **A** and B_j is a column of **B**. These can only be multiplied together if they have the same number of elements.

In general, as indicated in our definition, we can multiply an $m \times n$ matrix into an $n \times p$ matrix. The result will be an $m \times p$ matrix. As further examples, our matrix/vector product was an $m \times n$ matrix times an $n \times 1$ column vector, giving an $m \times 1$ column vector. And our general inner product was of a $1 \times n$ row vector times an $n \times 1$ column vector giving a 1×1 matrix! (i.e., a single number). Here are two other interesting possibilities.

EXAMPLE 7.7   A $1 \times n$ row vector times an $n \times p$ matrix gives a $1 \times p$ row vector. For example,

$$\begin{bmatrix} 1 & 2 & 3 \end{bmatrix} \begin{bmatrix} 1 & 2 \\ 0 & 2 \\ -1 & 3 \end{bmatrix} = \begin{bmatrix} -2 & 15 \end{bmatrix}.$$

An $n \times 1$ column vector times a $1 \times n$ row vector gives an $n \times n$ matrix. For example,

$$\begin{bmatrix} 4 \\ 5 \\ 6 \end{bmatrix} \begin{bmatrix} 1 & 2 & 3 \end{bmatrix} = \begin{bmatrix} 4 & 8 & 12 \\ 5 & 10 & 15 \\ 6 & 12 & 18 \end{bmatrix}.$$

Rather a contrast with the apparently similar inner product! This is an example of the fact that in matrix multiplication **AB** is not necessarily the same as **BA** (even if both exist).

EXERCISE 7.5

Evaluate the matrix products **AB** and **BA**, where possible, given that

(a) $A = \begin{bmatrix} 5 & -2 & 1 \\ 4 & 1 & -3 \end{bmatrix}$, $B = \begin{bmatrix} 2 & -1 \\ 3 & 4 \\ 5 & -2 \end{bmatrix}$

(b) $\mathbf{A} = \begin{bmatrix} 1 & 2 \\ 3 & -1 \end{bmatrix}$, $\qquad \mathbf{B} = \begin{bmatrix} 1 & 2 & 3 & 4 \\ 5 & 6 & 7 & 8 \end{bmatrix}$

(c) $\mathbf{A} = \begin{bmatrix} 1 & 2 & 3 \\ 4 & 5 & 6 \\ 7 & 8 & 9 \end{bmatrix}$, $\qquad \mathbf{B} = \begin{bmatrix} -3 & 4 & -5 \\ 6 & -7 & 8 \\ -9 & 10 & -11 \end{bmatrix}$

(d) $\mathbf{A} = \begin{bmatrix} 1 \\ 2 \\ 3 \end{bmatrix}$, $\qquad \mathbf{B} = \begin{bmatrix} 5 & 6 & 7 \end{bmatrix}$.

Corresponding to the zero matrix in addition and subtraction there is a unit matrix for multiplication. Consider for example

$$\begin{bmatrix} 1 & 2 & 3 \\ 4 & 5 & 6 \end{bmatrix} \begin{bmatrix} 1 & 0 & 0 \\ 0 & 1 & 0 \\ 0 & 0 & 1 \end{bmatrix} = \begin{bmatrix} 1 & 2 & 3 \\ 4 & 5 & 6 \end{bmatrix}.$$

(Check it for yourself.) Similarly

$$\begin{bmatrix} 1 & 0 \\ 0 & 1 \end{bmatrix} \begin{bmatrix} 1 & 2 & 3 \\ 4 & 5 & 6 \end{bmatrix} = \begin{bmatrix} 1 & 2 & 3 \\ 4 & 5 & 6 \end{bmatrix}.$$

$\begin{bmatrix} 1 & 0 & 0 \\ 0 & 1 & 0 \\ 0 & 0 & 1 \end{bmatrix}$ is called the 3×3 unit matrix and $\begin{bmatrix} 1 & 0 \\ 0 & 1 \end{bmatrix}$ the 2×2 unit matrix. They are both examples of *square* matrices. The general $n \times n$ unit matrix is an obvious extension. The unit matrices clearly play a role similar to the number 1 in ordinary algebra. In general if $\mathbf{I}_k$ indicates the $k \times k$ unit matrix for $k = 1, 2, 3, \ldots$ and $\mathbf{A}$ is an $m \times n$ matrix, then

$$\mathbf{I}_m \mathbf{A} = \mathbf{A} \mathbf{I}_n = \mathbf{A}. \qquad\qquad [7\text{-}7]$$

There are four other important properties of matrix multiplication to note.

(a) $\mathbf{A}(\mathbf{BC}) = (\mathbf{AB})\mathbf{C}$ whenever the products exist. This *associative* law allows us to multiply together more than two matrices.

(b) $\mathbf{A}(\mathbf{B} + \mathbf{C}) = \mathbf{AB} + \mathbf{AC}$ and

$\qquad (\mathbf{B} + \mathbf{C})\mathbf{A} = \mathbf{BA} + \mathbf{CA}$

whenever the matrix products exist. These *distributive* laws allow us to mix addition and multiplication.

(c) $\mathbf{AB}$ and $\mathbf{BA}$ are not necessarily equal. In fact one may exist while the other does not (if $\mathbf{A}$ is 3×2 and $\mathbf{B}$ is 2×2 for example — why?).

(d) **AB** = **CB** does not necessarily mean that **A** = **C**. Similarly **AB** = **0** does not necessarily mean that **A** = **0** or **B** = **0**.

Finally in this section we look at a bit of useful notation. Sometimes we let the element in the ith row and the jth column of **A** be $a_{i,j}$ or a_{ij} (often the comma is dropped out of sheer idleness!) The matrix **A** can be written as **A** = $[a_{ij}]$, and similarly **B** = $[b_{ij}]$. Looking at the matrix product **C** = $[c_{ij}]$ = **AB** we have $c_{ij} = A_i B_j$. This can be stated in words as 'the element in the ith row and the jth column of **AB** equals the inner product of the ith row of **A** with the jth column of **B**'. We can actually give a formula

$$c_{ij} = \sum_{k=1}^{n} a_{ik} b_{kj}$$

where n is the number of columns of **A** and rows of **B**.

Similarly if **C** = **A** + **B** then

$$c_{ij} = a_{ij} + b_{ij}$$

for all i and j, and if **C** = k**A**, then

$$c_{ij} = ka_{ij}$$

for all i and j. Lastly the equations [7-1] will be the same as **Ax** = **b** where

$$\mathbf{x} = \begin{bmatrix} x_1 \\ x_2 \\ \cdot \\ \cdot \\ \cdot \\ x_n \end{bmatrix} \quad \text{and} \quad \mathbf{b} = \begin{bmatrix} b_1 \\ b_2 \\ \cdot \\ \cdot \\ \cdot \\ b_n \end{bmatrix}.$$

7.7 TRANSPOSITION OF MATRICES

Writing down a column vector as $\begin{bmatrix} x_1 \\ x_2 \\ x_3 \end{bmatrix}$ is awkward in text and there is a useful notation which enables us to use a row vector instead. This is

$$\begin{bmatrix} x_1 & x_2 & x_3 \end{bmatrix}^T = \begin{bmatrix} x_1 \\ x_2 \\ x_3 \end{bmatrix}.$$

The T denotes that we are to *transpose* the vector.

There are other uses for transposition as well as this, particularly for the transposition of matrices. The rule for transposing a matrix is 'take the ith row and write it as the ith column, for each i'. This means that if A is $m \times n$ then A^T is $n \times m$.

EXAMPLE 7.8

$$\begin{bmatrix} 1 & 2 & 3 & 4 \\ 5 & 6 & 7 & 8 \\ 9 & 10 & 11 & 12 \end{bmatrix}^T = \begin{bmatrix} 1 & 5 & 9 \\ 2 & 6 & 10 \\ 3 & 7 & 11 \\ 4 & 8 & 12 \end{bmatrix}.$$

7.8 INVERSION OF MATRICES

When we have a single linear equation in one unknown, such as $7x = 3$, we solve it in effect by multiplying both sides by the reciprocal (inverse) of 7, i.e.

$$\tfrac{1}{7}(7x) = \tfrac{1}{7}(3)$$

$$\Rightarrow \qquad x = \tfrac{3}{7}.$$

We use the properties $(\tfrac{1}{7})(7) = 1$ and $(\tfrac{1}{7})(7x) = (\tfrac{1}{7} \times 7)x$.

This approach can be extended to matrices under certain circumstances. Suppose the matrix A is square ($n \times n$, say) and suppose that the system of linear equations $Ax = b$ has a unique solution for each and any b (this is not always the case as we shall see in Lesson 8). Then it is known that we can define the inverse of A, written A^{-1} and spoken 'A inverse', by the property

$$A^{-1}A = AA^{-1} = I_n \qquad\qquad [7\text{-}8]$$

where I_n is the $n \times n$ unit matrix. Then, premultiplying both sides of $Ax = b$ by A^{-1}, we have

$$A^{-1}(Ax) = A^{-1}b.$$

Using the associative law, we find that this becomes

$$(A^{-1}A)x = A^{-1}b$$

i.e. $\qquad I_n x = A^{-1}b$

i.e. $\qquad x = A^{-1}b \qquad\qquad [7\text{-}9]$

since I_n is the unit matrix.

There are various ways of finding inverse matrices but we must stress that we are not suggesting finding A^{-1} as a method for finding the solution x of the equation $Ax = b$. Rather Equation [7-9] is a theoretical expression for the solution, which enables us to do some further theoretical analysis.

EXAMPLE 7.9   Consider the equations

$$3x + 2y = 1$$
$$-x + 4y = -5.$$

They can be written as

$$\begin{bmatrix} 3 & 2 \\ -1 & 4 \end{bmatrix} \begin{bmatrix} x \\ y \end{bmatrix} = \begin{bmatrix} 1 \\ -5 \end{bmatrix}. \qquad [7\text{-}10]$$

Now in fact

$$\begin{bmatrix} \frac{4}{14} & -\frac{2}{14} \\ \frac{1}{14} & \frac{3}{14} \end{bmatrix} \begin{bmatrix} 3 & 2 \\ -1 & 4 \end{bmatrix} = \begin{bmatrix} 1 & 0 \\ 0 & 1 \end{bmatrix}$$

so that if we premultiply both sides of Equation [7-10] by $\begin{bmatrix} \frac{4}{14} & -\frac{2}{14} \\ \frac{1}{14} & \frac{3}{14} \end{bmatrix}$, we obtain

$$\begin{bmatrix} 1 & 0 \\ 0 & 1 \end{bmatrix} \begin{bmatrix} x \\ y \end{bmatrix} = \begin{bmatrix} \frac{4}{14} & -\frac{2}{14} \\ \frac{1}{14} & \frac{3}{14} \end{bmatrix} \begin{bmatrix} 1 \\ -5 \end{bmatrix} = \begin{bmatrix} \frac{14}{14} \\ -\frac{14}{14} \end{bmatrix}$$

i.e. $\begin{bmatrix} x & y \end{bmatrix}^T = \begin{bmatrix} 1 & -1 \end{bmatrix}^T$

which is the obvious solution to the equations. (Do check all this — it might be wrong!)

You may ask how the inverse matrix $\begin{bmatrix} \frac{4}{14} & -\frac{2}{14} \\ \frac{1}{14} & \frac{3}{14} \end{bmatrix}$ was found. One method for obtaining inverses is given in Lesson 8, others can be found in books on matrix algebra — see the Further Reading section

ANSWERS TO EXERCISES

7.1 (a) $\begin{bmatrix} 6 & 2 \\ 5 & 5 \\ -1 & -3 \end{bmatrix}$, $\begin{bmatrix} 2 & 4 \\ -1 & -3 \\ 13 & -7 \end{bmatrix}$.

(b) $\begin{bmatrix} -2 & -2 & -2 \\ -2 & -2 & -2 \\ -2 & 7 & 7 \end{bmatrix}$, $\begin{bmatrix} 4 & 6 & 8 \\ 10 & 12 & 14 \\ 16 & 9 & 11 \end{bmatrix}$

(c) $\begin{bmatrix} 7 \\ 5 \\ -4 \\ 7 \end{bmatrix}$, $\begin{bmatrix} -1 \\ -9 \\ 6 \\ 3 \end{bmatrix}$.

7.2 (a) $\begin{bmatrix} -17 & -18 \\ 25 & -7 \end{bmatrix}$ (b) $\begin{bmatrix} 23 & 0 \\ -1 & 19 \end{bmatrix}$ (c) $\begin{bmatrix} 11 & 13 \\ -18 & 4 \end{bmatrix}$.

7.3 (a) $\begin{bmatrix} 0 \end{bmatrix}$ (b) $\begin{bmatrix} 0 \\ 1 \end{bmatrix}$ (c) $\begin{bmatrix} 2 \\ 8 \\ 14 \end{bmatrix}$ (d) $\begin{bmatrix} 26 \\ 6 \\ -8 \end{bmatrix}$ (e) $\begin{bmatrix} 1 \\ 3 \\ -2 \\ -1 \end{bmatrix}$.

7.4 (a) $\begin{bmatrix} 2 & 10 \end{bmatrix} \begin{bmatrix} x \\ y \end{bmatrix} = \begin{bmatrix} 1 \end{bmatrix}$ (b) $\begin{bmatrix} 5 & -1 \\ -3 & 1 \end{bmatrix} \begin{bmatrix} x \\ y \end{bmatrix} = \begin{bmatrix} 0 \\ 4 \end{bmatrix}$

(c) $\begin{bmatrix} 1 & 2 & 3 \\ 4 & 5 & 6 \\ 7 & 8 & 9 \end{bmatrix} \begin{bmatrix} x_1 \\ x_2 \\ x_3 \end{bmatrix} = \begin{bmatrix} 2 \\ 8 \\ 14 \end{bmatrix}$ (d) $\begin{bmatrix} 5 & 1 \\ 2 & -1 \\ 1 & 3 \\ -1 & -3 \end{bmatrix} \begin{bmatrix} a \\ b \end{bmatrix} = \begin{bmatrix} 1 \\ 2 \\ 0 \\ -6 \end{bmatrix}$.

7.5 (a) $\mathbf{AB} = \begin{bmatrix} 10-6+5 & -5-8-2 \\ 8+3-15 & -4+4+6 \end{bmatrix} = \begin{bmatrix} 9 & -15 \\ -4 & 6 \end{bmatrix}$

$\mathbf{BA} = \begin{bmatrix} 10-4 & -4-1 & 2+3 \\ 15+16 & -6+4 & 3-12 \\ 25-8 & -10-2 & 5+6 \end{bmatrix} = \begin{bmatrix} 6 & -5 & 5 \\ 31 & -2 & -9 \\ 17 & -12 & 11 \end{bmatrix}$.

(b) $\mathbf{AB} = \begin{bmatrix} 11 & 14 & 17 & 20 \\ -2 & 0 & 2 & 4 \end{bmatrix}$

$\mathbf{BA}$ is not defined.

(c) $\mathbf{AB} = \begin{bmatrix} -18 & 20 & -22 \\ -36 & 41 & -46 \\ -54 & 62 & -70 \end{bmatrix}$ $\mathbf{BA} = \begin{bmatrix} -22 & -26 & -30 \\ 34 & 41 & 48 \\ -46 & -56 & -66 \end{bmatrix}$.

(d) $\mathbf{AB} = \begin{bmatrix} 5 & 6 & 7 \\ 10 & 12 & 14 \\ 15 & 18 & 21 \end{bmatrix}$, $\mathbf{BA} = \begin{bmatrix} 38 \end{bmatrix}$.

FURTHER READING

For a fuller account of elementary matrix algebra you could consult Brand and Sherlock, or Ayres.

LESSON 7 — COMPREHENSION TEST

1. Three matrices **A**, **B** and **C** are 2×3, 2×5 and 3×2 respectively. State all the possible products of two of **A**, **B** and **C**.

2. Is the transpose of a row vector (a) another row vector or (b) a column vector?

3. Let $\mathbf{Ax} = \mathbf{b}$ be a system of linear equations, where **A** is an $n \times m$ matrix. How many linear equations are there and in how many unknowns?

4. Describe the $n \times n$ unit matrix.

5. State the defining property of a matrix inverse.

6. (a) Can a non-square matrix have an inverse?

   (b) Do all square matrices have inverses? If not, give an example of one that has no inverse.

ANSWERS

1 AC, CA, CB.

2 (b).

3 n equations; m unknowns.

4 All diagonal elements = 1; all others = 0.

5 $\mathbf{AA}^{-1} = \mathbf{A}^{-1}\mathbf{A} = \mathbf{I}$.

6 (a) No, (b) No, e.g. any null matrix.

LESSON 7 — SUPPLEMENTARY EXERCISES

For practice at addition, subtraction, scalar and matrix multiplication see any book on elementary matrix algebra.

7.6. Given that

$$\mathbf{A} = \begin{bmatrix} 1 & 2 & -1 \\ 0 & 1 & 2 \\ 3 & -3 & 0 \end{bmatrix}$$

verify that

$$A^{-1} = \tfrac{1}{21} \begin{bmatrix} 6 & 3 & 5 \\ 6 & 3 & -2 \\ -3 & 9 & 1 \end{bmatrix}.$$

Also verify that

$$A \begin{bmatrix} 1 \\ 4 \\ -1 \end{bmatrix} = \begin{bmatrix} 10 \\ 2 \\ -9 \end{bmatrix} \quad \text{and} \quad A^{-1} \begin{bmatrix} 10 \\ 2 \\ -9 \end{bmatrix} = \begin{bmatrix} 1 \\ 4 \\ -1 \end{bmatrix}.$$

7.7. For the matrix A in Exercise 7.6, write down A^T, $(A^T)^T$, and $(A^{-1})^T$. Verify that

$$(A^T)(A^{-1})^T = I$$

which illustrates the general fact that

$$(A^{-1})^T = (A^T)^{-1}$$

for any invertible matrix A.

7.8. Given that $x = \begin{bmatrix} x_1 & x_2 & x_3 \end{bmatrix}^T$ write the equations

$$x_1 = (-2x_2 + 4x_3 + 1)/15$$
$$x_2 = (2x_1 - x_3 + 2)/10$$
$$x_3 = (x_1 + 2x_2 - 3)/12$$

in the matrix form

$$x = Mx + c$$

giving the matrix M and the vector c (both of which are independent of x).

Lesson 8 **Elimination Methods**

OBJECTIVES

At the end of this lesson you should

(a)   understand row equivalence of matrices and know the valid row operations;

(b)   be able to solve a system of linear equations by elimination and back-substitution in the augmented matrix;

(c)   understand that linear systems may have either a unique solution, or no solution, or an infinity of solutions;

(d)   understand the terms *consistent*, *inconsistent*, *dependent* and *independent*, as applied to linear systems;

(e)   be able to invert a matrix using row operations.

CONTENTS

8.1   Solving sets of linear equations
8.2   Row operations and row equivalent matrices
8.3   Elimination and back substitution
8.4   Inconsistent and dependent sets of equations
8.5   Matrix inversion

Answers to exercises. Further reading.
Comprehension test and answers. Supplementary exercises.

8.1   SOLVING SETS OF LINEAR EQUATIONS

One elementary method of solving simultaneous linear equations usually taught in schools consists essentially in solving one equation for one variable in terms of the others and substituting the resultant expression in the other equations. For the equations

$$3x + 2y = 1$$
$$-x + 4y = -5$$

[8-1]

we may find $x = 4y + 5$ from the second equation and hence replace x in the first equation by $4y + 5$, giving

$$3(4y + 5) + 2y = 1$$
$$\therefore \qquad 14y = -14$$
$$\therefore \qquad y = -1.$$

x can then be found from

$$x = 4y + 5 = -4 + 5 = 1.$$

The solution can then be checked by substitution in the original equations.

This approach has the disadvantage that it does not extend easily to the solution of more than two equations.

EXERCISE 8.1

Try to solve

$$x + y - z = -1$$
$$2x - 3y + 2z = 4$$
$$3x + 2y - 4z = -9$$

by the substitution method.

Another elementary method involves subtracting and adding multiples of the given equations to *eliminate* one variable at a time, repeating the process until an equation in only one variable is found. Finding the other variables is then fairly easy — if you can find the appropriate equation to use.

For the Equations [8-1] we proceed typically as follows:

$$3x + 2y = 1$$
$$-3x + 12y = -15.$$

[8-2]

Adding we have

$$14y = -14$$
$$\therefore \qquad y = -1.$$

From the first equation

$$3x - 2 = 1$$
$$\therefore \qquad 3x = 3$$
$$\therefore \qquad x = 1.$$

We ought to check the solution by substitution into the left-hand side of the second equation, giving

$$-1 - 4 = -5$$

as required.

In this lesson we will use this second method in a systematic way which will make it much easier to use than the first method, especially for the solution of large numbers of equations. It will tell us whether solutions exist (they don't always), and if so, what form they take. It also enables us to analyse whether there is anything peculiar about the original problem, or, as we say, to see whether the problem is *ill-conditioned* (see Lesson 9).

8.2 ROW OPERATIONS AND ROW-EQUIVALENT MATRICES

Reconsider equations [8-1]

$$\left. \begin{array}{r} 3x + 2y = 1 \\ -x + 4y = -5. \end{array} \right\}$$

Notice that the operations we performed on them (multiplying the second equation by 3 and adding the result to the first equation) were performed essentially on the coefficients only. Put another way, the equations when written in matrix form

$$\begin{bmatrix} 3 & 2 \\ -1 & 4 \end{bmatrix} \begin{bmatrix} x \\ y \end{bmatrix} = \begin{bmatrix} 1 \\ -5 \end{bmatrix}$$

are entirely represented by the *coefficient matrix* $\begin{bmatrix} 3 & 2 \\ -1 & 4 \end{bmatrix}$ and the *right-hand-side vector* $\begin{bmatrix} 1 \\ -5 \end{bmatrix}$. If we bring them together to form the so-called *augmented matrix*

$$\begin{bmatrix} 3 & 2 & \vdots & 1 \\ -1 & 4 & \vdots & -5 \end{bmatrix}$$

then we should be able to find the solutions by manipulating just the entries in this matrix.

In fact we could multiply the second row by 3, giving

$$\begin{bmatrix} 3 & 2 & \vdots & 1 \\ -3 & 12 & \vdots & -15 \end{bmatrix}$$

add row 1 to row 2, giving

$$\begin{bmatrix} 3 & 2 & \vdots & 1 \\ 0 & 14 & \vdots & -14 \end{bmatrix}$$

and divide row 2 by 14, giving

$$\begin{bmatrix} 3 & 2 & \vdots & 1 \\ 0 & 1 & \vdots & -1 \end{bmatrix}.$$

Interpreting this last augmented matrix as representing the equations

$$3x + 2y = 1$$
$$y = -1$$

gives the solution $y = -1$, $x = (1 - 2y)/3 = (1 + 2)/3 = 1$.

Now consider what we have done from a more general point of view. We have represented the original equations by an augmented matrix of coefficients. We have performed *row operations* on this matrix which are equivalent to the sort of operations that can be performed on linear equations *without changing the solutions*. The successive augmented matrices represent different but equivalent sets of equations, equivalent in the sense that they all have the same solutions. We call such matrices *row-equivalent* and denote the equivalence by $\sim$ ('twiddle'), thus

$$\begin{bmatrix} 3 & 2 & \vdots & 1 \\ -1 & 4 & \vdots & -5 \end{bmatrix} \sim \begin{bmatrix} 3 & 2 & \vdots & 1 \\ -3 & 12 & \vdots & -15 \end{bmatrix}$$

$$\sim \begin{bmatrix} 3 & 2 & \vdots & 1 \\ 0 & 14 & \vdots & -14 \end{bmatrix} \sim \begin{bmatrix} 3 & 2 & \vdots & 1 \\ 0 & 1 & \vdots & -1 \end{bmatrix}.$$

Having obtained a much simpler matrix we interpret it as a set of equations again, our aim being to find a set that can be solved easily.

More succinctly, given a set of linear equations $Ax = b$, write down the augmented matrix $[A \vdots b]$, and perform row operations until a much simpler, row-equivalent matrix is obtained. These questions remain: what row operations? how much simpler? Let's list the row operations we have used so far.

(a) Multiply (or divide) a row by a non-zero constant.

(b) Add (or subtract) one row to (or from) another.

Very often we use a hybrid of these two operations.

(b(i)) Add (or subtract) a multiple of one row to (or from) another.

One other row operation will be needed.

(c) Interchange two rows.

To check that (b(i)) and (c) are valid we should relate them to linear equations: we can add (or subtract) a non-zero multiple of one equation to (or from) another without changing the solutions; we can interchange two equations without changing the solutions.

It is a good idea when performing row operations to make a note of the operations performed. This is done in the following example, using a fairly obvious notation.

EXAMPLE 8.1   Use row operations to solve

$$3x + 3y = -1$$
$$4x - y = 9$$

The augmented matrix is

$$\begin{bmatrix} 3 & 3 & \vdots & -1 \\ 4 & -1 & \vdots & 9 \end{bmatrix} \quad R_1 \leftarrow \tfrac{1}{3}R_1$$

$$\sim \begin{bmatrix} 1 & 1 & \vdots & -\tfrac{1}{3} \\ 4 & -1 & \vdots & 9 \end{bmatrix} \quad R_2 \leftarrow R_2 - 4R_1$$

$$\sim \begin{bmatrix} 1 & 1 & \vdots & -\tfrac{1}{3} \\ 0 & -5 & \vdots & \tfrac{31}{3} \end{bmatrix}$$

$$\Rightarrow \quad x + y = -\tfrac{1}{3}$$
$$y = -31/15$$

i.e., $y = -31/15$ and $x = -\tfrac{1}{3} - y = 26/15$.

EXERCISE 8.2

Use the method of Example 8.1 to solve

$$2x + y = 4$$
$$3x - 2y = -1.$$

So far we have shown by simple examples that row operations on the augmented matrix seems a reasonable solution technique. We haven't given as yet any idea of how to tackle more general problems and, in particular, larger problems.

8.3   ELIMINATION AND BACK-SUBSTITUTION

To apply row operations to solving, say, 5 equations in 5 unknowns, we must have a logical strategy, otherwise we could go round in circles, never really obtaining a much simpler, equivalent system. The most popular strategy is as follows.

(a) Use row operations to obtain a row-equivalent augmented matrix in the form

$$\begin{bmatrix} \times & \times & \times & \cdots & \times & \times & \vdots & \times \\ 0 & \times & \times & \cdots & \times & \times & \vdots & \times \\ 0 & 0 & \times & \cdots & \times & \times & \vdots & \times \\ \cdots & \cdots & \cdots & \cdots & \cdots & \cdots & \vdots & \cdots \\ 0 & 0 & 0 & \cdots & \times & \times & \vdots & \times \\ 0 & 0 & 0 & \cdots & 0 & \times & \vdots & \times \end{bmatrix}$$

in which the left-hand part is in *upper-triangular form*. This is called the *elimination* stage. The zeros are obtained in one column at a time, starting at the left.

(b) This augmented matrix represents equations of the form

$$a_{11}x_1 + a_{12}x_2 + \ldots + a_{1n}x_n = b_1$$
$$a_{22}x_2 + \ldots + a_{2n}x_n = b_2$$
$$\vdots$$

[8-3]

$$a_{n-1,\,n-1}x_{n-1} + a_{n-1,\,n}x_n = b_{n-1}$$
$$a_{n,\,n}x_n = b_n.$$

Solve the last equation to find x_n. Substitute this value of x_n into the last but one equation to find x_{n-1} and so on. This process is called *back-substitution*. Note that it only works if all the diagonal entries $a_{i,i}$ are non-zero. More of this later.

The choice of row operations in the elimination stage is to some extent open. However, we shall see good reason in Lesson 9 to make the choice absolutely rigid, i.e., such that we always use exactly the same operations regardless of the system of equations being solved.

EXAMPLE 8.2   To solve the equations of Exercise 8.1

$$x + y - z = -1$$
$$2x - 3y + 2z = 4$$
$$3x + 2y - 4z = -9.$$

The augmented matrix is

$$\begin{bmatrix} 1 & 1 & -1 & \vdots & -1 \\ 2 & -3 & 2 & \vdots & 4 \\ 3 & 2 & -4 & \vdots & -9 \end{bmatrix} \begin{matrix} \\ R_2 \leftarrow R_2 - 2R_1 \\ R_3 \leftarrow R_3 - 3R_1 \end{matrix}$$

(to obtain zeros in the first column)

$$\sim \begin{bmatrix} 1 & 1 & -1 & \vdots & -1 \\ 0 & -5 & 4 & \vdots & 6 \\ 0 & -1 & -1 & \vdots & -6 \end{bmatrix} \begin{matrix} \\ \\ R_2 \leftrightarrow R_3 \end{matrix}$$

(to make the next elimination easy)

$$\sim \begin{bmatrix} 1 & 1 & -1 & \vdots & -1 \\ 0 & -1 & -1 & \vdots & -6 \\ 0 & -5 & 4 & \vdots & 6 \end{bmatrix} \begin{matrix} \\ \\ R_3 \leftarrow R_3 - 5R_2 \end{matrix}$$

(to obtain the zero in the second column)

$$\sim \begin{bmatrix} 1 & 1 & -1 & \vdots & -1 \\ 0 & -1 & -1 & \vdots & -6 \\ 0 & 0 & 9 & \vdots & 36 \end{bmatrix}.$$

This matrix is now in the form required for back-substitution. Hence

$$\begin{aligned} 9z &= 36 \quad \text{i.e.} \quad z = 4 \\ -y - z &= -6 \quad \text{i.e.} \quad y = 6 - z = 2 \\ x + y - z &= -1 \quad \text{i.e.} \quad x = -1 + z - y = 1. \end{aligned}$$

Notice the use of a row interchange to make the arithmetic of the elimination easier. Without the row interchange we would have been working with fractions. This is worth avoiding when the coefficients are integers but is irrelevant to a computer and is one of the choices that will be excluded in the formal algorithm given in Lesson 9.

EXERCISE 8.3

Use elimination and back-substitution to solve these equations:

$$\begin{aligned} 2x + 3y - z &= 5 \\ x + 4y + 2z &= 15 \\ 4x - y + 3z &= 11. \end{aligned}$$

The method that we have described above is by no means the only one. It is possible, for example, to continue row operations until the augmented matrix has the form

$$\begin{bmatrix} 1 & 0 & 0 & \ldots & 0 & \vdots & \times \\ 0 & 1 & 0 & \ldots & 0 & \vdots & \times \\ 0 & 0 & 1 & \ldots & 0 & \vdots & \times \\ \cdot & \cdot & \cdot & & \cdot & \vdots & \cdot \\ \cdot & \cdot & \cdot & & \cdot & \vdots & \cdot \\ \cdot & \cdot & \cdot & & \cdot & \vdots & \cdot \\ 0 & 0 & 0 & \ldots & 1 & \vdots & \times \end{bmatrix}$$

when the solutions can be read off immediately. We shall not consider this approach, but you may like to try it for yourself on the problems of Example 8.2 and Exercise 8.3.

The method of row operations has a further advantage. If we have two problems

$$\mathbf{Ax} = \mathbf{b} \quad \text{and} \quad \mathbf{Az} = \mathbf{c}$$

where the coefficient matrix $\mathbf{A}$ is the same in both equations we can collapse the two problems into one and solve it using row operations. We consider the augmented matrix

$$\begin{bmatrix} \mathbf{A} & \vdots & \mathbf{b} & \vdots & \mathbf{c} \end{bmatrix}$$

and work as before until $\mathbf{A}$ has been converted into upper triangular form.

EXAMPLE 8.3   If

$$\left. \begin{array}{r} x + y - z = -1 \\ 2x - 3y + 2z = 4 \\ 3x + 2y - 4z = -9 \end{array} \right\}$$

and

$$\left. \begin{array}{r} u + v - w = 3 \\ 2u - 3v + 2w = 19 \\ 3u + 2v - 4w = 8 \end{array} \right\}$$

then

$$\begin{bmatrix} 1 & 1 & -1 \\ 2 & -3 & 2 \\ 3 & 2 & -4 \end{bmatrix} \begin{bmatrix} x \\ y \\ z \end{bmatrix} = \begin{bmatrix} -1 \\ 4 \\ -9 \end{bmatrix}$$

and

$$\begin{bmatrix} 1 & 1 & -1 \\ 2 & -3 & 2 \\ 3 & 2 & -4 \end{bmatrix} \begin{bmatrix} u \\ v \\ w \end{bmatrix} = \begin{bmatrix} 3 \\ 19 \\ 8 \end{bmatrix}.$$

We can write these equations as one matrix equation

$$\begin{bmatrix} 1 & 1 & -1 \\ 2 & -3 & 2 \\ 3 & 2 & -4 \end{bmatrix} \begin{bmatrix} x & u \\ y & v \\ z & w \end{bmatrix} = \begin{bmatrix} -1 & 3 \\ 4 & 19 \\ -9 & 8 \end{bmatrix}.$$

Since the coefficient matrix is identical to that in Example 8.2 the row operations can be identical too, giving this sequence of augmented matrices:

$$\begin{bmatrix} 1 & 1 & -1 & \vdots & -1 & \vdots & 3 \\ 2 & -3 & 2 & \vdots & 4 & \vdots & 19 \\ 3 & 2 & -4 & \vdots & -9 & \vdots & 8 \end{bmatrix}$$

$$\sim \begin{bmatrix} 1 & 1 & -1 & \vdots & -1 & \vdots & 3 \\ 0 & -5 & 4 & \vdots & 6 & \vdots & 13 \\ 0 & -1 & -1 & \vdots & -6 & \vdots & -1 \end{bmatrix}$$

$$\sim \begin{bmatrix} 1 & 1 & -1 & \vdots & -1 & \vdots & 3 \\ 0 & -1 & -1 & \vdots & -6 & \vdots & -1 \\ 0 & -5 & 4 & \vdots & 6 & \vdots & 13 \end{bmatrix}$$

$$\sim \begin{bmatrix} 1 & 1 & -1 & \vdots & -1 & \vdots & 3 \\ 0 & -1 & -1 & \vdots & -6 & \vdots & -1 \\ 0 & 0 & 9 & \vdots & 36 & \vdots & 18 \end{bmatrix}.$$

Hence

$$z = 4 \qquad \text{and} \quad w = 2$$
$$y = 6 - 4 = 2 \qquad \text{and} \quad v = 1 - 2 = -1$$
$$x = -1 - 2 + 4 = 1 \quad \text{and} \quad u = 3 + 1 + 2 = 6.$$

If you don't see where these numbers come from write out the equations represented by the last augmented matrix!

This idea will be used in Lesson 9.

EXERCISE 8.4

Solve the equations $\mathbf{Ax} = \mathbf{b}$ and $\mathbf{Ay} = \mathbf{c}$ where

$$\mathbf{A} = \begin{bmatrix} 2 & 0 & 1 & 0 \\ 1 & 0 & -1 & 2 \\ 0 & 3 & 0 & -1 \\ 1 & 1 & -1 & 0 \end{bmatrix}, \quad \mathbf{b} = \begin{bmatrix} 1 \\ 0 \\ 0 \\ 0 \end{bmatrix}, \quad \mathbf{c} = \begin{bmatrix} 0 \\ 1 \\ 0 \\ 0 \end{bmatrix}.$$

8.4   INCONSISTENT AND DEPENDENT SETS OF EQUATIONS

So far we have chosen examples where nothing has 'gone wrong'. We have had n equations in n unknowns and we have been able to find unique solutions. For $n = 2$ this is the same as saying that the graphs of two linear equations in two unknowns (which are two straight lines of course) meet in a single (unique) point. But straight lines don't always do this. (What about parallel lines?) So perhaps we had better do some more general analysis.

To take the simplest case first, consider the single linear equation

$$ax = b$$

where a and b are constants and x is the unknown.

There are three cases.

CASE 1.1 $a \neq 0$

Then we have the *unique solution* $x = b/a$ for any $b \in \mathbb{R}$.

CASE 1.2 $a = b = 0$

Then x can take any real value since $0x = 0$ for any such x. In fact we have an *infinity of solutions*.

CASE 1.3 $a = 0, b \neq 0$

Then $x = b/a$ has no meaning, i.e., there is no $x \in \mathbb{R}$ such that $0x \neq 0$. There are *no solutions* to $ax = b$ in this case.

What happens in the case of 2 equations in 2 unknowns? We shall use examples here, and leave the general case to an exercise.

CASE 2.1

Consider

$$x + y = 3$$
$$x - y = 1.$$

These have the unique solution $x = 2$, $y = 1$. Using the augmented matrix and row operations gives

$$\begin{bmatrix} 1 & 1 & \vdots & 3 \\ 1 & -1 & \vdots & 1 \end{bmatrix} \quad R_2 \leftarrow R_2 - R_1 \qquad \sim \begin{bmatrix} 1 & 1 & \vdots & 3 \\ 0 & -2 & \vdots & -2 \end{bmatrix},$$

i.e. $-2y = -2$; hence $y = 1$; hence $x = 3 - y = 2$.

In coordinate geometry terms we have found the point of intersection of two non-parallel lines.

CASE 2.2

Consider

$$x + y = 3$$
$$2x + 2y = 6.$$

We observe that the second equation is a multiple of the first and hence adds no new information. We have in fact an infinity of solutions of the form $y = \alpha$, $x = 3 - \alpha$, where α can be any number in $\mathbb{R}$, i.e., all pairs (x, y) satisfying $x + y = 3$. If we use row operations on the augmented matrix the following is obtained:

$$\begin{bmatrix} 1 & 1 & \vdots & 3 \\ 2 & 2 & \vdots & 6 \end{bmatrix} \quad R_2 \leftarrow R_2 - 2R_1 \qquad \sim \begin{bmatrix} 1 & 1 & \vdots & 3 \\ 0 & 0 & \vdots & 0 \end{bmatrix},$$

i.e., $x + y = 3$ is all we know. In coordinate geometry terms the two given equations represent the same line and hence 'intersect' at all points of that line.

CASE 2.3

Consider

$$x + y = 3$$
$$2x + 2y = 7.$$

These equations have no solution, because $x + y = 3$ implies $2x + 2y = 6$ *for any x and y*; hence $2x + 2y = 7$ is impossible. Row operations on the augmented matrix reveal this as follows:

$$\begin{bmatrix} 1 & 1 & \vdots & 3 \\ 2 & 2 & \vdots & 7 \end{bmatrix} \quad \begin{matrix} \\ R_2 \leftarrow R_2 - 2R_1 \end{matrix} \quad \sim \begin{bmatrix} 1 & 1 & \vdots & 3 \\ 0 & 0 & \vdots & 1 \end{bmatrix},$$

i.e., $0x + 0y = 1$, which has no solutions (x, y).

In coordinate geometry terms the two equations represent parallel but non-coincident lines.

In this section we have given a brief introduction to the theory underlying systems of linear equations. The section is not essential to this text since we shall be concerned almost exclusively with n equations in n unknowns having a unique solution. Nevertheless the ideas above will be useful when we come to discuss (in Lesson 9) the intriguing idea of a system with a solution which is 'only just' unique.

It is also interesting to note that, given any system of linear equations, it is possible (in theory at least — i.e., with exact arithmetic) to discover whether there are solutions, and if so what they are, simply by applying row operations.

EXAMPLE 8.4   Describe the kinds of solutions to the system

$$\begin{bmatrix} 1 & 2 & 3 \\ -1 & 0 & 1 \\ \alpha & 2 & -3 \end{bmatrix} \begin{bmatrix} x_1 \\ x_2 \\ x_3 \end{bmatrix} = \begin{bmatrix} 4 \\ -1 \\ \beta \end{bmatrix}$$

that can be obtained by varying $\alpha, \beta \in \mathbb{R}$.

Using row operations gives

$$\begin{bmatrix} 1 & 2 & 3 & \vdots & 4 \\ -1 & 0 & 1 & \vdots & -1 \\ \alpha & 2 & -3 & \vdots & \beta \end{bmatrix} \quad \begin{matrix} \\ R_2 \leftarrow R_2 + R_1 \\ R_3 \leftarrow R_3 - \alpha R_1 \end{matrix}$$

$$\sim \begin{bmatrix} 1 & 2 & 3 & \vdots & 4 \\ 0 & 2 & 4 & \vdots & 3 \\ 0 & 2-2\alpha & -3-3\alpha & \vdots & \beta - 4\alpha \end{bmatrix} \quad \begin{matrix} \\ \\ R_3 \leftarrow R_3 - (1-\alpha)R_2 \end{matrix}$$

$$\sim \begin{bmatrix} 1 & 2 & 3 & \vdots & 4 \\ 0 & 2 & 4 & \vdots & 3 \\ 0 & 0 & -7+\alpha & \vdots & \beta - \alpha - 3 \end{bmatrix}.$$

(a) If $\alpha \neq 7$ the solution will be unique.

(b) If $\alpha = 7$ there will be an infinity of solutions if $\beta - \alpha - 3 = 0$ i.e., if $\beta = 10$.

(c) If $\alpha = 7$ and $\beta \neq 10$ there will be no solutions.

As examples: (a) If $\alpha = 0$ and $\beta = -4$ we have

$$\begin{bmatrix} 1 & 2 & 3 & \vdots & 4 \\ 0 & 2 & 4 & \vdots & 3 \\ 0 & 0 & -7 & \vdots & -7 \end{bmatrix};$$

i.e.,

$$-7x_3 = -7 \qquad \therefore x_3 = 1$$
$$2x_2 = 3 - 4x_3 \qquad \therefore x_2 = -\tfrac{1}{2}$$
$$x_1 = 4 - 3x_3 - 2x_2 \qquad \therefore x_1 = 2$$

the unique-solution case.

(b) If $\alpha = 7$ and $\beta = 10$ we have

$$\begin{bmatrix} 1 & 2 & 3 & \vdots & 4 \\ 0 & 2 & 4 & \vdots & 3 \\ 0 & 0 & 0 & \vdots & 0 \end{bmatrix};$$

i.e.,

$$2x_2 = 3 - 4x_3 \qquad \therefore x_2 = \tfrac{3}{2} - 2x_3$$
$$x_1 = 4 - 3x_3 - 2x_2 = 1 + x_3$$

where x_3 can be any real number; this is the infinity-of-solutions case.

(c) If $\alpha = 7$ and $\beta = 0$ we have

$$\begin{bmatrix} 1 & 2 & 3 & \vdots & 4 \\ 0 & 2 & 4 & \vdots & 3 \\ 0 & 0 & 0 & \vdots & -10 \end{bmatrix}$$

where the last equation cannot be satisfied by any x_1, x_2, x_3; i.e., the no-solutions case.

Introducing some terminology, we say that, if a set of linear equations has a solution or solutions, the set of equations is *consistent*. If no solution exists we say that the equations are *inconsistent*.

In Example 8.4, Case (c), for example, the equations are inconsistent.

In Cases (a) and (b) on the other hand the systems of equations are consistent.

However, Cases (a) and (b) are different. In Case (a) we have three 'quite different' equations, whereas in Case (b) one of the equations is effectively missing (or just represented by the other two equations anyway). In Case (b) we say that the three equations are *linearly dependent* because we can construct one of them by adding together suitable multiples of the other two. In Case (a) this cannot be done and we say that the original equations are *linearly independent*.

8.5   MATRIX INVERSION

Suppose A is an $n \times n$ matrix and has an inverse, and let

$$b_1 = \begin{bmatrix} 1 & 0 & 0 & \ldots & 0 \end{bmatrix}^T$$
$$b_2 = \begin{bmatrix} 0 & 1 & 0 & \ldots & 0 \end{bmatrix}^T$$

and so on to b_n where

$$b_n = \begin{bmatrix} 0 & 0 & 0 & \ldots & 1 \end{bmatrix}^T$$

(i.e., b_j is a column vector containing a 1 in the jth position and zeros everywhere else, for $j = 1(1)n$).

Then each system $Ax_j = b_j$ has a unique solution $x_j = A^{-1}b_j$ (see Section 7.8). Let X be the $n \times n$ matrix with columns x_j. Then

$$AX = A\begin{bmatrix} x_1 & x_2 & x_3 & \ldots & x_n \end{bmatrix}$$
$$= \begin{bmatrix} Ax_1 & Ax_2 & Ax_3 & \ldots & Ax_n \end{bmatrix}$$
$$= \begin{bmatrix} b_1 & b_2 & b_3 & \ldots & b_n \end{bmatrix}$$

by our definition of matrix multiplication. (See Equation [7-5] and its preamble.) But

$$\begin{bmatrix} b_1 & b_2 & b_3 & \ldots & b_n \end{bmatrix} = I_n$$

the $n \times n$ unit matrix. (We defined the b_j so that this would be the case.) Hence we have found a matrix X such that $AX = I_n$. It follows that $X = A^{-1}$.

Now we saw in Section 8.3 how we could solve, in one scheme, two systems of equations with the same coefficient matrix. This is equally true for any number of systems with the same coefficient matrix. Hence to solve $Ax_j = b_j$ for $j = 1(1)n$ we construct an augmented matrix of the form

$$\begin{bmatrix} A & \vdots & b_1 & b_2 & \ldots & b_n \end{bmatrix} = \begin{bmatrix} A & \vdots & I_n \end{bmatrix}.$$

We then apply the valid row operations until A is upper triangular, and back-substitute for each solution x_j. Alternatively we may continue with row operations until $\mathbf{A}$ becomes $\mathbf{I}_n$, when the old $\mathbf{I}_n$ will have become $\mathbf{A}^{-1}$ (rather like the alternative method for solving a single system as suggested in Section 8.3).

EXAMPLE 8.5 To invert $\begin{bmatrix} 3 & 2 \\ -1 & 4 \end{bmatrix}$ the augmented matrix is

$$\begin{bmatrix} 3 & 2 & \vdots & 1 & 0 \\ -1 & 4 & \vdots & 0 & 1 \end{bmatrix} \begin{matrix} R_2 \leftarrow -R_2 \\ R_1 \leftrightarrow R_2 \end{matrix}$$

$$\sim \begin{bmatrix} 1 & -4 & \vdots & 0 & -1 \\ 3 & 2 & \vdots & 1 & 0 \end{bmatrix} \quad R_2 \leftarrow R_2 - 3R_1$$

$$\sim \begin{bmatrix} 1 & -4 & \vdots & 0 & -1 \\ 0 & 14 & \vdots & 1 & 3 \end{bmatrix}.$$

Back-substitution: let $\mathbf{x}_1 = \begin{bmatrix} x_{11} & x_{21} \end{bmatrix}^T$ and $\mathbf{x}_2 = \begin{bmatrix} x_{12} & x_{22} \end{bmatrix}^T$.

Column 1: $x_{21} = \frac{1}{14}$, $x_{11} = 4x_{21} = \frac{4}{14}$.

Column 2: $x_{22} = \frac{3}{14}$, $x_{12} = -1 + 4x_{22} = -\frac{2}{14}$.

$$\therefore \quad \begin{bmatrix} 3 & 2 \\ -1 & 4 \end{bmatrix}^{-1} = \begin{bmatrix} \frac{4}{14} & -\frac{2}{14} \\ \frac{1}{14} & \frac{3}{14} \end{bmatrix} = \frac{1}{14} \begin{bmatrix} 4 & -2 \\ 1 & 3 \end{bmatrix}.$$

Alternatively we may continue row operations from

$$\begin{bmatrix} 1 & -4 & \vdots & 0 & -1 \\ 0 & 14 & \vdots & 1 & 3 \end{bmatrix} \quad R_2 \leftarrow R_2/14$$

$$\sim \begin{bmatrix} 1 & -4 & \vdots & 0 & -1 \\ 0 & 1 & \vdots & \frac{1}{14} & \frac{3}{14} \end{bmatrix} \quad R_1 \leftarrow R_1 + 4R_2$$

$$\sim \begin{bmatrix} 1 & 0 & \vdots & \frac{4}{14} & -\frac{2}{14} \\ 0 & 1 & \vdots & \frac{1}{14} & \frac{3}{14} \end{bmatrix}$$

with the same result.

EXAMPLE 8.6 To invert

$$\begin{bmatrix} 1 & 1 & -1 \\ 2 & -3 & 2 \\ 3 & 2 & -4 \end{bmatrix}.$$

Using the method which produces the unit matrix on the left gives

$$\begin{bmatrix} 1 & 1 & -1 & \vdots & 1 & 0 & 0 \\ 2 & -3 & 2 & \vdots & 0 & 1 & 0 \\ 3 & 2 & -4 & \vdots & 0 & 0 & 1 \end{bmatrix} \quad \begin{array}{l} \\ R_2 \leftarrow R_2 - 2R_1 \\ R_3 \leftarrow R_3 - 3R_1 \end{array}$$

$$\sim \begin{bmatrix} 1 & 1 & -1 & \vdots & 1 & 0 & 0 \\ 0 & -5 & 4 & \vdots & -2 & 1 & 0 \\ 0 & -1 & -1 & \vdots & -3 & 0 & 1 \end{bmatrix} \quad \begin{array}{l} \\ R_3 \leftarrow -R_3 \\ R_2 \leftrightarrow R_3 \end{array}$$

$$\sim \begin{bmatrix} 1 & 1 & -1 & \vdots & 1 & 0 & 0 \\ 0 & 1 & 1 & \vdots & 3 & 0 & -1 \\ 0 & -5 & 4 & \vdots & -2 & 1 & 0 \end{bmatrix} \quad \begin{array}{l} \\ R_1 \leftarrow R_1 - R_2 \\ R_3 \leftarrow R_3 + 5R_2 \end{array}$$

$$\sim \begin{bmatrix} 1 & 0 & -2 & \vdots & -2 & 0 & 1 \\ 0 & 1 & 1 & \vdots & 3 & 0 & -1 \\ 0 & 0 & 9 & \vdots & 13 & 1 & -5 \end{bmatrix} \quad \begin{array}{l} R_3 \leftarrow \frac{1}{9}R_3 \\ R_1 \leftarrow R_1 + 2R_3 \\ R_2 \leftarrow R_2 - R_3 \end{array}$$

$$\sim \begin{bmatrix} 1 & 0 & 0 & \vdots & \frac{8}{9} & \frac{2}{9} & -\frac{1}{9} \\ 0 & 1 & 0 & \vdots & \frac{14}{9} & -\frac{1}{9} & -\frac{4}{9} \\ 0 & 0 & 1 & \vdots & \frac{13}{9} & \frac{1}{9} & -\frac{5}{9} \end{bmatrix};$$

i.e., the inverse is $\frac{1}{9} \begin{bmatrix} 8 & 2 & -1 \\ 14 & -1 & -4 \\ 13 & 1 & -5 \end{bmatrix}$.

(Check that this matrix gives the unit matrix when multiplied by the original matrix.)

EXERCISE 8.5

Invert these matrices:

(a) $\begin{bmatrix} 3 & 2 \\ -1 & 1 \end{bmatrix}$ (b) $\begin{bmatrix} 2 & 3 & -1 \\ 1 & 4 & 2 \\ 4 & -1 & 3 \end{bmatrix}$.

ANSWERS TO EXERCISES

8.1   One possibility (of many) is to let $x = -1 - y + z$ from the first equation. Substituting this into the other two equations gives two equations in y and z only, which rearranged become

$$-5y + 4z = 6$$
$$-y - z = -6.$$

Substituting $y = 6 - z$ from the second of these equations into the first yields $z = 4$, hence $y = 2$, hence $x = 1$. This solution can be checked by substitution into the original equations.

This approach works if you know what you are doing, and if there are not too many equations. It does not lead, however, to a general algorithm suitable for solution by computer.

8.2
$$\begin{bmatrix} 2 & 1 & \vdots & 4 \\ 3 & -2 & \vdots & -1 \end{bmatrix} \quad R_1 \leftarrow \tfrac{1}{2}R_1$$

$$\sim \begin{bmatrix} 1 & \tfrac{1}{2} & \vdots & 2 \\ 3 & -2 & \vdots & -1 \end{bmatrix} \quad R_2 \leftarrow R_2 - 3R_1$$

$$\sim \begin{bmatrix} 1 & \tfrac{1}{2} & \vdots & 2 \\ 0 & -\tfrac{7}{2} & \vdots & -7 \end{bmatrix}$$

$$\Rightarrow \quad -\tfrac{7}{2}y = -7 \quad \text{i.e.,} \quad y = 2 \quad \text{and} \quad x = 2 - \tfrac{1}{2}y = 1.$$

8.3
$$\begin{bmatrix} 2 & 3 & -1 & \vdots & 5 \\ 1 & 4 & 2 & \vdots & 15 \\ 4 & -1 & 3 & \vdots & 11 \end{bmatrix} \quad R_1 \leftrightarrow R_2$$

(to make the elimination easier)

$$\sim \begin{bmatrix} 1 & 4 & 2 & \vdots & 15 \\ 2 & 3 & -1 & \vdots & 5 \\ 4 & -1 & 3 & \vdots & 11 \end{bmatrix} \quad \begin{array}{l} R_2 \leftarrow R_2 - 2R_1 \\ R_3 \leftarrow R_3 - 4R_1 \end{array}$$

(elimination in the first column)

$$\sim \begin{bmatrix} 1 & 4 & 2 & \vdots & 15 \\ 0 & -5 & -5 & \vdots & -25 \\ 0 & -17 & -5 & \vdots & -49 \end{bmatrix} \quad R_2 \leftarrow -\tfrac{1}{5}R_2$$

(to make elimination easier)

$$\sim \begin{bmatrix} 1 & 4 & 2 & \vdots & 15 \\ 0 & 1 & 1 & \vdots & 5 \\ 0 & -17 & -5 & \vdots & -49 \end{bmatrix} \quad R_3 \leftarrow R_3 + 17R_2$$

(elimination in the second column)

$$\sim \begin{bmatrix} 1 & 4 & 2 & \vdots & 15 \\ 0 & 1 & 1 & \vdots & 5 \\ 0 & 0 & 12 & \vdots & 36 \end{bmatrix}$$

$\Rightarrow \quad z = 3$

$\qquad y = 5 - z = 2$

$\qquad x = 15 - 4y - 2z = 1.$

Notice that the choice of row operations to achieve this result is not unique, but should be such that the zeros below the diagonal are found systematically, one column at a time, moving from left to right.

8.4
$$\begin{bmatrix} 2 & 0 & 1 & 0 & \vdots & 1 & 0 \\ 1 & 0 & -1 & 2 & \vdots & 0 & 1 \\ 0 & 3 & 0 & -1 & \vdots & 0 & 0 \\ 1 & 1 & -1 & 0 & \vdots & 0 & 0 \end{bmatrix} \quad R_1 \leftrightarrow R_2$$

$$\sim \begin{bmatrix} 1 & 0 & -1 & 2 & \vdots & 0 & 1 \\ 2 & 0 & 1 & 0 & \vdots & 1 & 0 \\ 0 & 3 & 0 & -1 & \vdots & 0 & 0 \\ 1 & 1 & -1 & 0 & \vdots & 0 & 0 \end{bmatrix} \quad \begin{matrix} R_2 \leftarrow R_2 - 2R_1 \\ R_4 \leftarrow R_4 - R_1 \end{matrix}$$

$$\sim \begin{bmatrix} 1 & 0 & -1 & 2 & \vdots & 0 & 1 \\ 0 & 0 & 3 & -4 & \vdots & 1 & -2 \\ 0 & 3 & 0 & -1 & \vdots & 0 & 0 \\ 0 & 1 & 0 & -2 & \vdots & 0 & -1 \end{bmatrix} \quad R_2 \leftrightarrow R_4$$

$$\sim \begin{bmatrix} 1 & 0 & -1 & 2 & \vdots & 0 & 1 \\ 0 & 1 & 0 & -2 & \vdots & 0 & -1 \\ 0 & 3 & 0 & -1 & \vdots & 0 & 0 \\ 0 & 0 & 3 & -4 & \vdots & 1 & -2 \end{bmatrix} \quad R_3 \leftarrow R_3 - 3R_2$$

$$\sim \begin{bmatrix} 1 & 0 & -1 & 2 & \vdots & 0 & 1 \\ 0 & 1 & 0 & -2 & \vdots & 0 & -1 \\ 0 & 0 & 0 & 5 & \vdots & 0 & 3 \\ 0 & 0 & 3 & -4 & \vdots & 1 & -2 \end{bmatrix} \quad R_3 \leftrightarrow R_4$$

$$\sim \begin{bmatrix} 1 & 0 & -1 & 2 & \vdots & 0 & 1 \\ 0 & 1 & 0 & -2 & \vdots & 0 & -1 \\ 0 & 0 & 3 & -4 & \vdots & 1 & -2 \\ 0 & 0 & 0 & 5 & \vdots & 0 & 3 \end{bmatrix}$$

$\Rightarrow \quad x_4 = 0, \qquad\qquad y_4 = \tfrac{3}{5}$

$\qquad\quad 3x_3 - 4x_4 = 1, \qquad 3y_3 - 4y_4 = -2$

$\therefore \quad x_3 = \tfrac{1}{3}, \qquad\qquad y_3 = \tfrac{2}{15}$

$\qquad\quad x_2 - 2x_4 = 0, \qquad y_2 - 2y_4 = -1$

$\therefore \quad x_2 = 0, \qquad\qquad y_2 = \tfrac{1}{5}$

$\qquad\quad x_1 - x_3 + 2x_4 = 0, \quad y_1 - y_3 + 2y_4 = 1$

$\therefore \quad x_1 = \tfrac{1}{3}, \qquad\qquad y_1 = -\tfrac{1}{15}.$

8.5 (a) $\tfrac{1}{5} \begin{bmatrix} 1 & -2 \\ 1 & 3 \end{bmatrix}$ (b) $\tfrac{1}{60} \begin{bmatrix} 14 & -8 & 10 \\ 5 & 10 & -5 \\ -17 & 14 & 5 \end{bmatrix}.$

FURTHER READING

For a well-written account of elimination, along the same lines as our own, see Phillips and Taylor, Section 8.3. You may need to read Section 8.2 beforehand, and in particular the subsection headed *Block Matrices*.

For a shorter account see Goult, Hoskins, Milner and Pratt, Section 3.2.1.

For a fuller treatment of the elementary theory of linear systems you could consult Cohn.

The books by Dixon and Hosking, Joyce and Turner contain accounts of elimination at a level similar to ours, but the approach taken makes them more suitable for consideration with Lesson 9.

LESSON 8 – COMPREHENSION TEST

1. Which of the following are true?
Row operations leave unchanged
(a) the rows of the augmented matrix;
(b) the columns of the augmented matrix;
(c) the solutions of the associated linear equations.

2. State the valid row operations (i.e., those that are allowed in the context of row equivalence).

3. Draw a diagram showing the structure of the augmented matrix after elimination.

4. State the three possible types of solution to a linear system.

5. Define the terms 'consistent' and 'linearly dependent' as they apply to linear systems of equations. Relate them to the three types of solution.

6. What is the solution X to the matrix equation $AX = I_n$, where A is $n \times n$ (assuming that the equation has a unique solution)?

ANSWERS

1   (c).

2   Multiplying a row by a non-zero scalar; adding one row (or a non-zero multiple of it) to another row; interchanging two rows.

3
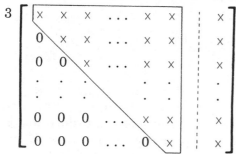

4   (a) A unique solution.
    (b) An infinity of solutions.
    (c) No solutions.

5   A consistent system has a solution, i.e. (a) or (b) in 4.

    A linearly dependent system has one or more equations that are combinations of the others.

    A linearly dependent system may have any of the three solution types.

6   $X = A^{-1}$.

LESSON 8 – SUPPLEMENTARY EXERCISES

8.6. Solve the systems of equations $Ax = b$ for the following A and b.

(a) $\mathbf{A} = \begin{bmatrix} 3 & 1 & 1 \\ 2 & 7 & 2 \\ 1 & 1 & 4 \end{bmatrix}$, $\quad \mathbf{b} = \begin{bmatrix} 4 \\ 9 \\ 5 \end{bmatrix}$

(b) $\mathbf{A} = \begin{bmatrix} 3 & 1 & -2 & -1 \\ 2 & -2 & 2 & 3 \\ 1 & 5 & -4 & -1 \\ 3 & 1 & 2 & 3 \end{bmatrix}$, $\quad \mathbf{b} = \begin{bmatrix} 3 \\ -8 \\ 3 \\ -1 \end{bmatrix}$

(c) $\mathbf{A} = \begin{bmatrix} 1 & 5 & 2 \\ 1 & 1 & 7 \\ 0 & -3 & 4 \end{bmatrix}$, $\quad \mathbf{b} = \begin{bmatrix} 9 \\ 6 \\ -2 \end{bmatrix}$ and $\begin{bmatrix} 2 \\ -3 \\ -4 \end{bmatrix}$.

8.7. Invert each of the matrices given in Exercise 8.6.

8.8. Solve the equations

$$x - 2y + 3z = 0$$
$$2x - 2y + 5z = 1$$
$$x \qquad + 2z = \alpha$$

for all real numbers α.

*8.9. Show that each of the valid row operations are equivalent to pre-multiplication by a matrix constructed by applying the row operation to the unit matrix. Show that these matrices are invertible and deduce that if the coefficient matrix $\mathbf{A}$ is reduced to the upper triangular matrix $\mathbf{U}$ by elimination, then there exists an invertible matrix $\mathbf{E}$ such that $\mathbf{U} = \mathbf{EA}$.

Lesson 9 **Instabilities and Pivoting**

OBJECTIVES

At the end of this lesson you should

(a)   understand the need for pivoting in elimination;

(b)   be able to perform Gaussian elimination with partial pivoting;

(c)   know how to use the standard checks for blunders;

(d)   understand the idea of ill-conditioning in linear equations and be able to recognise it.

CONTENTS

9.1   THE EFFECTS OF ROUNDING ERRORS

It will be remembered from Lesson 2 that, in using calculators or computers (or just using ordinary arithmetic if you have to), the results of arithmetic operations are usually rounded to fit storage space. It follows that, even if the numbers you start with can be given exactly (like 5 or -7, but not like π), after a few calculations you may have accumulated a certain amount of error due to the rounding processes. In Lesson 8 we saw how a set of equations can be represented by an augmented matrix and how equivalent matrices could be found. Back-substitution then gave the solutions. However in the problems considered there the elements of the matrices were chosen to be integers, the arithmetic was exact and no rounding occurred.

In practical problems rounding *will* occur and could grow to wreck the solution unless we are careful. Well, how? What about an example?

EXAMPLE 9.1   Consider the equations

$$-1.41x_1 + \quad 2x_2 \qquad\qquad = 1$$
$$x_1 - 1.41x_2 + \quad x_3 = 1$$
$$2x_2 - 1.41x_3 = 1.$$

We shall solve them using the augmented matrix and row equivalence method of Lesson 8, the only difference being that we shall not interchange any rows. To heighten the effect of the rounding errors we shall use 3-digit floating-point decimal arithmetic (like a primitive 3-digit decimal computer). See Sections 2.11 and 2.12 of Lesson 2. The effects that we shall observe would still occur with more accurate arithmetic, but not so noticeably.

$$\begin{bmatrix} -1.41 & 2 & 0 & \vdots & 1 \\ 1 & -1.41 & 1 & \vdots & 1 \\ 0 & 2 & -1.41 & \vdots & 1 \end{bmatrix} \quad R_2 \leftarrow R_2 + \dfrac{1}{1.41}R_1$$

$$= R_2 + 0.709R_1$$

$$\sim \begin{bmatrix} -1.41 & 2 & 0 & \vdots & 1 \\ 0 & 0.01 & 1 & \vdots & 1.71 \\ 0 & 2 & -1.41 & \vdots & 1 \end{bmatrix} \quad R_3 \leftarrow R_3 - 200R_2$$

$$\sim \begin{bmatrix} -1.41 & 2 & 0 & \vdots & 1 \\ 0 & 0.01 & 1 & \vdots & 1.71 \\ 0 & 0 & -201 & \vdots & -341 \end{bmatrix}.$$

Back-substitution, still working to 3S, gives

$$x_3 = 341/201 = 1.70$$
$$x_2 = (1.71 - 1.70)/0.01 = 1.00$$
$$x_1 = (1 - (2 \times 1.00))/(-1.41) = 0.710.$$

Substituting these values into the left-hand sides of the original equations gives

$$0.999, \quad 1, \quad -0.397$$

respectively. Some of the results for x_1, x_2, x_3 are plainly very inaccurate. (One might even be tempted to think that a numerical blunder had been made.)

Now consider what happens if we simply interchange the second and third equations after elimination in the first column. (Note that this would be necessary anyway if the 0.01 had in fact been zero.)

$$\begin{bmatrix} -1.41 & 2 & 0 & \vdots & 1 \\ 0 & 0.01 & 1 & \vdots & 1.71 \\ 0 & 2 & -1.41 & \vdots & 1 \end{bmatrix} \quad R_2 \leftrightarrow R_3$$

$$\sim \begin{bmatrix} -1.41 & 2 & 0 & \vdots & 1 \\ 0 & 2 & -1.41 & \vdots & 1 \\ 0 & 0.01 & 1 & \vdots & 1.71 \end{bmatrix} \quad R_3 \leftarrow R_3 - \frac{1}{200}R_2$$

$$= R_3 - 0.005R_2$$

$$\sim \begin{bmatrix} -1.41 & 2 & 0 & \vdots & 1 \\ 0 & 2 & -1.41 & \vdots & 1 \\ 0 & 0 & 1.01 & \vdots & 1.71 \end{bmatrix}$$

so that

$$x_3 = 1.71/1.01 = 1.69$$
$$x_2 = (1 - (-1.41 \times 1.69))/2 = 1.69$$
$$x_1 = (1 - (2 \times 1.69))/(-1.41) = 1.69.$$

The left-hand sides of the original equations with these values of x_1, x_2 and x_3 now equal

0.997,   0.997,   0.997

which are near enough to 1, 1, 1, considering that we are working with 3S. (In fact the new solutions $x_1 = x_2 = x_3 = 1.69$ are correct to 3S.)

Now what was the difference? Well, we incurred rounding errors in the first elimination operation. We then multiplied these errors in the second elimination by 200 no less. Moreover in the back-substitution we divided 1.71 − 1.70 = 0.01   with a large relative error (difference of two nearly equal quantities) by 0.01, i.e., multiplied the errors by 100. No wonder the answers would not satisfy the original equations! What about the second try? Well, the errors from the first elimination were multiplied by 0.005 at the second stage which reduces them drastically. Then in the back-substitution we divided by 2 instead of 0.01, hence further reducing the propagated errors.

Clearly the lessons to be learned from the above example are:

(a) avoid error magnification in elimination by using a *multiplier* less than 1 in magnitude (a multiplier is a number that multiplies a row in an elimination operation); and

(b) in back-substitution choose the equations that imply division by the largest possible numbers.

To incorporate these requirements we must use the row-interchange operation, although we shall not physically move the rows but rather consider them out of their natural order. This is done as part of the formal *Gaussian-elimination algorithm* described in the next section.

9.2 GAUSSIAN ELIMINATION

Writing down all the row-equivalent augmented matrices (as in Example 9.1) is very cumbersome, especially if there are more than three equations originally. The following tableau is easy to follow and gives the same information as the sequence of equivalent matrices, but without unnecessary repetitions.

Multipliers				
	-1.41	2	0	1
-0.709	1	-1.41	1	1
0	0	2	-1.41	1
0.005		0.01	1	1.71
		2	-1.41	1
			1.01	1.71

The only way in which it differs from its presentation in the previous section (apart from not repeating unaltered rows) is that, instead of writing, for example, $R_2 \leftarrow R_2 + 0.709R_1$ we have simply noted the elimination multiplier m. For consistency this is defined so that the elimination is always a subtraction; i.e., $R_i \leftarrow R_i - mR_j$, say. (We could equally well have made it always an addition.) Hence we interpret our example as $R_2 \leftarrow R_2 - (-0.709)R_1$ with multiplier -0.709. As promised we have not actually interchanged rows 2 and 3 at the second stage. It is sufficient to place the multiplier alongside the row in which the elimination is to be done so that the tableau says, in effect, $R_6 \leftarrow R_4 - 0.005R_5$. The rows that would appear in the final augmented matrix are those with the circled elements. The multipliers depend, of course, on which rows are used to do the elimination. In order to make all the multipliers less than 1 in magnitude we scan the column in question for the element which is largest in magnitude. This is called the *pivotal element*, or simply *pivot*, and the row in which it occurs is called the *pivotal row*. In the tableau above -1.41 was the pivot for elimination in the first column and 2 the pivot for elimination in the second column. In a trivial sense 1.01 is the pivot in the third column. The multipliers are then the other elements of the column divided by the pivot $(1/(-1.41) = -0.709, \ 0/(-1.41) = 0, \ 0.01/2 = 0.005)$.

Let us summarise the elimination stage. Suppose we have n linear equations in n unknowns. We write down the augmented matrix and proceed as follows.

(a) Find the element of largest magnitude in the first column — the pivot a_{p1} (in row p).

(b) Divide all the other elements in the first column by the pivot and write the results in the corresponding rows of the multiplier column; i.e., $m_i = a_{i1}/a_{p1}$, $i = 1(1)n$, $i \neq p$.

(c) Subtract from each row (except the pivotal row) the relevant multiple of the pivotal row; i.e., $R_i \leftarrow R_i - m_i R_p$, $i = 1(1)n$, $i \neq p$. These $n-1$ new rows are written down as the basis for the second stage of elimination.

(d) Repeat (a)—(c) for the second, third, ... , $(n-1)$th columns, thus producing the upper-triangular, row-equivalent system.

EXAMPLE 9.2   Consider

$$
\begin{aligned}
2x + 3y - z &= 5 \\
x + 4y + 2z &= 15 \\
4x - y + 3z &= 11
\end{aligned}
\qquad \text{i.e.} \qquad
\begin{bmatrix} 2 & 3 & -1 \\ 1 & 4 & 2 \\ 4 & -1 & 3 \end{bmatrix}
\begin{bmatrix} x \\ y \\ z \end{bmatrix}
=
\begin{bmatrix} 5 \\ 15 \\ 11 \end{bmatrix}.
$$

The Gaussian-elimination tableau is as follows.

Multipliers				
$\frac{1}{2}$	2	3	-1	5
$\frac{1}{4}$	1	4	2	15
	④	-1	3	11
$\frac{14}{17}$		$\frac{7}{2}$	$-\frac{5}{2}$	$-\frac{1}{2}$
		$\frac{17}{4}$	$\frac{5}{4}$	$\frac{49}{4}$
			$-\frac{120}{34}$	$-\frac{360}{34}$

The row operations implied are

$$R_4 \leftarrow R_1 - \tfrac{1}{2}R_3$$
$$R_5 \leftarrow R_2 - \tfrac{1}{4}R_3$$
$$R_6 \leftarrow R_4 - \tfrac{14}{17}R_5.$$

A 3×3 system of equations always requires at most 3 row operations to reach the upper triangular form. You should note the absolute rigidity of the method. There is no real choice in the row operations (unless two or more elements in a column are equally the largest in magnitude, in which case we can choose any one of them as pivot).

It is perfectly reasonable to do exact arithmetic in Example 9.2. The example merely illustrates the elimination process. But the whole point of the choice of pivot is to reduce the effect of rounding errors. Here is a more realistic example.

EXAMPLE 9.3   Consider

$$\begin{bmatrix} 2.26 & -8.89 & 1.76 & -1.84 \\ 2.11 & 4.03 & 1.70 & 9.04 \\ 10.18 & -2.19 & 3.63 & 5.54 \\ 1.19 & 2.48 & -8.31 & -2.02 \end{bmatrix} x = \begin{bmatrix} 5.19 \\ 5.05 \\ 8.16 \\ 3.36 \end{bmatrix}$$

where $x = \begin{bmatrix} x_1 & x_2 & x_3 & x_4 \end{bmatrix}^T$.

Here is the tableau for the elimination stage. Since the data are given to 2D we shall work to 3D.

Multipliers					
0.222	2.26	− 8.89	1.76	− 1.84	5.19
0.207	2.11	4.03	1.70	9.04	5.05
	(10.18)	− 2.19	3.63	5.54	8.16
0.117	1.19	2.48	− 8.31	− 2.02	3.36
		(− 8.404)	0.954	− 3.070	3.378
− 0.533		4.483	0.949	7.893	3.361
− 0.326		2.736	− 8.735	− 2.668	2.405
− 0.173			1.457	6.257	5.161
			(− 8.424)	− 3.669	3.506
				(5.662)	5.768

There are six row operations implied here, viz.

$$R_5 \leftarrow R_1 - 0.222R_3$$
$$R_6 \leftarrow R_2 - 0.207R_3$$
$$R_7 \leftarrow R_4 - 0.117R_3$$
$$R_8 \leftarrow R_6 + 0.533R_5$$
$$R_9 \leftarrow R_7 + 0.326R_5$$
$$R_{10} \leftarrow R_8 + 0.173R_9.$$

(In general an $n \times n$ system requires no more than $\frac{1}{2}n(n-1)$ elimination operations).

So much for the elimination stage. What about back-substitution? We require that we divide by the largest possible number, but that simply means sub-stituting back into the pivotal rows.

EXAMPLE 9.4   From the last pivotal row of the tableau in Example 9.3

$$x_4 = 5.768/5.622 = 1.026.$$

From the third pivotal row

$$x_3 = (3.506 + 3.669 \times 1.026)/(-8.424) = -0.863.$$

Here we have reduced the round off effects by a factor of $8\frac{1}{2}$ approximately. If we had used the other possible row defining x_3 we would get

$$x_3 = (5.161 - 6.257 \times 1.026)/1.457 = -0.864$$

in which the round off effects are diminished by a factor of only $1\frac{1}{2}$. The former is clearly preferable.

Similarly using the first two pivotal rows

$$x_2 = (3.378 - 0.954 \times (-0.863) + 3.070 \times 1.026)/(-8.404)$$
$$= -0.874_7$$

$$x_1 = (8.16 + 2.19 \times (-0.875) - 3.63 \times (-0.863) - 5.54 \times 1.026)/10.18$$
$$= 0.363.$$

Hence the solution vector is $\begin{bmatrix} 0.36 & -0.87 & -0.86 & 1.03 \end{bmatrix}^T$ when rounded to 2D. (The 2D data suggest that it is unlikely that we can quote more than 2D in the solutions. We shall see later that it could be much worse than that — rounding errors in the data may make the solutions anything from poor to meaningless. In this particular example our solutions are good.)

The algorithm that we have described in this section is known as *Gaussian elimination with partial pivoting* or *pivotal condensation*. The pivoting is called partial because it chooses pivots which are largest in their respective columns. Generally speaking this is sufficient to control the growth of rounding errors. Full pivoting, which means choosing the largest element in the whole matrix as pivot, is sometimes required, but we shall not describe how it is performed.

EXERCISE 9.1

Solve these equations by Gaussian elimination with partial pivoting. Use exact arithmetic.

$$x + y + z = 1$$
$$-3x - y + 4z = -10$$
$$-2x + 3y + z = -5.$$

9.3 CHECKING THE CALCULATIONS

The Gaussian elimination algorithm is well-suited to implementation on a computer. It is equally good for hand calculation. As a general rule it is a good idea to introduce checks into hand calculations if possible, and preferably not just by repetition as there is a tendency simply to repeat blunders.

To check the elimination stage it is usual to have a further column on the right which initially contains the sums of the elements in each of the original rows (apart from the multipliers in the first column). These numbers are treated in the elimination in exactly the same way as the other numbers in their rows. In theory the resulting elements in the check-sum column should be the same as the sums of the numbers in the new rows. In practice this will not occur because of rounding but the difference should be no more than 1 or 2 in the last recorded figure. If they differ then the number used for subsequent checking is the sum of the elements in the corresponding row, and not the number derived in the elimination. Thus all the eliminations can be checked.

The only part that remains to be checked is the back substitution. This is done in effect by adding the original equations together and substituting the solutions in the left-hand side. This will give (if correctly done) the sum of the right-hand sides (apart from rounding). It is usual to write the sum of the coefficients above the corresponding columns.

EXAMPLE 9.5   The result of incorporating these checks into the tableau at the beginning of Section 9.2 is to give the following.

Multipliers	-0.41	2.59	-0.41	3	Check sums
	-1.41	2	0	1	1.59
-0.709	1	-1.41	1	1	1.59
0	0	2	-1.41	1	1.59
0.005		0.01	1	1.71	2.72 (2.72)
		2	-1.41	1	1.59 (1.59)
			1.01	1.71	2.71 (2.72)

$x_3 = x_2 = x_1 = 1.69$ as before.

For the final check

$$-0.41 \times 1.69 + 2.59 \times 1.69 - 0.41 \times 1.69 = 2.99$$

which is quite satisfactorily near to 3 considering the 3-digit arithmetic.

Note that these checks are not infallible. It is possible for example to make two blunders that cancel as far as the row sums are concerned. Nevertheless the checks provide a worthwhile control over mistakes. There is nothing worse than making a mistake at the beginning of a long calculation and thereby having to repeat the whole process if it is not discovered until the end. The final check above is an alternative to substituting into the original equations, which might be regarded as the only safe check. Unfortunately it is possible for both of these checks to be misleading, as we shall see in the next section. What we *can* say is that if the left-hand sides are not close in value to the right-hand sides then a mistake is almost certainly implied. What we *cannot* say is that when they are close then the solutions are necessarily close to the exact solutions.

EXERCISE 9.2

Solve the following equations using Gaussian elimination with partial pivoting. Work with 4D and give your answers to 3D. Include checks in your calculations.

$$\begin{bmatrix} 4.193 & 1.274 & -0.851 \\ 1.085 & 5.329 & 1.542 \\ 0.968 & -1.437 & -4.616 \end{bmatrix} \begin{bmatrix} x_1 \\ x_2 \\ x_3 \end{bmatrix} = \begin{bmatrix} 6.730 \\ 5.828 \\ -4.207 \end{bmatrix}.$$

9.4 INDUCED INSTABILITY AND INHERENT INSTABILITY

The Gaussian elimination method is specifically designed to solve a set of simultaneous linear equations with a 'well-defined' solution, in such a way as to get the best possible solution within the limits of the accuracy of the data supplied and the precision of the arithmetic used.

If the multipliers chosen are not less than unity in magnitude the solution gained is very likely to be poor even if the checks show that the arithmetic has been performed correctly. We say that taking the multipliers greater than one in magnitude *induces* instability in the solution because such multipliers magnify the unavoidable rounding errors. *Induced instability* is a property of the *method* of solution used and may apply to methods other than the Gaussian-elimination method without pivoting, for example to some methods for the numerical solution of differential equations.

However, this discussion has been confined to a situation where the solution is 'well-defined'. So perhaps we now ought to go into what is meant by a 'well-defined' solution, or more precisely how we may be unable to find a good solution. We have already anticipated this in Exercise 1.4, which you should reconsider in the light of your new-found knowledge of linear equations.

In that exercise we saw that, given the equation

$$\begin{bmatrix} 1 & 1 \\ 1 & 1.01 \end{bmatrix} \begin{bmatrix} x \\ y \end{bmatrix} = \begin{bmatrix} 2 \\ 2.01 \end{bmatrix},$$

if the coefficients could be in error by as much as 0.01, then the solutions could be any of $(1, 1)$, $(2, 0)$, $(\alpha, 2 - \alpha)$, for all $\alpha \in \mathbb{R}$, or there might be no solutions at all. This is the situation that we have referred to before, where the apparent solution $(x = y = 1$ in this case) may be 'only just unique', and where errors in the data may severely affect the accuracy of the solutions.

There is a simple geometric interpretation of this phenomenon. If you try to draw the two straight lines

$$x + \quad y = 2$$
$$x + 1.01y = 2.01$$

you will find it very difficult to say where they intersect because they are virtually coincident. Corresponding to the effect of rounding error, a small drawing error will move the point of intersection a long way, or even make the lines parallel.

This example provides an excellent illustration of the difference between pure mathematics and numerical mathematics. As we saw in Lesson 8, pure mathematics indicates three possibilities — a unique solution, an infinity of solutions, or no solutions. The problem for the numerical mathematician is that the distinction between these three cases is not black and white in practice, but grey. It is possible for one case to be turned into another by rounding errors — for a unique solution to become non-unique or non-existent, and vice-versa.

When this occurs we say that the problem is *ill-conditioned* or *inherently unstable*. Such a problem is ultra-sensitive to errors in both the data and arithmetic. Note that this kind of instability is independent of any method of solution. In fact one way to check for it (if possible) is to solve the problem using exact (or very accurate) arithmetic, and then to solve the problem again, but using slightly changed data, and to study the resulting changes in the solution.

EXAMPLE 9.6   Consider the equations

$$\begin{bmatrix} \frac{1}{2} & \frac{1}{3} & \frac{1}{4} \\ \frac{1}{3} & \frac{1}{4} & \frac{1}{5} \\ \frac{1}{4} & \frac{1}{5} & \frac{1}{6} \end{bmatrix} \mathbf{x} = \begin{bmatrix} 0.95 \\ 0.67 \\ 0.52 \end{bmatrix}$$

Using exact arithmetic (check it!) the solution is $\mathbf{x} = \begin{bmatrix} 1.20 & 0.60 & 0.60 \end{bmatrix}^T$. If we change the right-hand-side vector to $\begin{bmatrix} 0.96 & 0.66 & 0.53 \end{bmatrix}^T$, the solution changes to $\mathbf{x} = \begin{bmatrix} 6.12 & -18.0 & 15.6 \end{bmatrix}^T$!!

This system is badly ill-conditioned. In practice this means that if the given right-hand sides are rounded data, then we can find the exact solution for the data given, but the exact solution for the true data could be completely different.

At this point we can return to a warning given in Example 9.3. If the problem contains rounded data then the accuracy of the solution will not be of the order of the accuracy of the data unless the equations are well-conditioned.

We can do a simple analysis of the effect of changing the right-hand-side vector which helps to explain why some systems are ill-conditioned while others are not. Consider the system $\mathbf{Ax} = \mathbf{b}$ and suppose it has a unique solution $\mathbf{x} = \mathbf{A}^{-1}\mathbf{b}$. Suppose that the data of the problem (which consists of $\mathbf{A}$ and $\mathbf{b}$) are perturbed (i.e., changed by 'small' amounts) in such a way that $\mathbf{b}$ becomes $\mathbf{b} + \delta\mathbf{b}$ but $\mathbf{A}$ remains unchanged. Then the solution $\mathbf{x}$ will be perturbed by some vector $\delta\mathbf{x}$ so that

$$\mathbf{A}(\mathbf{x} + \delta\mathbf{x}) = \mathbf{b} + \delta\mathbf{b}$$

$$\therefore \qquad \mathbf{x} + \delta\mathbf{x} = \mathbf{A}^{-1}(\mathbf{b} + \delta\mathbf{b}) = \mathbf{A}^{-1}\mathbf{b} + \mathbf{A}^{-1}\delta\mathbf{b}.$$

But $\mathbf{x} = \mathbf{A}^{-1}\mathbf{b}$ so that we have

$$\delta\mathbf{x} = \mathbf{A}^{-1}\delta\mathbf{b}.$$

We can now see that whether the entries of $\delta\mathbf{x}$ are significantly larger than those in $\delta\mathbf{b}$ will depend on $\mathbf{A}^{-1}$ and $\delta\mathbf{b}$.

EXAMPLE 9.7 In Example 9.6 $\delta\mathbf{b} = \begin{bmatrix} 0.01 & -0.01 & 0.01 \end{bmatrix}^T$. Also

$$\begin{bmatrix} \frac{1}{2} & \frac{1}{3} & \frac{1}{4} \\ \frac{1}{3} & \frac{1}{4} & \frac{1}{5} \\ \frac{1}{4} & \frac{1}{5} & \frac{1}{6} \end{bmatrix}^{-1} = \begin{bmatrix} 72 & -240 & 180 \\ -240 & 900 & -720 \\ 180 & -720 & 600 \end{bmatrix}.$$

(Check for yourself that the product of these two matrices is the unit matrix.) Hence

$$\delta\mathbf{x} = \begin{bmatrix} 72 & -240 & 180 \\ -240 & 900 & -720 \\ 180 & -720 & 600 \end{bmatrix} \begin{bmatrix} 0.01 \\ -0.01 \\ 0.01 \end{bmatrix} = \begin{bmatrix} 4.92 \\ -18.6 \\ 15.0 \end{bmatrix}$$

and this vector is, of course, precisely the difference between the two solution vectors given in Example 9.6. We observe that the cause of the ill-conditioning in the solutions is partly the large elements in A^{-1} and partly the sign patterns in A^{-1} and δb. If $\delta b = \begin{bmatrix} 0.01 & 0.01 & 0.01 \end{bmatrix}^T$, for example, then the elements of δx are not nearly so large (but are still too large for comfort).

In the example used above (the matrix appearing there is one of a famous type called Hilbert matrices, all of which produce ill-conditioning) we had an exact coefficient matrix A and used exact arithmetic. In reality we have rounded coefficients and make rounding errors in our arithmetic. An ill-conditioned system will cause magnification of errors in data and arithmetic, which tends to confuse the issue.

EXAMPLE 9.8   We shall solve the equation

$$\begin{bmatrix} 1.21 & -2.71 & 4.37 \\ 2.12 & -2.84 & 17.17 \\ -1.22 & 7.53 & 19.51 \end{bmatrix} x = b$$

with $b = \begin{bmatrix} 8.90 & 47.95 & 72.37 \end{bmatrix}^T$ and with $b = \begin{bmatrix} 8.91 & 47.96 & 72.38 \end{bmatrix}^T$ working with 4D and including the usual checks. We recall from Lesson 8 that we can solve for the two equations simultaneously since the choice of row operations depends only on the coefficient matrix, which is the same for both equations. Here is the tableau.

Multipliers	2.11	1.98	41.05	129.22	129.25	Checks
0.5708	1.21	−2.71	4.37	8.90	8.91	20.68
	(2.12)	−2.84	17.17	47.95	47.96	112.36
−0.5755	−1.22	7.53	19.51	72.37	72.38	170.57
−0.1847		−1.0889	−5.4306	−18.4699	−18.4656	−43.4551⁰
		(5.8956)	29.3913	99.9652	99.9610	235.2332¹
		(0.0020)	− 0.0063	− 0.0028	− 0.0074¹

Back-substitution in the pivotal rows yields

$$x_3 = -3.1500, \qquad y_3 = -1.4$$
$$x_2 = 32.6596, \qquad y_2 = 23.9346$$
$$x_1 = 91.8815, \qquad y_1 = 66.0247$$

where $\begin{bmatrix} x_1 & x_2 & x_3 \end{bmatrix}^T$ and $\begin{bmatrix} y_1 & y_2 & y_3 \end{bmatrix}^T$ are the solutions for the two b vectors respectively. The final checks give

$$2.11x_1 + 1.98x_2 + 41.05x_3 = 129.2285$$
$$2.11y_1 + 1.98y_2 + 41.05y_3 = 129.2326$$

which suggests that we have not made any blunders.

The two matrix equations are very similar, differing only by amounts of 0.01 in the b vectors. The solutions however are quite remarkably different. The true solutions for the first set are in fact $x = \begin{bmatrix} 1 & 2 & 3 \end{bmatrix}^T$. (Try it for yourself!). Clearly something very strange is going on and, moreover, it isn't obvious just by looking at the original equations.

What are the facts? Small changes in the data plus rounding errors in the arithmetic are producing vast changes in the solutions. (Neither solution is near the true solution, $x = \begin{bmatrix} 1 & 2 & 3 \end{bmatrix}^T$.) So the original equations must have been ill-conditioned.

How can we tell if this should happen again? We have implied that the answer could be — solve the equations twice using partial pivoting, changing the right-hand-side vector by a small amount to get the second problem. If the two sets of solutions are nowhere near the same the equations are ill-conditioned. But this means quite a lot more work. There is another way!

The last equation of the elimination sequence is

$$0.0020x_3 = -0.0063$$

so that $\qquad x_3 = -0.0063/0.0020$

which is the ratio of two small numbers, both of which have only 2 significant figures compared to five or six significant figures in the data that produced them. This loss of significance implies an increase in relative error by a factor of the order of 1000. As a general rule if the pivots in the Gaussian-elimination process show marked loss of significance then the system is likely to be ill-conditioned.

EXERCISE 9.3

Recalculate the solutions of the system of Example 9.8, working with 6D instead of 4D. Comment on the results.

9.5 SUMMARY

In this lesson we have considered numerical instabilities that can arise in solving linear equations. We have seen that if, in elimination, we take the equations in an unsuitable order then we may get small pivots and large multipliers, producing error magnification. This is called *induced instability*. By using partial pivoting we virtually guarantee no induced instability, which is entirely a property of a poor method.

On the other hand the system of equations itself may exhibit *inherent instability*, essentially because the system is near the borderline between having a unique solution and having a non-unique or non-existent solution.

We recognise this either by slightly perturbing the system and noting large changes in the solutions, or by noting loss of significance in the elimination pivots. Usually we have to accept that it is impossible to give good answers to an ill-conditioned (inherently unstable) system.

ANSWERS TO EXERCISES

9.1   Multipliers

$-\frac{1}{3}$	1 1 1			1
	$\boxed{-3}$ -1 4			-10
$\frac{2}{3}$	-2 3 1			-5
$\frac{2}{11}$	$\frac{2}{3}$ $\frac{7}{3}$			$-\frac{7}{3}$
	$\boxed{\frac{11}{3}}$ $-\frac{5}{3}$			$\frac{5}{3}$
	$\boxed{\frac{87}{33}}$			$-\frac{87}{33}$

$$\therefore \quad z = -1,$$
$$y = (\tfrac{5}{3} + \tfrac{5}{3}(-1))/(\tfrac{11}{3}) = 0$$
$$x = (-10 + 0 - 4(-1))/(-3) = 2$$

9.2

Multipliers	6.246	5.166	-3.925	8.351	Check sums
	$\boxed{4.193}$	1.274	-0.851	6.730	11.346
0.2588	1.085	5.329	1.542	5.828	13.784
0.2309	0.968	-1.437	-4.616	-4.207	$-\ 9.292$
		$\boxed{4.9993}$	1.7622	4.0863	$10.8477\sqrt{}$
-0.3463		-1.7312	-4.4195	-5.7610	-11.9118^7
			$\boxed{-3.8093}$	-4.3459	$-\ 8.1552\sqrt{}$

Back-substitution in the pivotal rows gives

$$x_3 = 1.1409$$
$$x_2 = 0.4152$$
$$x_1 = 1.7104$$

and the final check has

$$\dot{6}.246x_1 + 5.166x_2 - 3.925x_3 = 8.3501$$
$$\therefore \quad x_1 = 1.710, \quad x_2 = 0.415, \quad x_3 = 1.141$$

(when rounded to 3D) are good solutions as far as we can tell.

9.3

Multipliers	2.11	1.98	41.05	129.22	129.25	Check sums
0.570 755	1.21	− 2.71	4.37	8.90	8.91	20.68
− 0.575 472	(2.12)	− 2.84	17.17	47.95	47.96	112.36
− 0.184 722	− 1.22	7.53	19.51	72.37	72.38	170.57
		− 1.089 056	− 5.429 863	− 18.467 702	− 18.463 410	− 43.450 032[1]
		(5.895 660)	29.390 854	99.963 882	99.979 637	235.230 034[3]
			(− 0.000 726)	− 0.002 174	0.005 029	0.002 181[9]

Hence

$$x_3 = 2.994\,490 \qquad y_3 = -6.926\,997$$
$$x_2 = 2.027\,468 \qquad y_2 = 51.490\,417$$
$$x_1 = 1.081\,423 \qquad y_1 = 147.702\,511.$$

For the final checks

$$2.11x_1 + 1.98x_2 + 41.05x_3 = 129.220\,004$$

and

$$2.11y_1 + 1.98y_2 + 41.05y_3 = 129.250\,097.$$

We can still see that small changes in the data produce large changes in the solutions. The pivots still show the loss of significance. However, using 6D nearly gives 1D accuracy in the x solution (which we happen to know in this case). We can see (comparing with Example 9.8) that changing the working accuracy for one of the problems completely changes the solutions so that changing the working accuracy for a problem could also be used as a check on ill-conditioning. But this would entail reworking the problem.

FURTHER READING

For a concise but clear account of the need for and the technique of pivoting see Phillips and Taylor, Section 8.5.

For a simple summary of ill-conditioning and elimination with pivoting see Ribbans, Book 1, Sections 5.0—5.2.

A readable and comprehensive account can be found in Williams P. W., Sections 4.2.1.—4.2.6.

If you consulted the reference to it in Lesson 8 you may like to look at Goult, Hoskins, Milner and Pratt, Section 3.2.2. This book also contains a comprehensive treatment of condition in Chapter 5.

Both Dixon, Chapter 8 and Hosking, Joyce and Turner, 'Steps' 11, 12 and 14 contain accounts of the material in Lessons 8 and 9, with slightly different emphases and details to ours, but at a very similar level.

LESSON 9 – COMPREHENSION TEST

1. Given the augmented matrix

$$\begin{bmatrix} 1 & 2 & 1 & \vdots & 0 \\ -5 & 0 & 3 & \vdots & 2 \\ 4 & 1 & -2 & \vdots & 1 \end{bmatrix}$$

state

(a) the first pivot;

(b) the multipliers associated with the non-pivotal rows.

2. True or false?

(a) Partial pivoting is used to avoid inherent instability.

(b) If partial pivoting is applied all multipliers will be less than or equal to one in magnitude.

(c) Partial pivoting is equivalent to (selected) row interchanges.

(d) Back-substitution is made in the non-pivotal rows.

(e) 'Ill-conditioned' and 'inherently unstable' mean the same thing.

3. What is the number of row operations required for the elimination phase of the formal Gauss-elimination algorithm when applied to 5 equations in 5 unknowns? Can you generalise the result for n equations in n unknowns?

ANSWERS

1   (a) -5. (b) 1st row: $-1/5$; 3rd row: $-4/5$.

2   True: (b), (c), (e). False: (a), (d).

3   10 (4 at the first stage, 3 at the second, 2 at the third, 1 at the fourth). $\frac{1}{2} n(n-1)$ is the general result.

LESSON 9 – SUPPLEMENTARY EXERCISES

9.4. Solve the following systems of equations using Gauss-elimination with partial pivoting.

(a) $\begin{bmatrix} 2.2 & -4.5 & -2.0 \\ 3.0 & 2.6 & 4.3 \\ -6.0 & 3.5 & 2.6 \end{bmatrix} x = \begin{bmatrix} 19.07 \\ 3.21 \\ -18.25 \end{bmatrix}$

(b) $\begin{bmatrix} 1.1342 & 0.0432 & -0.0599 & 0.0202 \\ 0.0202 & 1.1342 & 0.0432 & -0.0599 \\ -0.0599 & 0.0202 & 1.1342 & 0.0432 \\ 0.0432 & -0.0599 & 0.0202 & 1.1342 \end{bmatrix} x = \begin{bmatrix} 1.941 \\ -0.230 \\ -1.941 \\ 0.230 \end{bmatrix}.$

9.5. Solve these equations both with and without partial pivoting, using 4-digit floating-point arithmetic.

$\begin{bmatrix} 0.3114 & 0.9438 & 0.4720 \\ 0.3808 & 0.7614 & 0.5028 \\ 0.5486 & 0.9051 & 0.6856 \end{bmatrix} x = \begin{bmatrix} 0.5883 \\ 0.5699 \\ 0.7474 \end{bmatrix}.$

Compare the solutions with the exact solution $x = \begin{bmatrix} \frac{1}{2} & \frac{1}{3} & \frac{1}{4} \end{bmatrix}^T$, commenting on any induced instability due to lack of pivoting.

9.6. Solve the system

$$\begin{bmatrix} 6 & 4 & 3 \\ 20 & 15 & 12 \\ 15 & 12 & 10 \end{bmatrix} x = b$$

for $b = \begin{bmatrix} 13 & 47 & 37 \end{bmatrix}^T$

and $b = \begin{bmatrix} 13.1 & 46.9 & 37.1 \end{bmatrix}^T$,

using *exact* arithmetic. Comment on the condition of the system.

9.7. Invert the matrix in Exercise 9.6 and use the inverse to explain the magnification of the differences between the elements of the two b vectors.

*9.8. Just as two linear equations represent two straight lines in a plane so three linear equations represent three planes in space. Describe all the different ways in which three planes may or may not intersect and relate an ill-conditioned system of three equations to these possibilities.

*9.9. Let x_0 be an approximate solution and x be the exact solution to the linear system $Ax = b$. Show that the *residual vector* $r_0 = b - Ax_0$ satisfies $A(x - x_0) = r_0$ and hence construct a method for iteratively improving x_0.

Lesson 10 Iterative Methods for Linear Systems

OBJECTIVES

At the end of this lesson you should

(a)   understand how simultaneous linear equations can be solved by iterative methods;

(b)   be able to apply the Gauss—Jacobi and Gauss—Seidel methods;

(c)   know a sufficient condition for the convergence of these two methods;

(d)   be able to apply the method of successive over-relaxation.

CONTENTS

0.1   INTRODUCTION

You will remember that in Lesson 5 non-linear algebraic equations were solved by one-point iterative methods. More precisely

(a)   an initial estimate, x_0, of the required root of the equation was found;

(b) the equation was rewritten in the form $x = f(x)$ implying the recurrence relation

$$x_{n+1} = f(x_n), \quad n = 0, 1, 2, 3, \ldots;$$

(c) a sequence of estimates of the root $\{x_0, x_1, x_2, x_3, \ldots\}$ was calculated;

(d) the process was terminated when two successive estimates of the root were found to differ by less than some prescribed tolerance.

This method of attack can be applied to the matrix equation

$$\mathbf{A}\mathbf{x} = \mathbf{b} \quad \text{where} \quad \mathbf{A} \text{ is } n \times n.$$

(We are assuming that a well-defined solution exists.)

The method for matrices is in general the same:

(a$'$) Choose some initial vector, $\mathbf{x}_0$. (This may or may not be a good approximation to the solution of $\mathbf{A}\mathbf{x} = \mathbf{b}$. Unlike the non-linear algebraic equation problem it is often not possible to choose a good first approximation. If this is so we may have to take an arbitrary vector, $\mathbf{x}_0$.)

(b$'$) Rearrange the equation $\mathbf{A}\mathbf{x} = \mathbf{b}$ in the form $\mathbf{x} = \mathbf{E}\mathbf{x} + \mathbf{c}$, where $\mathbf{E}$ and $\mathbf{c}$ do not depend on $\mathbf{x}$, implying the matrix recurrence relation

$$\mathbf{x}_{m+1} = \mathbf{E}\mathbf{x}_m + \mathbf{c}, \quad m = 0, 1, 2, 3, \ldots.$$

(c$'$) Calculate a sequence of vectors $\{\mathbf{x}_0, \mathbf{x}_1, \mathbf{x}_2, \ldots\}$, which are estimates of the solution $\mathbf{x}$.

(d$'$) Terminate the process when two successive estimates of the solution are found to differ, in some sense, by less than a given tolerance.

What then are the problems? How shall we do it in practice? For (a$'$) we may have to choose an arbitrary vector, say $\mathbf{0}$. (c$'$) is obvious when we have decided on (b$'$). In (d$'$) there are many ways of deciding on the termination criterion. We will use the criterion that the elements of the vector which is the difference between successive iterates must be less than $\frac{1}{2} \times 10^{-k-1}$ in modulus. If the rate of convergence is not very slow we shall almost certainly then have kD correct, and possibly $(k+1)D$ correct.

For (b$'$) there are infinitely many ways of rearranging $\mathbf{A}\mathbf{x} = \mathbf{b}$, only some of which give a convergent sequence of vectors, $\{\mathbf{x}_i\}$. We shall consider three methods, known as the Gauss—Jacobi, the Gauss—Seidel, and the Successive Over-Relaxation (SOR) methods.

10.2 NOTATION

So far we have denoted the sequence of estimates of $\mathbf{x}$ by $\{\mathbf{x}_0, \mathbf{x}_1, \mathbf{x}_2, \ldots\}$. But when we want to refer to an element of $\mathbf{x}$ we already use x_j for the

jth element. In consequence we will call the rth member of the sequence of vector estimates

$$^{(r)}\mathbf{x} \quad (\text{'the } r\text{th vector } \mathbf{x}').$$

Then the jth element of $^{(r)}\mathbf{x}$ will naturally be $^{(r)}x_j$.

The recurrence relation in (b') above becomes

$$^{(m+1)}\mathbf{x} = \mathbf{E}\,^{(m)}\mathbf{x} + \mathbf{c}, \quad m = 0, 1, 2, 3, \ldots \qquad [10\text{-}1]$$

and the sequence of estimates is

$$\{^{(0)}\mathbf{x},\ ^{(1)}\mathbf{x},\ ^{(2)}\mathbf{x},\ ^{(3)}\mathbf{x}, \ldots \}.$$

It is now easy to write down the convergence criterion which we put into words in Section 10.1. It becomes

$$|^{(m+1)}x_j - {}^{(m)}x_j| < \tfrac{1}{2} \times 10^{-k-1} \quad \text{for } j = 1(1)n.$$

$^{(m+1)}\mathbf{x}$ is then the accepted solution.

10.3 THE GAUSS–JACOBI ITERATIVE METHOD

Let us take the three equations

$$a_{11}x_1 + a_{12}x_2 + a_{13}x_3 = b_1$$
$$a_{21}x_1 + a_{22}x_2 + a_{23}x_3 = b_2$$
$$a_{31}x_1 + a_{32}x_2 + a_{33}x_3 = b_3,$$

i.e., $\mathbf{Ax} = \mathbf{b}$, where $\mathbf{A}$ is 3×3, and rearrange them in an obvious way as

$$\left.\begin{array}{l} x_1 = (b_1 - a_{12}x_2 - a_{13}x_3)/a_{11} \\ x_2 = (b_2 - a_{21}x_1 - a_{23}x_3)/a_{22} \\ x_3 = (b_3 - a_{31}x_1 - a_{32}x_2)/a_{33}. \end{array}\right\} \qquad [10\text{-}2]$$

Note that the development here is similar to that in Lesson 5 with $x = f(x)$.

We can represent this rearrangement in matrix form.

We first 'split' $\mathbf{A}$ into

$$\mathbf{A} = \mathbf{D} + \mathbf{B}$$

where

$$\mathbf{D} = \begin{bmatrix} a_{11} & 0 & 0 \\ 0 & a_{22} & 0 \\ 0 & 0 & a_{33} \end{bmatrix} \quad \text{and} \quad \mathbf{B} = \begin{bmatrix} 0 & a_{12} & a_{13} \\ a_{21} & 0 & a_{23} \\ a_{31} & a_{32} & 0 \end{bmatrix}$$

and then rewrite

$$(D + B)x = b$$

as $\quad\quad\quad Dx = -Bx + b.$

If $a_{ii} \neq 0$, $i = 1(1)3$, then D^{-1} exists and

$$D^{-1} = \begin{bmatrix} 1/a_{11} & 0 & 0 \\ 0 & 1/a_{22} & 0 \\ 0 & 0 & 1/a_{33} \end{bmatrix}$$

so that we can write

$$x = -D^{-1}Bx + D^{-1}b.$$

This matrix equation is the same as Equation [10-2]. (Multiply it all out if you don't believe us!) It has the form

$$x = Ex + c$$

required by (b'). Hence we can use the iterative scheme

$$^{(m+1)}x = E^{(m)}x + c$$

i.e. $\quad ^{(m+1)}x = (-D^{-1}B)^{(m)}x + (D^{-1}b)$ $\quad\quad\quad$ [10-3]

corresponding to Equation [10-1].

However, in practice the matrix form is *not* actually used to do the iteration. Equation [10-3] merely shows us that we have a process similar in essence to the one-point methods used for non-linear algebraic equations. Equation [10-3] can also be used for theoretical analysis of the convergence rate, but such analysis is beyond the scope of this text. Instead of using Equation [10-3] it is in fact more efficient to use the rearrangement of the individual equations [10-2]. Attaching the indices as in Equation [10-3] we get

$$\left.\begin{aligned} ^{(m+1)}x_1 &= (b_1 - a_{12}\,^{(m)}x_2 - a_{13}\,^{(m)}x_3)/a_{11} \\ ^{(m+1)}x_2 &= (b_2 - a_{21}\,^{(m)}x_1 - a_{23}\,^{(m)}x_3)/a_{22} \\ ^{(m+1)}x_3 &= (b_3 - a_{31}\,^{(m)}x_1 - a_{32}\,^{(m)}x_2)/a_{33}. \end{aligned}\right\} \quad [10\text{-}4]$$

In general (i.e., when A is $n \times n$) we have

$$\left.\begin{aligned} ^{(m+1)}x_1 &= (b_1 - a_{12}\,^{(m)}x_2 - a_{13}\,^{(m)}x_3 - \ldots - a_{1n}\,^{(m)}x_n)/a_{11} \\ ^{(m+1)}x_2 &= (b_2 - a_{21}\,^{(m)}x_1 - a_{23}\,^{(m)}x_3 - \ldots - a_{2n}\,^{(m)}x_n)/a_{22} \\ &\quad\cdot \\ &\quad\cdot \\ &\quad\cdot \\ ^{(m+1)}x_n &= (b_n - a_{n1}\,^{(m)}x_1 - a_{n2}\,^{(m)}x_2 - \ldots - a_{n,n-1}\,^{(m)}x_{n-1})/a_{nn}. \end{aligned}\right\}$$

$$[10\text{-}5]$$

This is the Gauss–Jacobi iterative method for solving $Ax = b$. Note that only values from $^{(m)}x$ are used on the right-hand sides.

EXAMPLE 10.1   Let us use the Gauss–Jacobi method to solve the equations

$$4x_1 - x_2 \qquad\qquad = 0.4$$
$$-x_1 + 4x_2 - x_3 \qquad = 0.8$$
$$- x_2 + 4x_3 - x_4 = 1.2$$
$$- x_3 + 2x_4 = 0.6.$$

The iterative scheme is

$$^{(m+1)}x_1 = 0.1 + {}^{(m)}x_2/4$$
$$^{(m+1)}x_2 = 0.2 + ({}^{(m)}x_1 + {}^{(m)}x_3)/4$$
$$^{(m+1)}x_3 = 0.3 + ({}^{(m)}x_2 + {}^{(m)}x_4)/4$$
$$^{(m+1)}x_4 = 0.3 + {}^{(m)}x_3/2.$$

It is not easy to see a *good* starting vector so let us take $^{(0)}x = \begin{bmatrix} 0 & 0 & 0 & 0 \end{bmatrix}^T$. We normally lay out the successive estimates in a vertical fashion so that we can easily see when successive values in each column are sufficiently close together. Let us work with 3D.

m	$^{(m)}x_1$	$^{(m)}x_2$	$^{(m)}x_3$	$^{(m)}x_4$
0	0	0	0	0
1	0.1	0.2	0.3	0.3
2	0.150	0.300	0.425	0.450
3	0.175	0.344	0.488	0.512
4	0.186	0.366	0.514	0.544
5	0.192	0.375	0.528	0.557
6	0.194	0.380	0.533	0.564
7	0.195	0.382	0.536	0.567
8	0.196	0.383	0.537	0.568
9	0.196	0.383	0.538	0.568_5
10	0.196	0.384	0.538	0.569
11	0.196	0.384	0.538	0.569

Working to 3D all the elements of Line 10 are the same as the corresponding elements of Line 11. Hence

$$x_1 = 0.20, \quad x_2 = 0.38, \quad x_3 = 0.54, \quad x_4 = 0.57$$

are almost certainly correct to 2D.

If we examine the table we can see that using Lines 8 and 9 x_1, x_2 and x_4 have 'converged'. However by Lines 10 and 11, when x_3 had also 'converged', we find a slight difference in x_2 and x_4.

If we had taken as a convergence criterion

$$|^{(m+1)}x_j - {}^{(m)}x_j| < \tfrac{1}{2} \times 10^{-2}$$

to give 2D accuracy we could have stopped at Lines 6 and 7, successfully in this case. But it is possible to imagine a situation where given a few more iterations the values may 'creep on' a little and change by 1 the value of the last place required. Hence we use what may appear a tough but safer criterion, i.e., with $\tfrac{1}{2} \times 10^{-k-1}$ on the right-hand side, and we study the resulting sequence in case very slow convergence makes even this insufficient to guarantee kD accuracy. (In fact we drop the $\tfrac{1}{2}$, so that the tolerance for the above problem is 10^{-3}, i.e., the calculation stops when successive values agree to 3D.)

EXERCISE 10.1

Use the Gauss—Jacobi method to solve the equations

$$
\begin{aligned}
5x_1 - x_2 - x_3 - x_4 &= -4 \\
-x_1 + 10x_2 - x_3 - x_4 &= 12 \\
-x_1 - x_2 + 5x_3 - x_4 &= 8 \\
-x_1 - x_2 - x_3 + 10x_4 &= 34
\end{aligned}
$$

starting with the trial vector $[0.9 \quad 1.9 \quad 2.9 \quad 3.9]^T$. Give your answers correct to 2D.

10.4 THE GAUSS—SEIDEL ITERATIVE METHOD

At some stages in the Gauss—Jacobi calculation of the vector iterates new values of some of the variables are available and could be used. For instance, at one stage in Example 10.1 the old line and part of the new line are:

m	$^{(m)}x_1$	$^{(m)}x_2$	$^{(m)}x_3$	$^{(m)}x_4$
4	0.186	0.366	0.514	0.544
5	0.192	0.375		

$^{(5)}x_3$ is about to be calculated as

$$^{(5)}x_3 = 0.3 + (^{(4)}x_2 + {}^{(4)}x_4)/4.$$

But $^{(5)}x_2$ is available and is hopefully a better estimate of x_2 than is $^{(4)}x_2$. The Gauss—Seidel method uses the latest values of all the unknowns at all times, i.e., would calculate $^{(5)}x_3$ from

$$^{(5)}x_3 = 0.3 + (^{(5)}x_2 + {^{(4)}}x_4)/4.$$

Applying this approach to Equations [10-2] gives

$$\left. \begin{array}{l} ^{(m+1)}x_1 = (b_1 - a_{12}\,^{(m)}x_2 - a_{13}\,^{(m)}x_3)/a_{11} \\[1mm] ^{(m+1)}x_2 = (b_2 - a_{21}\,^{(m+1)}x_1 - a_{23}\,^{(m)}x_3)/a_{22} \\[1mm] ^{(m+1)}x_3 = (b_3 - a_{31}\,^{(m+1)}x_1 - a_{32}\,^{(m+1)}x_2)/a_{33} \end{array} \right\} \qquad \text{[10-6]}$$

which should be carefully compared with Equations [10-4].

The general case is

$$\left. \begin{array}{l} ^{(m+1)}x_1 = (b_1 - a_{12}\,^{(m)}x_2 - a_{13}\,^{(m)}x_3 - \ldots - a_{1n}\,^{(m)}x_n)/a_{11} \\[1mm] ^{(m+1)}x_2 = (b_2 - a_{21}\,^{(m+1)}x_1 - a_{23}\,^{(m)}x_3 - \ldots - a_{2n}\,^{(m)}x_n)/a_{22} \\[2mm] \quad . \\ \quad . \\ \quad . \\[2mm] ^{(m+1)}x_n = (b_n - a_{n1}\,^{(m+1)}x_1 - a_{n2}\,^{(m+1)}x_2 - \ldots - a_{n,\,n-1}\,^{(m+1)}x_{n-1})/a_{nn} \end{array} \right\}$$

$$\text{[10-7]}$$

which should be compared with Equations [10-5].

EXAMPLE 10.2   Let us use the Gauss-Seidel method on the problem of Example 10.1 starting with the same $^{(0)}x$ and working with 3D.

The iterative scheme is

$$^{(m+1)}x_1 = 0.1 + {^{(m)}}x_2/4$$

$$^{(m+1)}x_2 = 0.2 + (^{(m+1)}x_1 + {^{(m)}}x_3)/4$$

$$^{(m+1)}x_3 = 0.3 + (^{(m+1)}x_2 + {^{(m)}}x_4)/4$$

$$^{(m+1)}x_4 = 0.3 + {^{(m+1)}}x_3/2.$$

This gives the table shown below.

m	$^{(m)}x_1$	$^{(m)}x_2$	$^{(m)}x_3$	$^{(m)}x_4$
0	0	0	0	0
1	0.1	0.225	0.356	0.478
2	0.156	0.328	0.502	0.551
3	0.182	0.371	0.530	0.565
4	0.193	0.381	0.536	0.568
5	0.195	0.383	0.538	0.569
6	0.196	0.384	0.538	0.569
7	0.196	0.384	0.538	0.569

We have Lines 6 and 7 the same and so we can quote the answers to 2D as before. But notice that the Gauss–Seidel method has needed only 7 iterations as against 11 iterations for the Gauss–Jacobi method.

This behaviour is typical. We would expect Gauss–Seidel to be faster than Gauss–Jacobi because it uses the latest information available. However it is possible (if unusual) for the Gauss–Jacobi method to converge when the Gauss–Seidel method diverges.

EXERCISE 10.2

Use the Gauss–Seidel method on the problem of Exercise 10.1.

To obtain the matrix representation of the Gauss–Seidel process (as for Gauss–Jacobi), we have to 'split' the matrix A, this time into an upper triangular matrix, U, a diagonal matrix, D, and a lower triangular matrix, L, so that

$$A = L + U + D$$

where

$$L = \begin{bmatrix} 0 & & & & & \\ a_{21} & 0 & & & & \\ a_{31} & a_{32} & 0 & & & \\ \cdot & \cdot & & \cdot & & \\ \cdot & \cdot & & & \cdot & \\ \cdot & \cdot & & & & 0 \\ a_{n1} & a_{n2} & \cdots & a_{n,n-1} & & 0 \end{bmatrix}$$

$$U = \begin{bmatrix} 0 & a_{12} & a_{13} & \cdots & a_{1n} \\ & 0 & a_{23} & \cdots & a_{2n} \\ & & 0 & \cdots & \cdot \\ & & & \cdot & \cdot \\ & & & & \cdot \\ & & & & a_{n-1,n} \\ & & & & 0 \end{bmatrix}$$

and

$$D = \begin{bmatrix} a_{11} & 0 & 0 & \cdots & 0 \\ 0 & a_{22} & 0 & \cdots & 0 \\ 0 & 0 & a_{33} & \cdots & 0 \\ \cdot & \cdot & \cdot & & \cdot \\ \cdot & \cdot & \cdot & & \cdot \\ \cdot & \cdot & \cdot & & \cdot \\ 0 & 0 & 0 & \cdots & a_{n,n} \end{bmatrix}$$

In each equation it will be the non-zero elements of L that will be associated with new values of the unknowns, i.e., elements of $^{(m+1)}x$, and non-zero elements of U that will be associated with elements of $^{(m)}x$. $Ax = b$ is rearranged thus:

$$(D + L + U)x = b$$

$\therefore$
$$Dx = b - Ux - Lx,$$

i.e.
$$x = D^{-1}(b - Ux - Lx).$$

This implies the iterative form

$$^{(m+1)}x = D^{-1}(b - U\,^{(m)}x - L\,^{(m+1)}x). \qquad [10\text{-}8]$$

This equation expresses the way the calculations are done, but is not of the form

$$^{(m+1)}x = E\,^{(m)}x + c.$$

It can be so arranged, however, giving

$$^{(m+1)}x = (L + D)^{-1}(b - U\,^{(m)}x)$$
$$= [-(L + D)^{-1}U]^{(m)}x + [(L + D)^{-1}b].$$

10.5 CONVERGENCE AND DIVERGENCE

No doubt the impression has grown up in the last two sections (and the examples and exercises) that nothing can go wrong with matrix iteration. However, let us consider the following trivial example. Let

$$\left. \begin{array}{l} 4x_1 - x_2 = 3 \\ -x_1 + 4x_2 = 3. \end{array} \right\} \qquad [10\text{-}9]$$

and

The (well-defined) solutions (row operations? what about trying it to check?) are $x_1 = 1$ and $x_2 = 1$. There's nothing like doing a very easy problem to see what is going on!

Using Gauss—Jacobi we can write

$$^{(m+1)}x_1 = (3 + {}^{(m)}x_2)/4$$
$$^{(m+1)}x_2 = (3 + {}^{(m)}x_1)/4.$$

Starting with $\begin{bmatrix} 0.9 & 0.9 \end{bmatrix}^T$ we get

m	$^{(m)}x_1$	$^{(m)}x_2$
0	0.9	0.9
1	0.975	0.975
2	0.994	0.994
3	0.999	0.999
4	1.000	1.000
5	1.000	1.000

The last two lines agree to 3D and so we can quote the answers as $x_1 = 1.00$, $x_2 = 1.00$ to 2D. The method converges nicely!

But what about rewriting the equations as

$$-x_1 + 4x_2 = 3$$
$$4x_1 - x_2 = 3$$

[10-10]

and using the scheme

$$^{(m+1)}x_1 = 4\,^{(m)}x_2 - 3$$
$$^{(m+1)}x_2 = 4\,^{(m)}x_1 - 3.$$

The table is now

m	$^{(m)}x_1$	$^{(m)}x_2$
0	0.9	0.9
1	0.6	0.6
2	— 0.6	— 0.6
3	— 5.4	— 5.4
4	— 24.6	— 24.6
5	— 101.4	— 101.4
6	— 408.6	— 408.6
7	— 1637.4	— 1637.4

Now let's stop for a minute! The vector estimates are nowhere near $\begin{bmatrix} 1 & 1 \end{bmatrix}^T$ and appear to be getting further away from the true solution.

The Gauss—Jacobi process is clearly diverging — and so also would the Gauss—Seidel (try it!)

How can we tell if a particular iterative scheme will give a divergent or convergent process? We will state a *sufficient* condition for the convergence of the Gauss—Jacobi and Gauss—Seidel schemes but we will not attempt to prove it as such a proof is beyond the scope of this text.

A *sufficient* condition for the convergence of the Gauss—Jacobi and Gauss—Siedel schemes is that the matrix A should be in such a form (by interchanging rows if necessary) that it (A) is *diagonally dominant*. When this is the case convergence occurs no matter what initial vector is chosen.

Diagonally dominant means that either

(a) for each row of A the diagonal element (a_{ii}) is bigger in modulus than the sum of the moduli of all the other elements in the row; or

(b) for each column of A the diagonal element (a_{jj}) is bigger in modulus than the sum of the moduli of all the other elements in the column.

We can state the condition for convergence now as either

(a) $|a_{ii}| > \sum\limits_{\substack{j=1 \\ j \neq i}}^{n} |a_{ij}|$ all $i = 1(1)n$,

or

(b) $|a_{jj}| > \sum\limits_{\substack{i=1 \\ i \neq j}}^{n} |a_{ij}|$ all $j = 1(1)n$,

and if this is so we can choose $^{(0)}x$ in any way we like.

EXAMPLE 10.3   For Equations [10-9] A is

$$\begin{bmatrix} 4 & -1 \\ -1 & 4 \end{bmatrix}$$

and $4 > |-1|$ for each row and each column. Hence the processes will converge.

For Equations [10-10] A is

$$\begin{bmatrix} -1 & 4 \\ 4 & -1 \end{bmatrix}$$

and $|-1| < |4|$ for each row and column. Hence the processes *may* not converge — and in fact they don't!

In Example 10.1

$$A = \begin{bmatrix} 4 & -1 & 0 & 0 \\ -1 & 4 & -1 & 0 \\ 0 & -1 & 4 & -1 \\ 0 & 0 & -1 & 2 \end{bmatrix}.$$

Here $|4| > |-1|$, $|4| > |-1| + |-1|$, $|4| > |-1| + |-1|$ and $|2| > |-1|$. The Gauss–Jacobi process converged nicely there, and so would the Gauss–Seidel.

EXERCISE 10.3

Write down A for the problem of Exercise 10.1 and confirm that A is diagonally dominant.

EXERCISE 10.4

Perform 10 iterations of the Gauss–Seidel method for the equations

$$\begin{bmatrix} 2 & -1 & 0 & 0 \\ -1 & 2 & -1 & 0 \\ 0 & -1 & 2 & -1 \\ 0 & 0 & -2 & 2 \end{bmatrix} \mathbf{x} = \begin{bmatrix} 0.4 \\ 0.8 \\ 1.2 \\ 0.6 \end{bmatrix}$$

starting with the zero vector. Comment on the rate of convergence. (The exact solution is $\mathbf{x} = \begin{bmatrix} 2.7 & 5.0 & 6.5 & 6.8 \end{bmatrix}^T$.)

10.6 SUCCESSIVE OVER-RELAXATION

The techniques of acceleration and relaxation described in Lesson 6 can be applied to matrix iteration too. We shall not describe the application of Aitken's Δ^2-process, although it is an important technique in this context, but shall concentrate on a particular method of relaxation that has become universally popular.

We could apply relaxation to any matrix iterative process. In fact we start with the Gauss–Seidel method [10-8] and, following the same course as in Lesson 6, rewrite it as

$$^{(m+1)}\mathbf{x} = {}^{(m)}\mathbf{x} + [\mathbf{D}^{-1}(\mathbf{b} - \mathbf{U}^{(m)}\mathbf{x} - \mathbf{D}^{(m)}\mathbf{x} - \mathbf{L}^{(m+1)}\mathbf{x})]$$

(since $\mathbf{D}\mathbf{D}^{-1} = \mathbf{I}$) so that the part in square brackets acts as a correction to $^{(m)}\mathbf{x}$, producing $^{(m+1)}\mathbf{x}$. Now we modify the correction by a relaxation parameter (the Greek letter omega $-\omega-$ is generally used in matrix iteration) to give

$$^{(m+1)}\mathbf{x} = {}^{(m)}\mathbf{x} + \omega[\mathbf{D}^{-1}(\mathbf{b} - \mathbf{U}^{(m)}\mathbf{x} - \mathbf{D}^{(m)}\mathbf{x} - \mathbf{L}^{(m+1)}\mathbf{x})]. \qquad [10\text{-}11]$$

This equation describes the method of *successive over-relaxation* (SOR for short). Equation [10-11] can be rearranged into the standard form

$$^{(m+1)}\mathbf{x} = \mathbf{E}^{(m)}\mathbf{x} + \mathbf{c}$$

giving

$$^{(m+1)}\mathbf{x} = (\mathbf{D} + \omega\mathbf{L})^{-1}[(1-\omega)\mathbf{D} - \omega\mathbf{U}]^{(m)}\mathbf{x} + \omega(\mathbf{D} + \omega\mathbf{L})^{-1}\mathbf{b}$$
$$[10\text{-}12]$$

Equation [10-12] allows us to analyse the convergence of the SOR method. An optimum value for ω can sometimes be found (rather as we made the relaxed processes second-order in Lesson 6), but this is difficult. We can, however, show that ω should normally satisfy $1 \leqslant \omega < 2$; in fact $\omega \geqslant 2$ causes divergence.

As we have previously pointed out, equations such as [10-12] are not used to perform the actual iteration, which is based on the individual (scalar) equations. For our three typical equations in their Gauss–Seidel form [10-6] we obtain

$$^{(m+1)}x_1 = {}^{(m)}x_1 + \omega[(b_1 - a_{11}\,{}^{(m)}x_1 - a_{12}\,{}^{(m)}x_2 - a_{13}\,{}^{(m)}x_3)/a_{11}]$$

$$^{(m+1)}x_2 = {}^{(m)}x_2 + \omega[(b_2 - a_{21}\,{}^{(m+1)}x_1 - a_{22}\,{}^{(m)}x_2 - a_{23}\,{}^{(m)}x_3)/a_{22}]$$

$$^{(m+1)}x_3 = {}^{(m)}x_3 + \omega[(b_3 - a_{31}\,{}^{(m+1)}x_1 - a_{32}\,{}^{(m+1)}x_2 - a_{33}\,{}^{(m)}x_3)/a_{33}]$$

$$[10\text{-}13]$$

with an obvious extension to the general case corresponding to Equations [10-7].

EXAMPLE 10.4   Applying SOR to the problem of Example 10.2 the iteration is defined by

$$^{(m+1)}x_1 = {}^{(m)}x_1 + \omega[0.1 - {}^{(m)}x_1 + {}^{(m)}x_2/4]$$

$$^{(m+1)}x_2 = {}^{(m)}x_2 + \omega[0.2 - {}^{(m)}x_2 + ({}^{(m+1)}x_1 + {}^{(m)}x_3)/4]$$

$$^{(m+1)}x_3 = {}^{(m)}x_3 + \omega[0.3 - {}^{(m)}x_3 + ({}^{(m+1)}x_2 + {}^{(m)}x_4)/4]$$

$$^{(m+1)}x_4 = {}^{(m)}x_4 + \omega[0.3 - {}^{(m)}x_4 + {}^{(m+1)}x_3/2]$$

with the following results, choosing $\omega = 1.2$.

m	$^{(m)}x_1$	$^{(m)}x_2$	$^{(m)}x_3$	$^{(m)}x_4$
0	0	0	0	0
1	0.12	0.276	0.443	0.628
2	0.262	0.396	0.579	0.582
3	0.186	0.390	0.536	0.565
4	0.200	0.383	0.537	0.569
5	0.195	0.383	0.538	0.569
6	0.196	0.384	0.538	0.569
7	0.196	0.384	0.538	0.569

We chose $\omega = 1.2$ because the results of Example 10.2 suggest that a 'small' overcorrection would be suitable. It appears that 20% is too much however, since we have caused an oscillation in the sequences, and have no quicker a convergence than with the Gauss—Seidel method. Let's try $\omega = 1.1$.

m	$^{(m)}x_1$	$^{(m)}x_2$	$^{(m)}x_3$	$^{(m)}x_4$
0	0	0	0	0
1	0.11	0.250	0.399	0.549
2	0.168	0.351	0.538	0.571
3	0.190	0.385	0.539	0.569
4	0.197	0.384	0.538	0.569
5	0.196	0.384	0.538	0.569

Clearly $\omega = 1.1$ is a better relaxation parameter value than 1.2. But with the Gauss—Seidel method converging in 7 iterations we would not expect a startling improvement. (Remember that we are not starting from a point particularly close to the solution.)

The SOR method has its most important application in problems where the Gauss—Seidel method is slow to converge, i.e., where the diagonal dominance is weak.

EXERCISE 10.5

Attempt to solve the problem of Exercise 10.4 by the SOR method, experimenting with the relaxation parameter to obtain reasonable convergence. (The best value is near 1.5. Try values both above and below this. To see what is happening you need to perform at least 10 iterations for each parameter value. An efficient way to do this would be by computer or by programmable electronic calculator.)

ANSWERS TO EXERCISES

10.1
$$^{(m+1)}x_1 = (-4 + {}^{(m)}x_2 + {}^{(m)}x_3 + {}^{(m)}x_4)/5$$
$$^{(m+1)}x_2 = (\ 12 + {}^{(m)}x_1 + {}^{(m)}x_3 + {}^{(m)}x_4)/10$$
$$^{(m+1)}x_3 = (\ \ 8 + {}^{(m)}x_1 + {}^{(m)}x_2 + {}^{(m)}x_4)/5$$
$$^{(m+1)}x_4 = (\ 34 + {}^{(m)}x_1 + {}^{(m)}x_2 + {}^{(m)}x_3)/10.$$

Working with 3D

m	$^{(m)}x_1$	$^{(m)}x_2$	$^{(m)}x_3$	$^{(m)}x_4$
0	0.9	1.9	2.9	3.9
1	0.94	1.97	2.94	3.97
2	0.976	1.985	2.976	3.985
3	0.989	1.994	2.989	3.994
4	0.995	1.997	2.995	3.997
5	0.998	1.999	2.999	3.999
6	0.999	2.000	2.999	4.000
7	1.000	2.000	3.000	4.000
8	1.000	2.000	3.000	4.000

Thus the answers are $\begin{bmatrix} 1.00 & 2.00 & 3.00 & 4.00 \end{bmatrix}^T$ all to 2D.

10.2
$$^{(m+1)}x_1 = (-4 + {}^{(m)}x_2 + {}^{(m)}x_3 + {}^{(m)}x_4)/5$$
$$^{(m+1)}x_2 = (\ 12 + {}^{(m+1)}x_1 + {}^{(m)}x_3 + {}^{(m)}x_4)/10$$

$$^{(m+1)}x_3 = (8 + {}^{(m+1)}x_1 + {}^{(m+1)}x_2 + {}^{(m)}x_4)/5$$
$$^{(m+1)}x_4 = (34 + {}^{(m+1)}x_1 + {}^{(m+1)}x_2 + {}^{(m+1)}x_3)/10.$$

The table is

m	$^{(m)}x_1$	$^{(m)}x_2$	$^{(m)}x_3$	$^{(m)}x_4$
0	0.9	1.9	2.9	3.9
1	0.94	1.974	2.963	3.988
2	0.985	1.994	2.993	3.997
3	0.997	1.999	2.999	4.000
4	1.000	2.000	3.000	4.000
5	1.000	2.000	3.000	4.000

The results are as before but are obtained in 5 iterations instead of 8.

10.3 $A = \begin{bmatrix} 5 & -1 & -1 & -1 \\ -1 & 10 & -1 & -1 \\ -1 & -1 & 5 & -1 \\ -1 & -1 & -1 & 10 \end{bmatrix}$

$$|5| > |-1| + |-1| + |-1| = 3$$
$$|10| > |-1| + |-1| + |-1| = 3$$
$$|5| > |-1| + |-1| + |-1| = 3$$
$$|10| > |-1| + |-1| + |-1| = 3.$$

10.4   After 10 iterations the process has reached

$$^{(10)}x = \begin{bmatrix} 2.0 & 3.9 & 5.1 & 5.4 \end{bmatrix}^T$$

to 1D, and is clearly converging very slowly to the solution. This is not at all surprising since the coefficient matrix is on the very borderline between being and not being diagonally dominant.

10.5   The optimum ω is near 1.475, which obtains 2D solutions in 10 iterations and 5D in about 20 iterations. For ω between 1 and about 1.45 convergence is monotonic (i.e., without overshoot and oscillation about the solution). Above 1.45 the process overshoots the solution and then returns. If ω is too large it overshoots again and continues to oscillate.

FURTHER READING

Goult, Hoskins, Milner and Pratt, Sections 4.1, 4.2 and 4.3.1, covers the same ground as us and in a straightforward manner. In other parts of Chapter 4 this book deals with the theory of the convergence of the iterative methods discussed, which is beyond the scope of our course.

For a slightly different approach see Conte, Section 5.8. (*NB Not* Conte and deBoor.)

For a short, elementary treatment of the Gauss—Seidel method see Hosking, Joyce and Turner, 'Step 13'.

LESSON 10 — COMPREHENSION TEST

1. Write in mathematical symbols the following convergence criterion for a sequence of vectors each with 4 elements.

'The magnitude of the difference between successive corresponding elements shall be less than $\frac{1}{2} \times 10^{-5}$ for all elements'.

2. For the following equations give the recurrence relations for each of the Gauss—Jacobi, Gauss—Seidel and SOR methods.

$$3x_1 - x_2 = 6$$
$$-2x_1 + 5x_2 = 5$$

3. What is the usual starting vector for the iterative processes of this lesson?

4. State a sufficient condition for the convergence of the Gauss—Jacobi and Gauss—Seidel methods.

5. Which of these matrices is diagonally dominant:

(a) $\begin{bmatrix} 3 & -1 \\ -2 & 5 \end{bmatrix}$; (b) $\begin{bmatrix} 2 & -1 & 0 \\ -1 & 2 & -1 \\ 0 & -2 & 2 \end{bmatrix}$; (c) $\begin{bmatrix} 10 & 1 & -2 \\ 0 & 8 & 3 \\ -2 & 3 & 4 \end{bmatrix}$?

6. (a) What is the normal range of the relaxation parameter ω for the SOR method?

(b) What method does the SOR method reduce to when $\omega = 1$?

ANSWERS

1 $|^{(m+1)}x_j - {}^{(m)}x_j| < \frac{1}{2} \times 10^{-5}$, $j = 1(1)4$, for some m.

2 G—J ${}^{(m+1)}x_1 = (6 + {}^{(m)}x_2)/3$

  ${}^{(m+1)}x_2 = (5 + 2\,{}^{(m)}x_1)/5$

  G—S ${}^{(m+1)}x_1 = (6 + {}^{(m)}x_2)/3$

  ${}^{(m+1)}x_2 = (5 + 2\,{}^{(m+1)}x_1)/5$

SOR $\quad {}^{(m+1)}x_1 = {}^{(m)}x_1 + \omega[(6 - 3\,{}^{(m)}x_1 + {}^{(m)}x_2)/3]$

$\qquad\quad {}^{(m+1)}x_2 = {}^{(m)}x_2 + \omega[(5 + 2\,{}^{(m+1)}x_1 - 5\,{}^{(m)}x_2)/5].$

3　Zero vector.

4　That the coefficient matrix $\mathbf{A}$ in the equation $\mathbf{Ax} = \mathbf{b}$ shall be diagonally dominant (as defined in Section 10.5).

5　(a) only. (b) fails the column test in column 2, and just fails the row test on rows 2 and 3. In fact this is a borderline case and the corresponding iteration may well converge although the matrix fails the test. It is the kind of matrix for which SOR is effective. (c) fails both row and column tests on row and column 3. Again it may still provide a convergent method this time because of the dominance of the first two diagonal elements.

6　(a) $1 < \omega < 2$;　(b) Gauss-Seidel.

LESSON 10 – SUPPLEMENTARY EXERCISES

10.6. Apply the Gauss—Jacobi and Gauss—Seidel methods to

$$\begin{bmatrix} 1 & 2 & -2 \\ 1 & 1 & 1 \\ 2 & 2 & 1 \end{bmatrix} \mathbf{x} = \begin{bmatrix} 1 \\ 0 \\ 1 \end{bmatrix}.$$

Comment on your results. Repeat the calculations with the first and third equations interchanged. Comment again!

10.7. Solve the equations

$$3.04x_1 + 6.18x_2 + 1.22x_3 = 8.20$$
$$-2.44x_1 + 1.22x_2 + 8.44x_3 = 3.93$$
$$9.37x_1 + 3.04x_2 - 2.44x_3 = 9.23$$

using both the Gauss—Jacobi and the Gauss—Seidel methods and starting with the same initial vector. Give your answers correct to 2D. Comment on the effectiveness of the two methods.

10.8. Apply the Gauss—Seidel method to solving the system

$$\begin{bmatrix} 7.2 & 0.1 & 0.3 & 0.4 & -0.5 \\ 0.5 & 6.4 & 0.1 & 0.2 & 0.4 \\ 0 & 0.4 & 7.3 & 0.6 & 0.2 \\ 0 & 0 & 0.4 & 6.3 & 0.5 \\ 0 & 0 & 0 & 0.1 & 7.2 \end{bmatrix} \mathbf{x} = \begin{bmatrix} 0.1 \\ 0.4 \\ -0.3 \\ 1.1 \\ -0.1 \end{bmatrix}$$

giving solutions correct to 2D.

10.9. Apply SOR to the following system, experimenting with the relaxation parameter until a much-improved convergence rate is achieved.

$$\begin{bmatrix} -3 & 1 & 0 & 2 \\ 1 & -3 & 1 & 0 \\ 0 & 1 & -3 & 1 \\ 2 & 0 & 1 & -3 \end{bmatrix} \mathbf{x} = \begin{bmatrix} 1 \\ 2 \\ 3 \\ 4 \end{bmatrix}.$$

10.10. Solve the system of equations given in Example 1.3 when $a_i = 1$, $i = 1(1)8$. Use any suitable iterative method and obtain the answers correct to 2D.

PART D INTERPOLATION AND APPROXIMATION BY POLYNOMIALS

Lesson 11 Taylor Polynomials

OBJECTIVES

At the end of this lesson you should

(a) understand and be able to use the abbreviated notations f_k and $f_k^{(n)}$;

(b) understand what Taylor polynomials are and be able to construct them in simple cases.

CONTENTS

11.1 An abbreviated function notation

11.2 Taylor polynomials

Answers to exercises. Further reading.
Comprehension test and answers. Supplementary exercises.

11.1 AN ABBREVIATED FUNCTION NOTATION

There is one bit of notation to look at before we get to the meat of this lesson. We shall often need to evaluate $f, f^{(1)}, f^{(2)}, \ldots$ for a given function f at some fixed point say x_0. Instead of writing $f(x_0), f^{(1)}(x_0), f^{(2)}(x_0), \ldots$ we shall abbreviate to $f_0, f_0^{(1)}, f_0^{(2)}, \ldots$. In general we may have a sequence of values $x_0, x_1, x_2, \ldots$ and the values of $f, f^{(1)}, f^{(2)}, \ldots$ at a general point x_k may be written $f_k, f_k^{(1)}, f_k^{(2)}, \ldots$. In other words the subscript k denotes the point x_k at which the functions are evaluated.

EXAMPLE 11.1 (a) Let $f(x) = x^3 - 3x^2 + 7x - 11$, $x \in \mathbb{R}$. Then, supposing that $x_0 = 1$, we find that

$$f_0 = (1)^3 - 3(1)^2 + 7(1) - 11 = -6$$

$$f_0^{(1)} = 3(1)^2 - 6(1) + 7 = 4$$

$$f_0^{(2)} = 0$$

$$f_0^{(3)} = 6$$

$$f_0^{(n)} = 0, \quad n = 4, 5, 6, \dots.$$

(Are you checking?!)

(b) Let $h(x) = \log_e(1 + x)$, $x > -1$. Then

$$h^{(1)}(x) = 1/(1 + x)$$

$$h^{(2)}(x) = -1/(1 + x)^2$$

$$h^{(3)}(x) = 2/(1 + x)^3$$

. . . .

$$h^{(n)}(x) = (-1)^{n-1}(n-1)!/(1+x)^n, \quad n \in \mathbb{N}.$$

Now let $x_0 = 0$, $x_1 = \frac{1}{2}$, $x_2 = 1, \dots$.

Then

$$h_0 = 0, h_0^{(1)} = 1, h_0^{(2)} = -1, h_0^{(3)} = 2, \dots, h_0^{(n)} = (-1)^{n-1}(n-1)!$$

$$h_1 = \log_e \tfrac{3}{2}, h_1^{(1)} = \tfrac{2}{3}, h_1^{(2)} = -\tfrac{4}{9}, h_1^{(3)} = \tfrac{16}{27}, \dots,$$

$$h_1^{(n)} = (-1)^{n-1}(n-1)!(\tfrac{2}{3})^n$$

etc.

(Are you still checking?)

EXERCISE 11.1

(a) Let $f(x) = x^4 - 3x^2 + 5x + 11$, $x \in \mathbb{R}$, and $x_1 = 2$. Evaluate f_1, $f_1^{(1)}, f_1^{(2)}, f_1^{(3)}$.

(b) Let $g(x) = \sin x \cos x$, $x \in \mathbb{R}$, and $x_0 = \pi/2$. Evaluate $g_0, g_0^{(1)}$, $g_0^{(2)}, g_0^{(3)}$.

We are now ready to tackle the main part of this lesson which is about approximating functions by polynomials. As a recap on polynomials try this exercise.

EXERCISE 11.2

(a) Starting from the general polynomial function

$$p_n(x) = a_0 + a_1 x + \dots + a_n x^n, \quad x \in \mathbb{R}$$

write down the functions

$$p_n^{(1)}, p_n^{(2)}, p_n^{(3)}, \ldots.$$

Hence deduce that $p_n^{(n)}$ is a constant function and that $p_n^{(k)}(x) = 0$, $k > n$.

(b) Given any $x_0 \in \mathbb{R}$,

$$q_n(x) = b_0 + b_1(x - x_0) + b_2(x - x_0)^2 + \ldots + b_n(x - x_0)^n$$

is also a perfectly general nth degree polynomial. Show that

$$q_n(x_0) = b_0$$
$$q_n^{(1)}(x_0) = b_1 = 1!\, b_1$$
$$q_n^{(2)}(x_0) = 2b_2 = 2!\, b_2$$
$$q_n^{(3)}(x_0) = 6b_3 = 3!\, b_3$$

etc. until finally

$$q_n^{(n)}(x_0) = n!\, b_n.$$

.2 TAYLOR POLYNOMIALS

In numerical mathematics polynomials are very important if only because the operations we have available are addition, subtraction, multiplication and division. Polynomials can be evaluated using these operations alone.

Suppose that we wish to evaluate $\sin x$ for a particular value of x using a computer. The machine cannot draw right-angled triangles and measure lengths, and it would be inefficient to store sufficiently accurate tables of values. For various reasons we wish to leave $\sin x$ for the moment.

Consider instead the equally difficult problem of evaluating e^x for some small value of x, i.e., at some point of its graph near the origin.

If we draw the tangent to $y = e^x$ at $x = 0$, $y = 1$, we can see from Figure 11.1 that, for small values of x, $p_1(x)$ is close to e^x. In particular for $x = \alpha$ the point B is near to the point A. (We have called the tangent at $x = 0$, $y = p_1(x)$ because the tangent is the graph of a first degree polynomial.)

The conditions which determine $p_1(x)$ are that it has the same value and gradient as e^x at $x = 0$, i.e.

$$p_1(0) = e^0 \quad (= 1)$$

and

$$p_1^{(1)}(0) = e^0 \quad (= 1).$$

If $p_1(x) = a_0 + a_1 x$ then $p_1(0) = a_0$ and $p_1^{(1)}(0) = a_1$, so $a_0 = a_1 = 1$ and

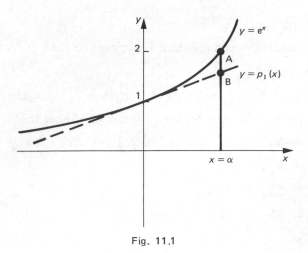

Fig. 11.1

$p_1(x) = 1 + x$. This is our first *Taylor-polynomial approximation*, a first degree approximation to e^x, based on $x = 0$.

Of course if α were large, $p_1(\alpha)$ would not be very effective in approximating e^α. Thus we need to generalise the approach in such a way that the results will be good for larger values of x. The analysis of how good the results are is left to Lesson 12.

Before making the generalisation we should note that there are two processes involved:

(a) the approximation of the function (be it the sine function, the exponential function or whatever) by some more easily managed function (like a polynomial);

(b) the evaluation of the second function.

In this lesson we are only concerned with (a), specifically with *constructing* polynomial approximations to functions. We choose to use polynomials because they are, amongst other things, easy to evaluate by hand or inside computer programs. Hence (b) will be straightforward.

In our example we put

$$p_1(0) = f(0) \quad \text{and} \quad p_1^{(1)}(0) = f^{(1)}(0)$$

(replacing e^x by $f(x)$ for generality); i.e., we made the polynomial and $f(x)$ equal at $x = 0$ and we made the derivatives of the polynomial and $f(x)$ equal at $x = 0$. To generalise further, take $p_n(x)$ as the polynomial (of general degree n), and make the polynomial and *all* its derivatives equal to the function and its corresponding derivatives at the general point $x = x_0$.

More tidily let $f(x)$ be defined in $[a, b]$ and let $x_0 \in [a, b]$.

Let

$$p_n(x) = a_0 + a_1 x + a_2 x^2 + \ldots + a_n x^n, \quad x \in [a, b].$$

Then we require

$$
\begin{aligned}
p_n(x_0) &= f(x_0) \\
p_n^{(1)}(x_0) &= f^{(1)}(x_0) \\
p_n^{(2)}(x_0) &= f^{(2)}(x_0) \\
&\cdots\cdots\cdots\cdots \\
p_n^{(n)}(x_0) &= f^{(n)}(x_0)
\end{aligned}
\qquad \text{[11-1]}
$$

which gives explicitly the $n + 1$ conditions which we need to define the $n + 1$ coefficients $a_0, a_1, a_2, \ldots, a_n$.

To be able to do this we have to assume that $f(x)$ can be differentiated n times. More precisely we shall assume that $f^{(n)}(x_0)$ exists. For many common functions this is true for all n and all x_0 in the interval where the function is defined but you should be aware that this is not always so. For example, if $f(x) = \sqrt{x}$ what is $f^{(1)}(0)$?

To see more clearly what the conditions mean let us write them out in full. The left-hand sides are the answers to Exercise 11.2(a) evaluated at $x = x_0$.

$$
\begin{aligned}
a_0 + a_1 x_0 + a_2 x_0^2 + \ldots + \quad & a_{n-1} x_0^{n-1} + \quad a_n x_0^n = f(x_0) \\
a_1 + 2a_2 x_0 + \ldots + (n-1)\, & a_{n-1} x_0^{n-2} + n\, a_n x_0^{n-1} = f^{(1)}(x_0) \\
& \cdots\cdots\cdots\cdots\cdots\cdots\cdots\cdots \\
(n-1)!\, a_{n-1} \quad & + n!\, a_n x_0 = f^{(n-1)}(x_0) \\
& n!\, a_n = f^{(n)}(x_0).
\end{aligned}
$$

$$\text{[11-2]}$$

The last equation defines a_n immediately. Knowing a_n, we can use the next to last equation to find a_{n-1}. Working backwards through the set of $n + 1$ equations in $n + 1$ unknowns we can clearly evaluate $a_n, a_{n-1}, a_{n-2}, \ldots, a_2, a_1, a_0$, uniquely, i.e., given values for $f(x_0), f^{(1)}(x_0), \ldots, f^{(n)}(x_0)$ there is just one value for each of the a_i, $i = 0, 1, 2, \ldots, n$. (This process is of course back-substitution.)

EXERCISE 11.3

Write Equations [11-2] in matrix form $\mathbf{Ba} = \mathbf{c}$ where $\mathbf{a} = \begin{bmatrix} a_0 & a_1 & a_2 \cdots a_n \end{bmatrix}^T$:

(a) when $n = 4$,

(b) for the general case.

Although we can see that we could find the a_i in a particular case it would be nice to be able to give an answer for the general case which requires no solution process. We can do this by choosing a different definition of p_n, which is nevertheless still a perfectly general nth degree polynomial.

Let

$$p_n(x) = b_0 + b_1(x - x_0) + b_2(x - x_0)^2 + \ldots + b_n(x - x_0)^n$$

and apply the same conditions as before. We get (see Exercise 11.2(b))

$$
\begin{aligned}
b_0 &= f(x_0) \\
b_1 &= f^{(1)}(x_0) \\
2b_2 &= f^{(2)}(x_0) \\
&\cdots\cdots\cdots\cdots\cdots \\
(n-1)!\, b_{n-1} &= f^{(n-1)}(x_0) \\
n!\, b_n &= f^{(n)}(x_0)
\end{aligned}
\qquad \text{[11-3]}
$$

and the b_i are all given immediately without any back-substitution.

EXERCISE 11.4

Write Equation [11-3] in matrix form $\mathbf{Ab} = \mathbf{c}$ where $\mathbf{b} = \begin{bmatrix} b_0 & b_1 & b_2 & \ldots & b_n \end{bmatrix}^T$:

(a) when $n = 4$,

(b) for the general case.

From Equation [11-3] we can write down p_n explicitly:

$$p_n(x) = f(x_0) + (x - x_0) f^{(1)}(x_0) + \frac{(x - x_0)^2}{2!} f^{(2)}(x_0) + \ldots$$

$$+ \frac{(x - x_0)^{n-1}}{(n-1)!} f^{(n-1)}(x_0) + \frac{(x - x_0)^n}{n!} f^{(n)}(x_0). \qquad \text{[11-4]}$$

Using the abbreviated notation described in Section 11.1 we often write Equation [11-4] as

$$p_n(x) = f_0 + (x - x_0) f_0^{(1)} + \frac{(x - x_0)^2}{2!} f_0^{(2)} + \ldots + \frac{(x - x_0)^n}{n!} f_0^{(n)}.$$

$$\text{[11-5]}$$

The right hand side of Equation [11-4] or Equation [11-5] is known as the *Taylor polynomial of degree n for $f(x)$ constructed at the point $x = x_0$.*

EXAMPLE 11.2 To find the Taylor polynomial of degree 7 for $\sin x$ constructed at the point $x = 0$.

$$f(x) \quad = \sin x \qquad\qquad f(0) \quad = 0$$
$$f^{(1)}(x) \quad = \cos x \qquad\qquad f^{(1)}(0) = 1$$
$$f^{(2)}(x) \quad = -\sin x \qquad\qquad f^{(2)}(0) = 0$$
$$f^{(3)}(x) \quad = -\cos x \qquad\qquad f^{(3)}(0) = -1$$
$$f^{(4)}(x) \quad = \sin x \qquad\qquad f^{(4)}(0) = 0$$
$$f^{(5)}(x) \quad = \cos x \qquad\qquad f^{(5)}(0) = 1$$
$$f^{(6)}(x) \quad = -\sin x \qquad\qquad f^{(6)}(0) = 0$$
$$f^{(7)}(x) \quad = -\cos x \qquad\qquad f^{(7)}(0) = -1.$$

$$\therefore \quad p_7(x) = x - \frac{x^3}{3!} + \frac{x^5}{5!} - \frac{x^7}{7!}$$

is the required Taylor polynomial for $\sin x$ constructed at $x = 0$.

Notice that p_7 contains p_1 to p_6. In fact

$$p_1(x) \quad = p_2(x) = x$$
$$p_3(x) \quad = p_4(x) = x - x^3/3!$$
$$p_5(x) \quad = p_6(x) = x - x^3/3! + x^5/5!$$

In general $p_n(x)$ contains $p_1(x)$ through $p_{n-1}(x)$. For the sine function we can draw the graphs of the successive Taylor polynomial approximations to see how they improve as the degree increases. This is shown in Fig. 11.2.

EXAMPLE 11.3 To find the Taylor polynomial of degree n for e^x constructed at the point $x = a$.

$$f(x) \quad = e^x \qquad\qquad f(a) \quad = e^a$$
$$f^{(1)}(x) = e^x \qquad\qquad f^{(1)}(a) = e^a.$$

In fact

$$f^{(n)}(x) = e^x \qquad\qquad f^{(n)}(a) = e^a \quad \text{for all } n \in \mathbb{N}.$$

$$\therefore \quad p_n(x) = e^a + (x - a) e^a + \frac{(x - a)^2}{2!} e^a + \ldots + \frac{(x - a)^n}{n!} e^a$$

is the required Taylor polynomial for e^x constructed at $x = a$. We can tidy up this equation by writing

$$p_n(x) = e^a(1 + (x - a) + (x - a)^2/2! + \ldots + (x - a)^n/n!).$$

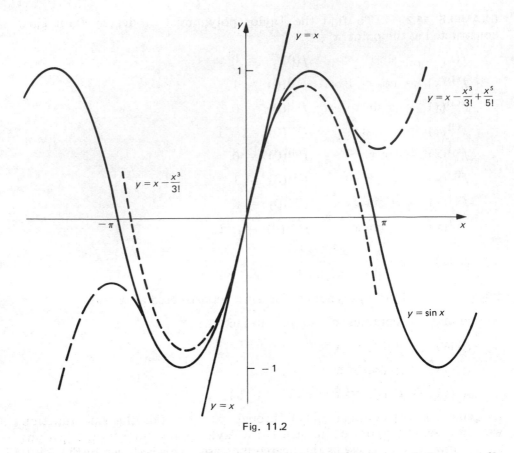

$y = x$

$y = x - \dfrac{x^3}{3!} + \dfrac{x^5}{5!}$

$y = x - \dfrac{x^3}{3!}$

$y = \sin x$

$y = x$

Fig. 11.2

When the Taylor polynomial is to be constructed at $x = 0$ it is usually called the *Maclaurin* polynomial. $p_7(x)$ in Example 11.2 is the Maclaurin polynomial of degree 7 for $\sin x$.

EXAMPLE 11.4   To find the Maclaurin polynomial of degree n for $(1 + x)^6$.

$$f(x) \quad = (1 + x)^6 \qquad\qquad f(0) \quad = 1$$
$$f^{(1)}(x) = 6(1 + x)^5 \qquad\qquad f^{(1)}(0) = 6$$
$$f^{(2)}(x) = 6 \times 5 \times (1 + x)^4 \qquad f^{(2)}(0) = 6 \times 5$$
$$f^{(3)}(x) = 6 \times 5 \times 4 \times (1 + x)^3 \qquad f^{(3)}(0) = 6 \times 5 \times 4$$

$$\cdots\cdots\cdots\cdots\cdots\cdots\cdots\cdots\cdots\cdots\cdots\cdots\cdots\cdots$$

$$f^{(6)}(x) = 6! \qquad\qquad f^{(6)}(0) = 6!$$
$$f^{(7)}(x) = f^{(8)}(x) = \ldots = 0 \qquad f^{(7)}(0) = f^{(8)}(0) = \ldots = 0.$$

Hence we can write down Maclaurin polynomials of degree 0 (a constant) through 6. Those of degree 7 or higher are the same as that of degree 6. Since p_6 contains p_0 through p_5 let us write p_6:

$$p_6(x) = 1 + 6x + \frac{6 \cdot 5}{2!}x^2 + \frac{6 \cdot 5 \cdot 4}{3!}x^3 + \frac{6 \cdot 5 \cdot 4 \cdot 3}{4!}x^4$$

$$+ \frac{6 \cdot 5 \cdot 4 \cdot 3 \cdot 2}{5!}x^5 + \frac{6!}{6!}x^6$$

$$= 1 + 6x + 15x^2 + 20x^3 + 15x^4 + 6x^5 + x^6.$$

This is exactly equal to $(1 + x)^6$ — multiply it out if you need convincing! Intuitively we would expect that a 6th degree polynomial approximation to a 6th degree polynomial would be the original polynomial.

If you have seen binomial expansions before you will recognise p_6 as one. Example 11.5 is a brief account of the general case.

EXAMPLE 11.5   To find the Maclaurin polynomial of degree n for $(1 + x)^m$, for any $m \in \mathbb{R}$.

$$\begin{aligned}
f(x) &= (1 + x)^m & f(0) &= 1 \\
f^{(1)}(x) &= m(1 + x)^{m-1} & f^{(1)}(0) &= m \\
f^{(2)}(x) &= m(m - 1)(1 + x)^{m-2} & f^{(2)}(0) &= m(m - 1)
\end{aligned}$$

$$\cdots \cdots \cdots \cdots \cdots \cdots \cdots \cdots \cdots \cdots$$

$$f^{(k)}(x) = m(m - 1) \ldots (m - k + 1)(1 + x)^{m-k}, \quad k = 1, 2, 3, \ldots$$

$$\therefore \quad f^{(k)}(0) = m(m - 1) \ldots (m - k + 1), \quad k = 1, 2, 3, \ldots$$

and

$$p_n(x) = 1 + mx + \frac{m(m - 1)}{2!}x^2 + \ldots + \frac{m(m - 1) \ldots (m - k + 1)}{k!}x^k$$

$$+ \ldots + \frac{m(m - 1) \ldots (m - n + 1)}{n!}x^n.$$

If $m \in \mathbb{N}$ and we take $n = m$ we have the well-known binomial expansions

$$(1 + x)^1 = 1 + x$$
$$(1 + x)^2 = 1 + 2x + x^2$$
$$(1 + x)^3 = 1 + 3x + 3x^2 + x^3$$

etc.,

the coefficients of which form 'Pascal's Triangle':

$$
\begin{array}{ccccccc}
 & & 1 & & 1 & & \\
 & 1 & & 2 & & 1 & \\
1 & & 3 & & 3 & & 1 \\
\end{array}
$$

1   1

1   2   1

1   3   3   1

1   4   6   4   1

1   5   10   10   5   1

so that, for example

$$(1 + x)^4 \ = \ 1 + 4x + 6x^2 + 4x^3 + x^4.$$

The general expansion is often written in terms of the *binomial coefficients*

$$\binom{m}{k} \ = \ \frac{m(m-1)\ldots(m-k+1)}{k!}.$$

When $m \in \mathbb{N}$ this is sometimes written as

$$^mC_k \ = \ \frac{m!}{k!\,(m-k)!}.$$

Hence we have

$$(1 + x)^m \ = \ \sum_{k=0}^{m} {}^mC_k x^k \quad \text{for } m \in \mathbb{N}$$

and

$$p_n(x) \ = \ \sum_{k=0}^{n} \binom{m}{k} x^k \quad \text{otherwise.}$$

To summarise, the Taylor polynomial of degree n for $f(x)$ constructed at x_0 is an approximation to $f(x)$ which will be very good near x_0 (because the polynomial and its derivatives (up to the nth) have the same values as f and its derivatives at x_0) but may not be very good away from x_0. As we have said before, we shall return in Lesson 12 to how good the approximation is.

EXERCISE 11.5

Find the Taylor (or Maclaurin) polynomials for the following functions with the degrees indicated, and constructed at the points indicated.

	Function	Degree	Point
(a)	$\cos x$	7	0
(b)	$\sin x$	4	$\pi/4$
(c)	e^x	n	0
(d)	$\ln(1+x)$	7	0
(e)	$\ln(1-x)$	n	0
(f)	$\ln x$	5	1
(g)	$\ln x$	n	0
(h)	x^4	3	1
(i)	x^4	4	1
(j)	x^4	3	0

ANSWERS TO EXERCISES

11.1 (a) $f_1 = 25$, $f_1^{(1)} = 25$, $f_1^{(2)} = 42$, $f_1^{(3)} = 48$

(b) $g_0 = 0$, $g_0^{(1)} = -1$, $g_0^{(2)} = 0$, $g_0^{(3)} = 4$.

11.2 (a) $p_n^{(1)}(x) = a_1 + 2a_2 x + 3a_3 x^2 + \ldots + na_n x^{n-1}$

$p_n^{(2)}(x) = 2a_2 + 6a_3 x + \ldots + n(n-1)a_n x^{n-2}$

$p_n^{(3)}(x) = 6a_3 + \ldots + n(n-1)(n-2)a_n x^{n-3}$

and so on. The degree drops by one at each differentiation and reaches zero at the nth derivative which is

$$p_n^{(n)}(x) = n!\, a_n = \text{constant}.$$

Clearly further differentiations give only the zero function.

11.3 (a)
$$\begin{bmatrix} 1 & x_0 & x_0^2 & x_0^3 & x_0^4 \\ 0 & 1 & 2x_0 & 3x_0^2 & 4x_0^3 \\ 0 & 0 & 2 & 6x_0 & 12x_0^2 \\ 0 & 0 & 0 & 6 & 24x_0 \\ 0 & 0 & 0 & 0 & 24 \end{bmatrix} \begin{bmatrix} a_0 \\ a_1 \\ a_2 \\ a_3 \\ a_4 \end{bmatrix} = \begin{bmatrix} f_0 \\ f_0^{(1)} \\ f_0^{(2)} \\ f_0^{(3)} \\ f_0^{(4)} \end{bmatrix}$$

(b)
$$\begin{bmatrix} 1 & x_0 & x_0^2 & \cdots & x_0^{n-1} & x_0^n \\ & 1 & 2x_0 & \cdots & (n-1)x_0^{n-2} & nx_0^{n-1} \\ & & 2 & \cdots & & n(n-1)x_0^{n-2} \\ & & & \ddots & & \vdots \\ & & & & (n-1)! & n!\,x_0 \\ & & & & & n! \end{bmatrix} \begin{bmatrix} a_0 \\ a_1 \\ a_2 \\ \vdots \\ a_{n-1} \\ a_n \end{bmatrix} = \begin{bmatrix} f_0 \\ f_0^{(1)} \\ f_0^{(2)} \\ \vdots \\ f_0^{(n-1)} \\ f_0^{(n)} \end{bmatrix}$$

11.4 (a)
$$\begin{bmatrix} 1 & & & & \\ & 1 & & & \\ & & 2 & & \\ & & & 6 & \\ & & & & 24 \end{bmatrix} \begin{bmatrix} b_0 \\ b_1 \\ b_2 \\ b_3 \\ b_4 \end{bmatrix} = \begin{bmatrix} f_0 \\ f_0^{(1)} \\ f_0^{(2)} \\ f_0^{(3)} \\ f_0^{(4)} \end{bmatrix}.$$

(b)
$$\begin{bmatrix} 1 & & & & & \\ & 1 & & & & \\ & & 2 & & & \\ & & & \ddots & & \\ & & & & (n-1)! & \\ & & & & & n! \end{bmatrix} \begin{bmatrix} b_0 \\ b_1 \\ b_2 \\ \vdots \\ b_{n-1} \\ b_n \end{bmatrix} = \begin{bmatrix} f_0 \\ f_0^{(1)} \\ f_0^{(2)} \\ \vdots \\ f_0^{(n-1)} \\ f_0^{(n)} \end{bmatrix}.$$

11.5 (a) $p_7(x) = 1 - \dfrac{x^2}{2!} + \dfrac{x^4}{4!} - \dfrac{x^6}{6!}$.

(b) $p_4(x) = \dfrac{1}{\sqrt{2}} \left(1 + \dfrac{(x - \pi/4)}{1!} - \dfrac{(x - \pi/4)^2}{2!} - \dfrac{(x - \pi/4)^3}{3!} \right.$

$\left. + \dfrac{(x - \pi/4)^4}{4!} \right)$.

(c) $p_n(x) = 1 + x + x^2/2! + x^3/3! + \ldots + x^n/n!$

(d) $p_7(x) = x - x^2/2 + x^3/3 - x^4/4 + x^5/5 - x^6/6 + x^7/7$.

(e) $p_n(x) = -x - x^2/2 - x^3/3 - \ldots - x^n/n$.

(f) $p_5(x) = (x - 1) - (x - 1)^2/2 + (x - 1)^3/3 - (x - 1)^4/4$

$+ (x - 1)^5/5$.

(g) Not possible.

(h) $1 + 4(x - 1) + 6(x - 1)^2 + 4(x - 1)^3$.

(i) $1 + 4(x - 1) + 6(x - 1)^2 + 4(x - 1)^3 + (x - 1)^4 \quad (= x^4?)$.

(j) 0.

FURTHER READING

Most books develop Taylor *series* rather than Taylor polynomials and we are leaving consideration of series until Lesson 12. If you can accept for the moment the idea of a Taylor polynomial of degree n where n can be as large as you like:

$$f(x) \simeq f_0 + (x - x_0) f_0^{(1)} + \frac{(x - x_0)^2}{2!} f_0^{(2)} + \dots$$

then the following reference may suit you: Stark, Chapter 2, pp. 30–36 and (optionally) pp. 37–38.

LESSON 11 – COMPREHENSION TEST

1. Let $p_2(x)$ be the Taylor quadratic polynomial for $f(x)$ constructed at $x = -3$. What are the three conditions that define $p_2(x)$?

2. State the mth degree Taylor polynomial for a general function $f(x)$ constructed at a general point $x = x_0$.

3. State (without calculation) the Maclaurin cubic polynomial for $f(x) = 2x^3 + 3x^2 + x - 4$.

4. Obtain the Taylor cubic polynomial for $f(x) = 2x^3 + 3x^2 + x - 4$ constructed at $x = -2$. How is the answer related to the answer to 3?

5. Obtain the linear Taylor polynomials for $f(x) = \tan x$ constructed: (a) at $x = 0$, and (b) at $x = \pi/4$.
Give a geometric description of these polynomials.

6. Let $p_n(x)$ be the Taylor polynomial for $f(x)$ constructed at $x = x_0$ where $x_0 \geqslant 0$. When are the following true:
(a) $p_n(x) = f(x), \quad x \in \mathbb{R}$
(b) $p_n(x) = f(x), \quad 0 \leqslant x \leqslant x_0$
(c) $p_n(x_0) = f(x_0)$
(d) $p_n^{(r)}(x_0) = f^{(r)}(x_0), \quad r \in \mathbb{N}$?

ANSWERS

1 $p_2(-3) = f(-3), \quad p_2^{(1)}(-3) = f^{(1)}(-3), \quad p_2^{(2)}(-3) = f^{(2)}(-3)$.

2 $p_m(x) = f(x_0) + (x - x_0) f^{(1)}(x_0) + \frac{1}{2!}(x - x_0)^2 f^{(2)}(x_0)$

$$+ \dots + \frac{(x - x_0)^m}{m!} f^{(m)}(x_0).$$

3 $-4 + x + 3x^2 + 2x^3$.

4   $-10 + 13(x + 2) - 9(x + 2)^2 + 2(x + 2)^3$. They are identical (in value).

5   (a) x (b) $1 + 2x$.   They are tangents to $y = \tan x$ at $x = 0$ and $x = \pi/4$.

6   (a) When f is a polynomial of degree n or less.
    (b) Ditto for $x \in [0, x_0]$.
    (c) Always.
    (d) When $r \leqslant n$.

LESSON 11 – SUPPLEMENTARY EXERCISES

11.6. Obtain the Taylor polynomials of the given degree n for the given functions $f(x)$, constructed at the given points $x = x_0$.

	$f(x)$	n	x_0
(a)	$\cosh x$	6	0
(b)	$\cosh x$	m	0
(c)	$\sinh x$	7	0
(d)	$\sinh x$	m	0
(e)	$\arctan x$	3	0
(f)	e^{-x^2}	4	0
(g)	$\sec x$	4	$\pi/3$

*11.7. Show that the Taylor polynomial of degree $m - 1$ for $f^{(1)}(x)$, constructed at $x = x_0$, is the derivative of the Taylor polynomial of degree m for $f(x)$, also constructed at $x = x_0$. Generalise this fact.

*11.8. By replacing x by $x_0 + h$ in the general form of Equation [11-5] obtain

$$p_n(x_0 + h) = f_0 + hf_0^{(1)} + \frac{h^2}{2!}f_0^{(2)} + \ldots + \frac{h^n}{n!}f_0^{(n)}.$$

Hence deduce that, if h is small,

$$f_0^{(1)} \simeq \{f(x_0 + h) - f(x_0 - h)\}/2h$$

$$f_0^{(2)} \simeq \{f(x_0 + h) - 2f(x_0) + f(x_0 - h)\}/h^2.$$

Lesson 12 Taylor Series

OBJECTIVES

By the end of this lesson you should be able to

(a)   understand the concept of a remainder term;

(b)   find remainder terms in simple cases;

(c)   appreciate the ideas of sequence, convergence, infinite series, truncation error, order of approximation;

(d)   say what Taylor and Maclaurin series are;

(e)   (i) estimate a bound on $R_n(x_0 + h)$ given n and h, or
(ii) given a maximum requirement on $|R_n(x_0 + h)|$ estimate n given h, or h given n (if possible!).

CONTENTS

2.1   THE ERROR TERM

In Lesson 11 we discovered Taylor polynomials, which enable us to calculate approximations to the values of functions which we would not otherwise be able to obtain. Let f be the function approximated and p_n be a Taylor polynomial for f. In this lesson we shall consider the error in the approximation, i.e.,

$$\text{true} - \text{approximate} = f - p_n.$$

This *error function* is called the *remainder* and is denoted by R_n; i.e.,

$$R_n = f - p_n$$

or $\quad f = p_n + R_n.$

When we use a value $p_n(x)$ to approximate $f(x)$, for some x, the error is now the number $R_n(x)$ defined by

$$R_n(x) = f(x) - p_n(x).$$

First of all we shall consider the case $n = 0$ denoting approximation by polynomials of degree zero, i.e., constants. From Lesson 11 we know that

$$p_0(x_0) = f(x_0)$$

where x_0 is the point at which the Taylor polynomial is being constructed. Hence we need to find an expression for

$$R_0(x) = f(x) - f(x_0).$$

You may remember that the First Mean Value Theorem (Appendix 2) gives us an expression for this difference provided that f is continuous on the interval with endpoints at x and x_0, and has a derivative everywhere inside that interval. In that case we have

$$f(x) - f(x_0) = (x - x_0)f^{(1)}(\xi) \qquad [12\text{-}1]$$

where ξ lies somewhere between x and x_0 but is otherwise unknown. In other words $x_0 < \xi < x$ or $x < \xi < x_0$, depending on whether $x_0 < x$ or $x < x_0$. To make sure that f has the required properties throughout the interval $[x_0, x]$ (or $[x, x_0]$), when x may need to vary, we usually ensure that the properties hold on some larger interval $[a, b]$ such that $x_0 \in [a, b]$ and all $x \in [a, b]$.

Equation [12-1] gives us our first remainder term

$$R_0(x) = (x - x_0)f^{(1)}(\xi).$$

Note that it looks like the next term in a Taylor polynomial of degree one which would be $(x - x_0)f^{(1)}(x_0)$, except that the derivative is evaluated at an unknown point ξ and not at x_0. It is this that makes it difficult in practice to actually evaluate these errors. Later on we shall show how we may sometimes be able to estimate such errors, or upper bounds on them.

EXAMPLE 12.1   Let $f(x) = e^x$ and $x_0 = 0$. e^x is in fact continuous and differentiable everywhere in $(-\infty, \infty)$. Applying Equation [12-1] we have

$$e^x - 1 = x\, e^\xi$$

since $f(x_0) = e^0 = 1$ and $f^{(1)}(x) = e^x$ so that $f^{(1)}(\xi) = e^\xi$. Here e^x is being approximated by the constant function $p_0(x) = 1$, with error $x\, e^\xi$ where $0 < \xi < x$ or $x < \xi < 0$, but ξ is otherwise unknown. This is represented in Fig. 12.1 where AB is the error.

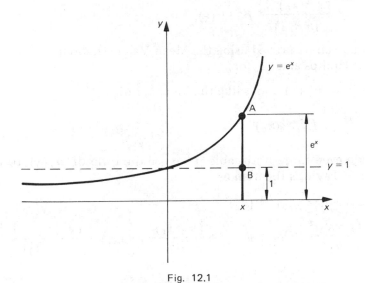

Fig. 12.1

Clearly in this case the errors are going to be too large to be generally acceptable, but if we wished to estimate the value of $x\, e^\xi$ we would be in trouble, because, after all, we are trying to approximate e^x in the first place. This is our first indication of the practical difficulties in using the remainder.

We have already indicated in Lesson 11 that, in general, the way to improve Taylor-polynomial approximations is to increase their degree. What we need therefore is a result extending the Mean Value Theorem to give us an expression for $R_n(x)$ for any n. This result is given by *Taylor's Theorem*.

2.2   TAYLOR'S THEOREM

Here is a precise statement of Taylor's theorem. You should notice that the conditions are very similar to those of the Mean Value Theorem. Basically they require f to be a smooth, 'well-behaved' function over an interval large enough to contain all the values of x in which we are interested.

(a) Let $f, f^{(1)}, f^{(2)}, \ldots, f^{(n)}$ exist and be continuous in an interval $[a, b]$.

(b) Let $f^{(n+1)}$ exist in (a, b).

(c) Let p_n be the Taylor polynomial of degree n for f constructed at some point $x_0 \in (a, b)$.

(d) Let $x \in (a, b)$. Then there exists a number $\xi \in (x_0, x)$ or (x, x_0), with ξ depending on x, such that

$$R_n(x) = f(x) - p_n(x)$$

$$= \frac{(x - x_0)^{n+1}}{(n+1)!} f^{(n+1)}(\xi).$$

This theorem can be proved using the Mean Value Theorem (see for example the book of Phillips and Taylor).

Note again how $R_n(x)$ looks like the 'next term'

$$\frac{(x - x_0)^{n+1}}{(n+1)!} f^{(n+1)}(x_0).$$

This is made even more noticeable if we use the form of $p_n(x)$ to write the conclusion of Taylor's theorem as

$$f(x) = f(x_0) + (x - x_0)f^{(1)}(x_0)$$

$$+ \frac{(x - x_0)^2}{2!} f^{(2)}(x_0) + \ldots + \frac{(x - x_0)^n}{n!} f^{(n)}(x_0) + \frac{(x - x_0)^{n+1}}{(n+1)!} f^{(n+1)}(\xi)$$

[12-2]

or

$$f(x) = f_0 + (x - x_0)f_0^{(1)}$$

$$+ \frac{(x - x_0)^2}{2!} f_0^{(2)} + \ldots + \frac{(x - x_0)^n}{n!} f_0^{(n)} + \frac{(x - x_0)^{n+1}}{(n+1)!} f^{(n+1)}(\xi)$$

[12-3]

using the notation $f^{(r)}(x_0) \equiv f_0^{(r)}$.

EXAMPLE 12.2 Let $f(x) = e^x$. The Taylor quadratic for f constructed at $x = 0$ is

$$p_2(x) = 1 + x + \tfrac{1}{2}x^2$$

and

$$R_2(x) = x^3 \, e^\xi / 6$$

where $0 < \xi < x$ or $x < \xi < 0$. (Derive these for yourself.) Figs. 12.2 and 12.3 may help you to understand what the remainder term represents.

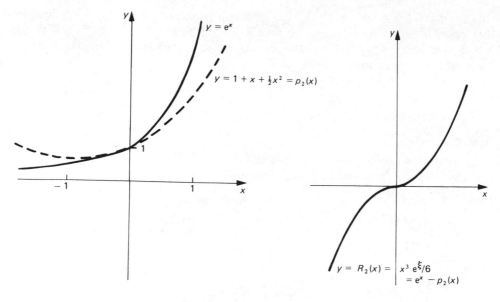

Fig. 12.2

Fig. 12.3

From Fig. 12.3 we can see that the error $R_2(x) = x^3 e^{\xi}/6$ is zero at $x = 0$ (the point at which $p_2(x)$ is constructed) and is very small near $x = 0$. However, further away from $x = 0$, $R_2(x)$ grows rapidly so that the approximation becomes rapidly worse. This is generally true. The expression

$$\frac{(x - x_0)^{n+1} f^{(n+1)}(\xi)}{(n+1)!}$$

behaves very roughly like $(x - x_0)^{n+1}$ in many cases (when $f^{(n+1)}(\xi)$ stays roughly constant as x and hence ξ vary). $(x - x_0)^{n+1}$ increases very rapidly as x moves away from x_0, even for quite small values of n.

EXERCISE 12.1

Sketch the graphs of $(x - 2)^4$, $(x - 2)^5$, $(x - 2)^6$, etc. (until you see the pattern that is developing).

12.2.1 ALTERNATIVE STATEMENTS OF THE CONCLUSION OF TAYLOR'S THEOREM

If $x = x_0 + h$ (h is the distance between x and x_0) the conclusion of Taylor's Theorem (under the same conditions of course) is

$$f(x_0 + h) = f_0 + \frac{h}{1!} f_0^{(1)} + \frac{h^2}{2!} f_0^{(2)} + \ldots + \frac{h^n}{n!} f_0^{(n)} + \frac{h^{n+1}}{(n+1)!} f^{(n+1)}(\xi)$$

[12-4]

where $x_0 < \xi < x$ or $x < \xi < x_0$. We can replace ξ by $x_0 + \theta h$ and in this case θ must lie between 0 and 1 $(0 < \theta < 1)$ so that $x_0 < \xi < x$ (if $h > 0$) or $x < \xi < x_0$ (if $h < 0$). In consequence we have another equivalent form of the conclusion of Taylor's Theorem

$$f(x_0 + h) = f_0 + \frac{h}{1!}f_0^{(1)} + \frac{h^2}{2!}f_0^{(2)} + \ldots + \frac{h^n}{n!}f_0^{(n)} + \frac{h^{n+1}}{(n+1)!}f^{(n+1)}(x_0 + \theta h)$$

[12-5]

where $0 < \theta < 1$.

This form has the advantage that it applies to $h > 0$ *and* $h < 0$. We may use any of these forms as they are equivalent to each other.

Hence the remainder term R_n can be expressed as

$$\frac{(x - x_0)^{n+1}}{(n+1)!}f^{(n+1)}(\xi) \qquad \text{where } x_0 < \xi < x \text{ or } x < \xi < x_0$$

or as

$$\frac{h^{n+1}}{(n+1)!}f^{(n+1)}(\xi) \qquad \text{where } x_0 < \xi < x \text{ or } x < \xi < x_0$$

or as

$$\frac{h^{n+1}}{(n+1)!}f^{(n+1)}(x_0 + \theta h) \qquad \text{where } 0 < \theta < 1.$$

EXAMPLE 12.3   In Exercise 11.5(f) you should have found that for $f(x) = \ln x$ and $x_0 = 1$

$$p_5(x) = (x-1) - \frac{(x-1)^2}{2} + \frac{(x-1)^3}{3} - \frac{(x-1)^4}{4} + \frac{(x-1)^5}{5}.$$

Going on we find that

$$f^{(n)}(x) = \frac{(-1)^{n+1}(n-1)!}{x^n}.$$

$$\therefore \quad p_n(x) = (x-1) - \frac{(x-1)^2}{2} + \frac{(x-1)^3}{3} + \ldots + \frac{(-1)^{n+1}(x-1)^n}{n}$$

and

$$R_n(x) = \frac{(x-1)^{n+1}}{(n+1)!}\frac{(-1)^{n+2}n!}{\xi^{n+1}}$$

[12-6]

where $1 < \xi < x$ or $0 < x < \xi < 1$,   i.e., ξ lies between 1 and x.

(Remember that $\ln x$ is not even defined for $x \leqslant 0$.)

If we write $h = x - 1$ and $\xi = 1 + \theta h$ we get

$$p_n(1 + h) = h - h^2/2 + h^3/3 - \ldots + (-1)^{n+1} h^n/n$$

and

$$R_n(1 + h) = \frac{(-1)^n h^{n+1}}{(n+1)(1+\theta h)^{n+1}}$$

$$\left. \right\} \quad [12\text{-}7]$$

where $0 < \theta < 1$. You should compare Equations [12-6] and [12-7].

EXERCISE 12.2

Find the appropriate remainder terms for the Taylor polynomials found in Exercise 11, namely:

	Function	Point	Remainder
(a)	$\cos x$	0	R_7
(b)	$\sin x$	$\pi/4$	R_4
(c)	e^x	0	R_n
(d)	$\ln(1+x)$	0	R_7
(e)	$\ln(1-x)$	0	R_n
(f)	$\ln x$	1	R_5
(g)	$\ln x$	0	R_n
(h)	x^4	1	R_3
(i)	x^4	1	R_4
(j)	x^4	0	R_3

12.3 CONVERGENCE

If $h = x - x_0$ is fixed and finite we may ask 'do we get smaller remainders by increasing the degree of the Taylor polynomial (which is the approximation to $f(x)$)?' More precisely, 'can we get as small a remainder as we like by making the degree large enough?'

EXAMPLE 12.4   Consider $f(x) = 1/(1-x)$, $x \in (0, b)$ for some $b > 0$. The Maclaurin polynomial of degree n for f is

$$p_n(x) = 1 + x + x^2 + \ldots + x^n.$$

If $x = 0.1$, $f(x) = 1/0.9 = 1.1111 \ldots$.

But

$$p_1(0.1) = 1 + 0.1 \qquad\qquad = 1.1$$

$$p_2(0.1) = 1 + 0.1 + (0.1)^2 \qquad = 1.11$$
$$p_3(0.1) = 1 + 0.1 + (0.1)^2 + (0.1)^3 = 1.111$$

and so on, showing how well the p_n approximate f for $x = 0.1$.

Here we must reintroduce the idea of a *sequence*, which is a set (usually of numbers) considered in a particular order. From Example 12.4, the ordered set

$$\{p_1(0.1), p_2(0.1), p_3(0.1), \ldots\} = \{1.1, 1.11, 1.111, \ldots\}$$

which we usually abbreviate to $\{p_n(0.1)\}$, is an *infinite sequence* with the property that the elements of the sequence get nearer and nearer to the predicted value $10/9$, the value of $f(0.1)$.

This sort of behaviour in a sequence is important and has the name *convergence*. We say that a sequence $\{a_n\} = \{a_1, a_2, a_3, \ldots\}$ *converges* to a particular number, a, called *the limit of the sequence*, when choosing n to be sufficiently large makes a_n as near in value to a as we please. If this is so we write $a_n \to a$ as $n \to \infty$ and read it 'a_n tends to a as n tends to infinity'. This is an important idea in pure mathematics and a detailed discussion of convergence is beyond the scope of this book, but the idea itself is also important in numerical mathematics.

To get back to our example the elements of the sequence $\{p_n(0.1)\}$ clearly get closer to $f(0.1)$ as n gets bigger and bigger and so we may feel fairly happy to say that $\{p_n(0.1)\}$ converges to $f(0.1)$.

EXAMPLE 12.5   Consider the problem of Example 12.4 and the sequence $\{p_n(1)\}$ instead of $\{p_n(0.1)\}$.

$$p_1(1) = 2$$
$$p_2(1) = 3$$
$$p_3(1) = 4$$

and in general

$$p_n(1) = n + 1 \text{ for all } n \in \mathbb{N}.$$

Clearly the sequence $\{p_n(1)\}$ does not converge. In fact the values of the p_n's get larger and larger without limit. We say that the sequence *diverges*.

From these two examples we can see that the convergence or otherwise of $\{p_n(x)\}$ to $f(x)$ depends on the value of x, and so the answer to the question posed at the start of this section is 'sometimes'. Each case has to be examined separately. This examination usually requires some fairly difficult mathematics. The example which follows is typical of what has to be done, so do not expect to understand the details at the first reading.

EXAMPLE 12.6   Consider

$$f(x) = \frac{1}{1+x}, \quad x \in [0, b).$$

The Maclaurin polynomial of degree n for f is

$$p_n(x) = 1 - x + x^2 + \ldots + (-x)^n$$

and

$$R_n(x) = \frac{(-x)^{n+1}}{(1+\xi)^{n+2}}$$

where $0 < \xi < x < b$.

(Check that these statements are correct!)

Now since $\xi > 0$, $1/(1 + \xi) < 1$.

$$\therefore \quad |R_n(x)| < x^{n+1}.$$

If $\{p_n(x)\}$ is to converge to $f(x)$ then $\{R_n(x)\}$ must converge to zero.

If $\{R_n(x)\}$ is to converge to zero ($R_n(x) \to 0$ as $n \to \infty$) x must be less than one in magnitude, and this will be true (in this example) if $b \leqslant 1$. (Remember that $0 \leqslant x < b$.)

So in this case we have the desired behaviour (i.e., the approximations $p_n(x)$ to $f(x)$ become better and better as n increases) for $x \in [0, b)$ where $0 < b \leqslant 1$.

To complete the analysis we ought to consider the case $b > 1$. Since $0 \leqslant x < b$ we could now have $x \geqslant 1$, e.g. at $x = 1$,

$$p_1(1) = 1$$

$$p_2(1) = 0$$

$$p_3(1) = 1$$

$$p_4(1) = 0 \quad \text{etc.},$$

i.e., $p_{2n}(1) = 0$ and $p_{2n-1}(1) = 1$, for all $n \in \mathbb{N}$.

The sequence oscillates indefinitely and does not converge ($f(1) = \frac{1}{2}$ in fact.) So convergence is clearly not guaranteed for $x \geqslant 1$. (We have not proved that convergence will never occur for $x \geqslant 1$. This is true, however, and follows from the identity

$$p_n(x) = 1 - x + x^2 - \ldots + (-x)^n = (1 - (-x)^{n+1})/(1 + x).$$

You should be able to see that $|p_n(x)| \to \infty$ as $n \to \infty$ if $|x| > 1$, and we have already seen that the sequence does not converge for $x = 1$.)

To summarise: the remainder term

$$R_n(x) = \frac{(x - x_0)^{n+1} f^{(n+1)}(\xi)}{(n + 1)!}$$

where $x_0 < \xi < x$ or $x < \xi < x_0$ (ξ otherwise unknown) satisfies $R_n(x) \to 0$ as $n \to \infty$ in some cases only. We have seen in the examples that this behaviour may depend on the values of x (and hence ξ) and that we may have to restrict $x - x_0$ (and hence ξ) to get the required convergence.

12.4 INFINITE SERIES

Let us return to the equation defining R_n

$$f(x) = p_n(x) + R_n(x)$$

and assume now that $R_n(x) \to 0$ as $n \to \infty$ for some suitable interval containing x. Then if we take larger and larger values for n the number of terms in $p_n(x)$ becomes larger and larger and the value of $p_n(x)$ becomes nearer and nearer to that of $f(x)$.

For example, let $f(x) = 1/(1 + x)$, $x \in [0, 1)$, then

$$p_n(x) = 1 + x + x^2 + \ldots + x^n.$$

If n gets larger and larger we can write

$$p(x) = 1 + x + x^2 + \ldots + x^n + \ldots$$

where the last three dots indicate that the number of terms in this series is infinite. We call such an expression, $p(x)$, an *infinite series*. The idea of an infinite series is a difficult one. We shall restrict ourselves to saying that the value of $p(x)$ is defined to be the limit of the sequence $\{p_n(x)\}$, as n tends to infinity, if the sequence converges. We can then write

$$p_n(x) \to p(x) \quad \text{as} \quad n \to \infty.$$

Even when $\{p_n(x)\}$ converges and the infinite series $p(x)$ is defined $p(x)$ is not necessarily equal to $f(x)$. But if in addition $R_n(x) \to 0$ as $n \to \infty$ then $f(x)$ and $p(x)$ have the same value and we write, e.g.,

$$\frac{1}{1 - x} = 1 + x + x^2 + x^3 + \ldots \quad \text{for} \quad x \in [0, 1).$$

We shall only be interested in this course in infinite series for which this is true, i.e., they converge to $f(x)$:

$$p_n(x) \to f(x) \quad \text{as} \quad n \to \infty$$

for all x in some given interval.

In general $p(x)$ is called the *Taylor series* for $f(x)$ constructed as $x = x_0$. If $x_0 = 0$ it is usual to call $p(x)$ the *Maclaurin series* for $f(x)$.

TRUNCATION ERROR AND THE ORDER OF APPROXIMATION

We started this lesson with Taylor's theorem giving the error in Taylor polynomial approximation, and developed the idea of Taylor series. We finish now by returning to the remainder term as an error and discussing the problems of actually calculating its value.

We can think of $p_n(x)$ now as the Taylor series for $f(x)$ *truncated* at degree n. The remainder term

$$R_n(x_0 + h) = \frac{h^{n+1}}{(n+1)!} f^{(n+1)}(x_0 + \theta h), \quad 0 < \theta < 1$$

is hence sometimes called the *truncation error*.

Calculating values of $R_n(x)$ is generally impossible. (If we knew the error we would not need to approximate!) The problem is that we do not know the value of θ, only that it satisfies $0 < \theta < 1$. What we can often do is estimate an error bound, i.e. a number B such that

$$|R_n(x)| < B$$

for all x values of interest, say $x \in [a, b]$.

In particular if $f^{(n+1)}$ is a bounded function for $x \in [a, b]$ (one that cannot get as large as it likes in that interval) we can write that

$$|f^{(n+1)}(x)| < M, \quad x \in [a, b]$$

for some number M, and then

$$|R_n(x)| < \frac{M h^{n+1}}{(n+1)!}$$

The right-hand side will be close to the maximum value of $|R_n(x)|$ only if M is close to the maximum value of $|f^{(n+1)}(x_0 + \theta h)|$. In other words if you have to be pessimistic in estimating M then you will get a pessimistic bound on $|R_n(x)|$.

In practice we can use this idea in one of three ways:

(a) given n and h, estimate a bound on R_n;

(b) given n and a maximum requirement for R_n, find h;

(c) given h and a maximum requirement for R_n, find n.

EXAMPLE 12.7   Let $f(x) = e^x$, $x_0 = 0$ (so $h = x$ in what follows), and restrict x to $x \in [0, \frac{1}{2}]$. Then

$$p_n(x_0 + h) = p_n(h) = 1 + h + h^2/2! + h^3/3! + \ldots + h^n/n!$$
$$R_n(x_0 + h) = R_n(h) = h^{n+1} e^\xi /(n+1)!$$

where $0 < \xi < h$ since $h > 0$.

Since e^x is an increasing function (sketch its graph) $|e^\xi| < e^{1/2}$, $(0 < \xi < h \leqslant \frac{1}{2})$, so

$$|R_n(h)| < h^{n+1} e^{1/2}/(n+1)! < 2h^{n+1}/(n+1)!$$

since $e^{1/2} < 2$ ($e^{1/2} = 1.6 \ldots$, but 2 is a nice round number to work with. This crude working is typical of error estimate calculations). We shall now illustrate the three possible uses for our error bound.

(a) Suppose $n = 3$ and $h = 0.4$, then

$$|R_3(0.4)| < 2(0.4)^4/4! \simeq 0.02$$

so $|e^{0.4} - p_3(0.4)| < 0.02$, i.e.

$$e^{0.4} = 1 + 0.4 + (0.4)^2/2! + (0.4)^3/3! \pm 0.02 \quad \text{(at most)}.$$

In fact

$$e^{0.4} = 1.491\,82\ldots \quad \text{and} \quad p_3(0.4) = 1.490\,66\ldots$$

and so $R_3(0.4) = 0.0011$ (to 2S), showing our error bound to be 20 times too pessimistic, which is not unusual. (It's better than nothing!)

(b) Now suppose $n = 3$ and that we require $|R_n(h)| < 0.5 \times 10^{-4}$ (so that $p_3(h)$ equals e^h to 4D). Then we need $2h^4/4! < 0.5 \times 10^{-4}$

i.e. $h^4 < 0.0006$

i.e. $h < 0.16$.

For example, if $h = 0.1$, $p_3(0.1) = 1.1052$ to 4D (check this) which *is* $e^{0.1}$ to 4D.

(c) On the other hand, if $h = 0.4$ and we need 4D accuracy, then we must have

$$2(0.4)^{n+1}/(n+1)! < 0.5 \times 10^{-4}.$$

This is more difficult to solve. Since $n \in \mathbb{N}$ we can do it by trial and error. Let $g_n = 2(0.4)^{n+1}/(n+1)!$ Then

$g_3 = 0.0021$ (to 2S)   — no good

$g_4 = 0.000\,17$ (to 2S)   — much better

$g_5 = 0.000\,011$ (to 2S)   — satisfactory.

So $p_5(h)$ approximates e^h correct to 4D for $h = 0.4$ (and hence for all $h \in [0, 0.4]$ since we know that $R_n(h)$ decreases as h decreases.)

Apart from allowing us to estimate a bound on $R_n(x_0 + h)$ the expression $Mh^{n+1}/(n + 1)!$ also gives us a qualitative feeling for how good p_n is as an approximation to f. It tells us that, as $h \to 0$, (for 'small' h if you like) $R_n(x_0 + h)$ behaves like h^{n+1}. A common notation is

$$R_n(x_0 + h) = O(h^{n+1})$$

which is read as '$R_n(x_0 + h)$ is of order h^{n+1}'. We sometimes say, as an alternative, that the truncation error is of order $n + 1$. This kind of description of an error will occur in other parts of this course.

EXERCISE 12.3

(a) Find h_0 such that, for $0 \leqslant h \leqslant h_0$, the seventh degree Maclaurin polynomial $p_7(h)$ for $\ln(1 + h)$, where $h \geqslant 0$, approximates $\ln(1 + h)$ correctly to 3D. (See Exercise 12.2(d) for $R_7(h)$).

(b) What degree of polynomial is required if the 3D accuracy is to be extended to

(i) $0 \leqslant h \leqslant 1$ (ii) $0 \leqslant h \leqslant 2$?

EXERCISE 12.4

Using the error bound expression $g_n = 2(0.4)^{n+1}/(n + 1)!$ found in Example 12.7 estimate the degrees of Taylor polynomial required for $p_n(h)$ to approximate e^h, for $h \in [0, 0.4]$, correct to

(a) 2D   (b) 6D   (c) 10D   (d) 4S.

EXERCISE 12.5

Use the answer to Exercise 12.2(b) to estimate a bound on the error $|\sin(\pi/6) - p_4(\pi/6)|$. Calculate the actual error and compare it with your error bound.

12.6   CONVERGENCE OF ITERATIVE PROCESSES

Taylor's Theorem gives us an expression for the remainder or error term when $f(x)$ is approximated by the appropriate Taylor polynomial. Usually we can at best find a bound for $|R_n(x_0 + h)|$, or alternatively if a required accuracy is given either find a maximum value for h for a given n, or a minimum value of n for a given h.

In practice the remainder term is rarely used numerically; we find other ways of estimating errors. But it is of considerable use in developing error theory for many other processes in numerical mathematics.

As an example of this we shall prove the statements made in Section 6.2 about the errors $\xi_1, \xi_2, \ldots$ of the approximations $x_1, x_2, \ldots$ to the root X of $x = f(x)$ generated by the iteration $x_{n+1} = f(x_n)$. By definition $\xi_n = X - x_n$ and $\xi_{n+1} = X - x_{n+1}$, hence $x_{n+1} = f(x_n)$ is the same as

$$X - \xi_{n+1} = f(X - \xi_n).$$

Replacing the right-hand side of this by the Taylor polynomial of degree m for f, constructed at $x = X$, and evaluated at $x = X - \xi_n$, plus the associated remainder,

$$X - \xi_{n+1} = f(X) - \xi_n f^{(1)}(X) + \frac{\xi_n^2}{2!} f^{(2)}(X) - \ldots$$

$$+ \frac{(-\xi_n)^m}{m!} f^{(m)}(X) + \frac{(-\xi_n)^{m+1}}{(m+1)!} f^{(m+1)}(\eta_n)$$

where η_n lies between x_n and X.

Now $X = f(X)$, since X is a root of $x = f(x)$. Hence

$$\xi_{n+1} = \xi_n f^{(1)}(X) - \frac{\xi_n^2}{2!} f^{(2)}(X) + \ldots - \frac{(-\xi_n)^m}{m!} f^{(m)}(X)$$

$$- \frac{(-\xi_n)^{m+1}}{(m+1)!} f^{(m+1)}(\eta_n).$$

If $f^{(1)}(X) \neq 0$ and ξ_n is small then ξ_{n+1} is dominated by the first term. Hence take $m = 0$ and obtain

$$\xi_{n+1} = \xi_n f^{(1)}(\eta_n).$$

This characterises a first-order process. If $f^{(1)}(X) = 0$ but $f^{(2)}(X) \neq 0$ then ξ_{n+1} is dominated by the second term. Hence take $m = 1$, giving

$$\xi_{n+1} = -\tfrac{1}{2} \xi_n^2 f^{(2)}(\eta_n)$$

which characterises a second-order process. In general if $f^{(1)}(X) = \ldots = f^{(r)}(X) = 0$ but $f^{(r+1)}(X) \neq 0$ then the process is rth order and

$$\xi_{n+1} = -\frac{(-\xi_n)^{r+1}}{(r+1)!} f^{(r+1)}(\eta_n).$$

ANSWERS TO EXERCISES

12.2   (a) $x^8 \cos \xi/8!$ $0 < \xi < x$ or $x < \xi < 0$,

        or $h^8 \cos (\theta h)/8!$ $0 < \theta < 1$ $(h = x)$.

(b) $(x - \pi/4)^5 \cos \xi/5!$

or $\quad h^5 \cos(\pi/4 + \theta h)/5!$

$\pi/4 < \xi < x$ or $x < \xi < \pi/4$,

$0 < \theta < 1$ $(h = x - \pi/4)$.

(c) $x^{n+1} e^\xi /(n+1)!$

or $\quad h^{n+1} e^{\theta h}/(n+1)!$

$0 < \xi < x$ or $x < \xi < 0$,

$0 < \theta < 1$ $(h = x)$.

(d) $-x^8/8(1+\xi)^8$

or $\quad -h^8/8(1+\theta h)^8$

$0 < \xi < x$ or $-1 < x < \xi < 0$,

$0 < \theta < 1$ $(h = x, \; h > -1)$.

(e) $-x^{n+1}/(n+1)(1-\xi)^{n+1}$

or $\quad -h^{n+1}/(n+1)(1-\theta h)^{n+1}$

$0 < \xi < x < 1$ or $x < \xi < 0$,

$0 < \theta < 1$ $(h = x, \; h < 1)$.

(f) $-(x-1)^6/6\xi^6$

or $\quad -h^6/6(1+\theta h)^6$

$1 < \xi < x$ or $0 < x < \xi < 1$,

$0 < \theta < 1$ $(h = x - 1, \; h > -1)$.

(g) No polynomial, therefore no remainder.

(h) $(x-1)^4$.

(i) 0.

(j) x^4.

12.3 (a) $R_7(x_0 + h) = R_7(h) = -h^8/8(1+\theta h)^8$; $0 < \theta < 1$, and $h > 0$.
Therefore $|R_7(h)| < h^8/8$ (because $\theta h > 0$) and this is the best
bound we can achieve on R_7 (independent of θ). So for 3D
accuracy we need $h^8/8 < \frac{1}{2} \times 10^{-3}$,

i.e. $\quad h^8 < 4 \times 10^{-3}$

i.e. $\quad h_0 < 0.5$ approx.

So choose $h_0 = 0.5$. (For absolute security choose h_0 rather less,
e.g. 0.2.)

(b) (i) To get 3D for all $x \in [0, 1]$ we need $|R_n(1)| < \frac{1}{2} \times 10^{-3}$.
Now

$$|R_n(1)| = \frac{1^{n+1}}{(n+1)(1+\theta h)^{n+1}} < \frac{1}{n+1}.$$

So $\quad \dfrac{1}{n+1} < \frac{1}{2} \times 10^{-3}$ will suffice,

i.e. $n + 1 > 2000$,

i.e. the 2000th-degree Taylor polynomial!!!

(ii) For 3D on $[0, 2]$ we need $2^{n+1}/(n+1) < \frac{1}{2} \times 10^{-3}$.

This is impossible. So

$$p_n(2) = 2 - \frac{2^2}{2} + \frac{2^3}{3} - \frac{2^4}{4} + \dots + (-1)^{n-1}\frac{2^n}{n}$$

is such that $|p_n(2)| \to \infty$ as $n \to \infty$, i.e., the Maclaurin series for $\ln(1+h)$ diverges for $h = 2$.

(To show the impossibility prove that $2^{x+1}/(x+1)$ increases as x increases from zero, i.e., has a positive derivative for positive x. *Hint*: take logs first.)

12.4   (a) $n = 3$ ($g_3 = 0.0021$).

      (b) $n = 7$ ($g_6 = 0.65 \times 10^{-6}$, $g_7 = 0.32 \times 10^{-7}$).

      (c) $n = 10$ ($g_9 = 0.58 \times 10^{-10}$, $g_{10} = 0.21 \times 10^{-11}$).

      (d) Two approaches:

(i) since for $0 \leqslant x \leqslant \frac{1}{2}$, $1 \leqslant e^x \leqslant \sqrt{e} \simeq 1.6$, 4S in this case implies 3D. So $n = 4$ since $g_4 = 0.000\,17$.

(ii) A number quoted to kS has a *relative* error bound of 5×10^{-k} (*not* 0.5×10^{-k}) — see Section 2.11. So we need

$$\frac{2h^{n+1}/(n+1)!}{e^h} < 5 \times 10^{-4}.$$

Since $e^h > 1$ for $h > 0$ it will be sufficient if

$$\frac{2(0.4)^{n+1}}{(n+1)!} = g_n < 5 \times 10^{-4} = 0.5 \times 10^{-3},$$

i.e., the 3D requirement, as before.

12.5 $R_4\left(\dfrac{\pi}{4} + h\right) = h^5 \cos\left(\dfrac{\pi}{4} + \theta h\right) \Big/ 5! \quad 0 < \theta < 1.$

$$\therefore \quad R_4\left(\frac{\pi}{6}\right) = R_4\left(\frac{\pi}{4} - \frac{\pi}{12}\right) = \left(-\frac{\pi}{12}\right)^5 \cos\left(\frac{\pi}{4} - \theta\,\frac{\pi}{12}\right)\Big/5!$$

Since $0 < \theta < 1$,

$$\cos\frac{\pi}{4} < \cos\left(\frac{\pi}{4} - \theta\,\frac{\pi}{12}\right) < \cos\frac{\pi}{6}$$

i.e. $0.7 < \cos\left(\dfrac{\pi}{4} - \theta\,\dfrac{\pi}{12}\right) < 0.9$ (approx.).

So $\left| R_4\left(\dfrac{\pi}{6}\right) \right| < \left(\dfrac{\pi}{12}\right)^5 (0.9)/120 = 0.92 \times 10^{-5}$ (to 2S).

In this case since $0.7 < \cos(\pi/4 + \theta h) < 0.9$ we can get quite close to the actual error. Suppose $\cos(\pi/4 + \theta h) \simeq 0.8$. Then

$$R_4\left(\frac{\pi}{6}\right) \simeq -\left(\frac{\pi}{12}\right)^5 (0.8)/120 = -0.82 \times 10^{-5} \text{ (to 2S)}$$

whereas

$$\sin \frac{\pi}{6} = 0.5$$

and

$$p_4\left(\frac{\pi}{6}\right) = \frac{1}{\sqrt{2}}\left(1 + \left(\frac{\pi}{6} - \frac{\pi}{4}\right) - \frac{1}{2!}\left(\frac{\pi}{6} - \frac{\pi}{4}\right)^2 - \frac{1}{3!}\left(\frac{\pi}{6} - \frac{\pi}{4}\right)^3\right.$$
$$\left. + \frac{1}{4!}\left(\frac{\pi}{6} - \frac{\pi}{4}\right)^4\right)$$

$$= 0.500\,007\,55\ldots$$

(see Exercise 11.5(b)) with error $= -0.76 \times 10^{-5}$ (to 2S).

FURTHER READING

At the end of Lesson 3 we recommended Stark, Chapter 2, pp. 30—38 for further reading. On pp. 39—43 Stark develops the remainder term R_n and from p. 43 onwards gives some examples including computer programs for evaluating Taylor polynomials. The approach is rather different to ours, but it's a good thing to get a second point of view.

Another alternative is Unit M100 14 Sequences and Limits II of the original Open University foundation course in mathematics. This contains some excellent pictures and examples if you are prepared to accept that it is part of another course and therefore contains references to material that you won't understand. Simply skip over parts that aren't familiar to you, and look closely at those that are specifically about Taylor polynomials and Taylor's theorem.

LESSON 12 – COMPREHENSION TEST

1. Which of these equations defineṣ the Taylor remainder R_n:
(a) $R_n = f + p_n$ (b) $R_n = f - p_n$ (c) $R_n = p_n - f$?

2. State the remainder term $R_n(x)$ for the Taylor polynomial of degree n for the function $f(x)$ constructed at $x = x_0$.

3. State the Taylor polynomial for $f(x)$ constructed at $x = x_0$, with remainder, when x is replaced by $x_0 + h$.

4. (a) Obtain the remainder term $R_2(x)$ for the linear Maclaurin polynomial for $f(x) = \tan x$.
(b) Estimate an upper bound for $|R_2(x)|$ for $0 \leqslant x \leqslant \pi/4$.

(c) How large can x ($\geqslant 0$) be while $|R_2(x)| \leqslant \frac{1}{2} \times 10^{-2}$?

5. Which of these describes best the region where Taylor polynomials are, in general, accurate:

(a) everywhere;   (b) near x_0;   (c) far from x_0;   (d) near the origin?

6. State the condition on $|R_n(x)|$ for the associated Taylor series to converge for all $x \in [a, b]$.

ANSWERS

1   (b).

2   $R_n(x) = \dfrac{(x - x_0)^{n+1}}{(n+1)!} f^{(n+1)}(\xi), \quad x_0 < \xi < x \text{ or } x < \xi < x_0.$

3   $f(x_0) + hf^{(1)}(x_0) + \dfrac{h^2}{2!} f^{(2)}(x_0) + \cdots + \dfrac{h^m}{m!} f^{(m)}(x_0)$

$+ \dfrac{h^{m+1}}{(m+1)!} f^{(m)}(x_0 + \theta h), \quad 0 < \theta < 1.$

4   (a) $x^2 \sec^2 \xi \tan \xi, \ 0 < \xi < x \text{ or } x < \xi < 0.$

(b) Since x^2, $\sec^2 \xi$ and $\tan \xi$ all increase for $x > 0$ and $\xi > 0$ (up to $\pi/2$) the maximum value of $|R_2(x)|$ will be $(\pi/4)^2 \sec^2(\pi/4) \tan \pi/4 = \frac{1}{8}\pi^2.$

(c) We require $|x^2 \sec^2 \xi \tan \xi| < \frac{1}{2} \times 10^{-2}$. Using the worst case again this means that $x^2 < \frac{1}{2} \times 10^{-2} \cos^2 \pi/4 \cot \pi/4 = \frac{1}{4} \times 10^{-2} \Rightarrow x < \frac{1}{2} \times 10^{-1} = 0.05.$

5   (b).

6   $|R_n(x)| \to 0$ as $n \to \infty$ for all $x \in [a, b]$.

LESSON 12 – SUPPLEMENTARY EXERCISES

12.6. Obtain remainder terms for the Taylor polynomials constructed in Exercise 11.6.

12.7. Using the remainder term found in Exercise 12.2(e) show

(a) that the Maclaurin series for $\ln(1-x)$ converges for $-1 < x < 0$;

(b) that $|R_4(x)| \leqslant 0.2$ for $-1 < x \leqslant 0$;

(c) that we require $|x| \leqslant 0.3$ (approx.) for $|R_4(x)| < \frac{1}{2} \times 10^{-3}$; and

(d) that we require $n > 2000$ so that $|R_n(x)| < \frac{1}{2} \times 10^{-3}$ for all x such that $-1 < x \leqslant 0$.

12.8. Using the remainder term found in Exercise 12.6(a) estimate a bound on the error $|\cosh(\tfrac{1}{2}) - p_6(\tfrac{1}{2})|$.

Calculate the actual error (using Exercise 11.6(a)) and compare it with the estimated bound.

*12.9. Using integration by parts show that

$$\int_{x_0}^{x} \frac{(x-t)^n}{n!} f^{(n+1)}(t) \, dt \; = \; f(x) - f(x_0) - (x-x_0)f^{(1)}(x_0) - \ldots$$

$$- \frac{(x-x_0)^n}{n!} f^{(n)}(x_0).$$

(This is 'the integral form' of the remainder $R_n(x)$.)

Lesson 13 Interpolating Polynomials

OBJECTIVES

By the end of this lesson you should

(a) understand the concept of an interpolating polynomial;

(b) be able to write down the Lagrange and divided-difference interpolating polynomials for a given set of points (particular or
general);

(c) appreciate the advantages and disadvantages of the Lagrange and
the divided-difference forms of the interpolating polynomials.

CONTENTS

13.1 INTRODUCTION

In Lessons 11 and 12 we have seen how, given a particular function $f(x)$,
defined for *all* values of x in a given interval, we can approximate $f(x)$ by
a Taylor polynomial $p_n(x)$ with remainder (error) $R_n(x)$. In that case

we were discussing the *approximation of the explicit function* f by a polynomial p_n.

However, there is another important situation that can arise. In a practical experiment the experimenter may only be able to measure the value of the function $f(x)$ for particular (or *discrete*) values of x. He (or she) certainly cannot do it for all the (infinite) values of x in the interval for which $f(x)$ is defined. But it would be very useful if, from the measured data above, values of $f(x)$ could be found for any required x in this interval. Put another way the problem is: can we find a reasonable approximation to the function f given a finite set of points on the graph of $y = f(x)$?

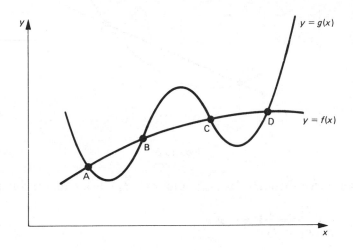

Fig. 13.1

In Fig. 13.1 the curve $y = f(x)$ (unknown) has four known points on it — A, B, C and D. One obvious way to approximate f is to choose a function whose graph includes A, B, C, D. There are, of course, many curves through A, B, C and D, of which one $(y = g(x))$ is shown in Fig. 13.1.

The most usual way of solving this problem is to construct a *polynomial* through the given points. Such a polynomial is called an *interpolating polynomial*. It should be noted that this is not by any means the only way of solving the problem, but other solutions are beyond the scope of this course.

Most of the material in this lesson is concerned with the possible forms of interpolating polynomials. The problem of their evaluation and use is left until Lesson 14.

13.2 LINEAR INTERPOLATION

Let us consider a very simple case where we know $f(x)$ only for $x = x_0$ and $x = x_1$ (where $x_0 \neq x_1$). Fig. 13.2 shows us that we know the position of the two points whose coordinates are (x_0, f_0) and (x_1, f_1). (We shall consistently use the notation f_i as an abbreviation for $f(x_i)$.)

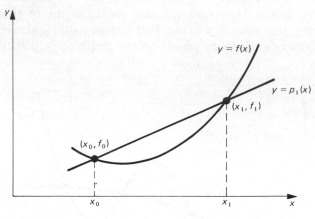

Fig. 13.2

The simplest curve through (x_0, f_0) and (x_1, f_1) is a straight line given by

$$y - f_0 = \left(\frac{f_1 - f_0}{x_1 - x_0}\right)(x - x_0).$$

(Check that this line does go through (x_0, f_0) and (x_1, f_1).)

Hence we get

$$y = f_0 + \left(\frac{f_1 - f_0}{x_1 - x_0}\right)(x - x_0)$$

or equivalently,

$$y = \left(\frac{x - x_1}{x_0 - x_1}\right)f_0 + \left(\frac{x - x_0}{x_1 - x_0}\right)f_1.$$

Now this equation gives the graph of a polynomial of degree 1, so we shall call it $p_1(x)$, i.e.

$$p_1(x) = \left(\frac{x - x_1}{x_0 - x_1}\right)f_0 + \left(\frac{x - x_0}{x_1 - x_0}\right)f_1. \qquad [13\text{-}1]$$

This is known as Lagrange's form of the interpolating polynomial of degree 1 through (x_0, f_0) and (x_1, f_1). Notice that in this form it is very easy to check that setting $x = x_0$ gives $p_1(x_0) = f_0$ and setting $x = x_1$ gives $p_1(x_1) = f_1$, i.e., the graph of the polynomial does indeed pass through the two given points.

It is also clear that $p_1(x)$ is unique — there is only one straight line through two distinct points.

However, we can write $p_1(x)$ in other ways. We already have in effect

$$p_1(x) = f_0 + \left(\frac{f_1 - f_0}{x_1 - x_0}\right)(x - x_0) \qquad [13\text{-}2]$$

which is *Newton's* form of the interpolating polynomial of degree 1 through (x_0, f_0) and (x_1, f_1). (In future the given points will not be mentioned but rather understood by implication). This form (Newton's) is also known as Newton's *divided-difference* form or simply the divided-difference form of $p_1(x)$. It is of the form

$$p_1(x) = a_0 + a_1(x - x_0)$$

where a_0 and a_1 are constants depending on the data. Frequently a special 'divided-difference' notation is used. $a_0 = f_0$ is written as $f[x_0]$ and $a_1 = (f_1 - f_0)/(x_1 - x_0)$ as $f[x_0, x_1]$ so that

$$p_1(x) = f[x_0] + f[x_0, x_1](x - x_0)$$

This notation and the idea of divided differences will be extended later, but we can see already in $(f_1 - f_0)/(x_1 - x_0)$ the reason for calling Equation [13-2] the divided-difference form.

There is one further form of $p_1(x)$ which we shall need. Recall that Lagrange's form of $p_1(x)$ is

$$p_1(x) = \left(\frac{x - x_1}{x_0 - x_1}\right)f_0 + \left(\frac{x - x_0}{x_1 - x_0}\right)f_1.$$

This can be converted into

$$p_1(x) = \frac{(x - x_0)f_1 - (x - x_1)f_0}{x_1 - x_0}. \qquad [13\text{-}3]$$

(Check this!) This form will be used in constructing the Neville—Aitken method in Lesson 14.

SUMMARY   For the interpolating polynomial of degree 1 we have:

Lagrange form: $\qquad\qquad p_1(x) = \left(\dfrac{x - x_1}{x_0 - x_1}\right)f_0 + \left(\dfrac{x - x_0}{x_1 - x_0}\right)f_1.$

Divided-difference form: $\quad p_1(x) = f[x_0] + (x - x_0)f[x_0, x_1].$

Neville—Aitken form: $\qquad p_1(x) = \dfrac{(x-x_0)f_1 - (x-x_1)f_0}{x_1 - x_0}.$

In the following sections we will see how Lagrange's form and the divided-difference form can be extended. The extension of the Neville—Aitken form is left to Lesson 14.

EXAMPLE 13.1   To find the linear interpolating polynomial using both Lagrange's method and the divided-difference method for the points (2, 6), (5, 12), or equivalently the data

x	2	5
$f(x)$	6	12

Using Lagrange's interpolating polynomial we get

$$\begin{aligned}
p_1(x) &= \frac{6(x-5)}{2-5} + \frac{12(x-2)}{5-2} \\
       &= \frac{6(x-5)}{-3} + \frac{12(x-2)}{3} \\
       &= -2(x-5) + 4(x-2) \\
       &= 2x + 2.
\end{aligned}$$

Using the divided-difference form we get

$$p_1(x) = 6 + \left(\frac{12-6}{5-2}\right)(x-2)$$

(see Equation [13-2]), i.e.

$$p_1(x) = 2x + 2$$

as before. (The polynomials must of course be the same.) In divided-difference notation

$$f[x_0] = 6$$

$$f[x_0, x_1] = \frac{12-6}{5-2} = 2$$

so that $p_1(x) = 6 + 2(x-2) = 2x + 2.$

EXERCISE 13.1

Write down and simplify the Lagrange interpolating polynomials for each of these sets of points.

(a) (0, 1), (2, 3)   (b) (3, 4), (2, 1)

EXERCISE 13.2

From the following tables find the interpolating polynomial for f using the divided-difference form.

(a)

x	0	2
$f(x)$	1	3

(b)

x	3	2
$f(x)$	4	1

13.3 QUADRATIC INTERPOLATION – THE LAGRANGE FORM

So far we have constructed the linear interpolating polynomial for the two points (x_0, f_0) and (x_1, f_1) and it is clearly unique if the points are distinct. (Can you prove this statement? Hint: let $y = a_1 x + b_1$ and $y = a_2 x + b_2$ be two different linear interpolating polynomials for the same data!) However, we could have chosen

$$p_2(x) = ax^2 + bx + c$$

instead of $p_1(x)$. In this case we would require that

$$f_0 = ax_0^2 + bx_0 + c$$

and

$$f_1 = ax_1^2 + bx_1 + c.$$

These equations could be solved for a and b in terms of c and this means that by choosing c to be any number we please we can get an infinite number of parabolas passing through (x_0, f_0) and (x_1, f_1).

However, if we increase the number of known points to three by adding the point (x_2, f_2) then clearly a unique parabola will pass through all three points (unless all three points are collinear, i.e., lie on a straight line). With three points the interpolating polynomial will be called $p_2(x)$ (even if the coefficient of x^2 turns out to be zero because the three points are collinear).

To see what form $p_2(x)$ might take let us return to the first-degree Lagrange form

$$p_1(x) = \left(\frac{x - x_1}{x_0 - x_1} \right) f_0 + \left(\frac{x - x_0}{x_1 - x_0} \right) f_1.$$

For the points $(x_0, f_0), (x_1, f_1)$ and (x_2, f_2) it seems likely that the form of the interpolating polynomial of degree 2, $p_2(x)$, should be similar to that of $p_1(x)$. More specifically we might guess at

$$p_2(x) = L_0(x)f_0 + L_1(x)f_1 + L_2(x)f_2$$

where $L_0(x), L_1(x)$ and $L_2(x)$ are *quadratic* polynomials which do not depend on f_0, f_1, f_2.

At $x = x_0$ we require $p_2(x_0) = f_0$, so we must have

$$L_0(x_0) = 1$$
$$L_1(x_0) = 0$$
$$L_2(x_0) = 0.$$

The second and third of these imply that $x - x_0$ must be a factor of L_1 and L_2. Similarly for $p_2(x_1) = f_1$ we need

$$L_1(x_1) = 1$$
$$L_0(x_1) = 0$$
$$L_2(x_1) = 0$$

so $x - x_1$ must be a factor of L_0 and L_2. Finally for $p_2(x_2) = f_2$ we need

$$L_2(x_2) = 1$$
$$L_0(x_2) = 0$$
$$L_1(x_2) = 0$$

so $x - x_2$ must be a factor of L_0 and L_1. But L_0, L_1 and L_2 are quadratic polynomials. Hence

$$L_0(x) = \alpha(x - x_1)(x - x_2)$$
$$L_1(x) = \beta(x - x_0)(x - x_2)$$
$$L_2(x) = \gamma(x - x_0)(x - x_1)$$

where α, β and γ are constants. Now imposing $L_0(x_0) = L_1(x_1) = L_2(x_2) = 1$ from above, we get

$$\alpha = 1/(x_0 - x_1)(x_0 - x_2)$$
$$\beta = 1/(x_1 - x_0)(x_1 - x_2)$$
$$\gamma = 1/(x_2 - x_0)(x_2 - x_1).$$

Now what does $p_2(x)$ look like?

$$p_2(x) = \frac{(x - x_1)(x - x_2)}{(x_0 - x_1)(x_0 - x_2)} f_0 + \frac{(x - x_0)(x - x_2)}{(x_1 - x_0)(x_1 - x_2)} f_1$$
$$+ \frac{(x - x_0)(x - x_1)}{(x_2 - x_0)(x_2 - x_1)} f_2 \qquad [13\text{-}4]$$

which has the same sort of shape as $p_1(x)$ in Lagrange form.

Before doing some exercises check that you can see why $p_2(x)$ must be like this when written in Lagrange form. To get $p_2(x_i) = f_i$ for $i = 0, 1, 2$ the expression multiplying f_0 must have the value unity when $x = x_0$, and the value zero when $x = x_1$ and $x = x_2$; similar remarks apply to the expressions multiplying f_1 and f_2.

EXAMPLE 13.2 To find the Lagrangian interpolating polynomial for the three points $(0, -3), (1, 4), (2, 15)$.

$$p_2(x) = -3 \frac{(x-1)(x-2)}{(0-1)(0-2)} + 4 \frac{(x-0)(x-2)}{(1-0)(1-2)} + 15 \frac{(x-0)(x-1)}{(2-0)(2-1)}$$

$$= -\frac{3}{2}(x^2 - 3x + 2) - 4(x^2 - 2x) + \frac{15}{2}(x^2 - x)$$

$$= \frac{x^2}{2}(-3 - 8 + 15) + \frac{x}{2}(9 + 16 - 15) + \frac{1}{2}(-6)$$

$$= 2x^2 + 5x - 3.$$

Normally we don't simplify the Lagrange form but use it in its original state. We have simplified it here for comparison with later examples.

EXERCISE 13.3

Write down and simplify the Lagrange interpolating polynomials for each of these sets of points:

(a) $(1, 2), (3, 7), (5, 11)$;

(b) $(-1, 1), (1, 4), (3, 2)$;

(c) $(0, -1), (2, 1), (4, 3)$.

13.4 QUADRATIC INTERPOLATION – THE DIVIDED–DIFFERENCE FORM

In Section 13.2 we noted that the divided-difference form of $p_1(x)$ is

$$p_1(x) = a_0 + a_1(x - x_0)$$

where a_0 and a_1 are constants depending on the data $(x_0, f_0), (x_1, f_1)$.

Suppose that we again add the third data point (x_2, f_2), but this time guess the interpolating polynomial has the form

$$p_2(x) = a_0 + a_1(x - x_0) + a_2(x - x_0)(x - x_1).$$

Notice that $p_2(x)$ is gained from $p_1(x)$ by adding a quadratic term. The question is: what are a_0, a_1 and a_2 now? Making the graph of $y = p_2(x)$ pass through $(x_0, f_0), (x_1, f_1)$ and (x_2, f_2) gives three conditions which should be enough to find the three unknowns a_0, a_1 and a_2; i.e., at $x = x_0$, $f_0 = a_0$

at $x = x_1$, $f_1 = a_0 + a_1(x_1 - x_0)$

and

at $x = x_2$, $f_2 = a_0 + a_1(x_2 - x_0) + a_2(x_2 - x_0)(x_2 - x_1)$.

$\therefore \quad a_0 = f_0 \quad$ (as in Section 13.2)

$$a_1 = \frac{f_1 - f_0}{x_1 - x_0} \quad \text{(as in Section 13.2)}$$

and

$$a_2 = (f_2 - a_0 - a_1(x_2 - x_0))/(x_2 - x_0)(x_2 - x_1).$$

The form of p_2 that we have chosen is a perfectly general quadratic polynomial, but it has the great advantage that a_0 and a_1 are exactly the same as for the linear polynomial interpolating at x_0 and x_1. Adding the extra data at x_2 to give an interpolating parabola involves calculating just one extra coefficient a_2. (Later on we shall see how this idea can be extended to polynomials of any degree n for $n + 1$ data points.)

Although the idea of extending the polynomial one degree at a time is simple, the calculation of a_2 (and later of a_3, a_4, etc.) is cumbersome by the method above. What follows may appear to be a rather odd way of going about things compared to the direct attack on the interpolation equations attempted above, but it does lead to an extremely simple way of constructing the divided-difference interpolating polynomial, whether quadratic as in this section, or of higher degree as in a later section.

We can easily see that a_0 involves only f_0, a_1 involves only f_0 and f_1, and a_2 involves only f_0, f_1 and f_2. In consequence we write

$$a_0 = f[x_0]$$
$$a_1 = f[x_0, x_1]$$

and

$$a_2 = f[x_0, x_1, x_2]$$

and we would like a simple method for calculating these values.

Let us compare the Newton formula with the Lagrange formula for $p_2(x)$:

$$p_2(x) = f[x_0] + (x - x_0)f[x_0, x_1] + (x - x_0)(x - x_1)f[x_0, x_1, x_2]$$

and

$$p_2(x) = \frac{f_0(x - x_1)(x - x_2)}{(x_0 - x_1)(x_0 - x_2)} + \frac{f_1(x - x_0)(x - x_2)}{(x_1 - x_0)(x_1 - x_2)} + \frac{f_2(x - x_0)(x - x_1)}{(x_2 - x_0)(x_2 - x_1)}.$$

These must be equivalent! Comparing the coefficients of x^2 in these two expressions we get

$$f[x_0, x_1, x_2] = \frac{f_0}{(x_0 - x_1)(x_0 - x_2)} + \frac{f_1}{(x_1 - x_0)(x_1 - x_2)}$$

$$+ \frac{f_2}{(x_2 - x_0)(x_2 - x_1)}. \qquad [13\text{-}5]$$

We could calculate $f[x_0, x_1, x_2]$ from this, but there is a more convenient method resulting from it.

Reconsider the calculation of $f[x_0, x_1]$:

$$f[x_0, x_1] = \frac{f[x_1] - f[x_0]}{x_1 - x_0} = \frac{f_1 - f_0}{x_1 - x_0}. \qquad [13\text{-}6]$$

Similarly

$$f[x_1, x_2] = \frac{f[x_2] - f[x_1]}{x_2 - x_1} = \frac{f_2 - f_1}{x_2 - x_1}. \qquad [13\text{-}7]$$

Perhaps we can write $f[x_0, x_1, x_2]$ in a similar form; e.g.,

$$f[x_0, x_1, x_2] = \alpha f[x_0, x_1] + \beta f[x_1, x_2] \qquad [13\text{-}8]$$

where α and β depend on x_0, x_1, x_2. Substituting Equations [13-6] and [13-7] into Equation [13-8] gives

$$f[x_0, x_1, x_2] = \left(-\frac{\alpha}{x_1 - x_0}\right)f_0 + \left(\frac{\alpha}{x_1 - x_0} - \frac{\beta}{x_2 - x_1}\right)f_1 + \left(\frac{\beta}{x_2 - x_1}\right)f_2.$$

Comparing the coefficients of f_0 and f_2 with those in Equation [13-5] gives

$$\alpha = \frac{1}{x_0 - x_2} = -\beta$$

so that

$$f[x_0, x_1, x_2] = \frac{f[x_1, x_2] - f[x_0, x_1]}{x_2 - x_0}. \qquad [13\text{-}9]$$

We have not actually proved that this identity holds for any three points $(x_0, f_0), (x_1, f_1)$ and (x_2, f_2) because we have only equated coefficients of f_0 and f_2. To complete the argument you should check that the coefficients of f_1 also agree.

Summarising: to calculate a_0, a_1, a_2 we need to calculate $f[x_0, x_1]$, $f[x_1, x_2]$ and $f[x_0, x_1, x_2]$. Each of these is obtained by a similar process (see Equations [13-6], [13-7] and [13-9]), and the results are usually tabulated in a *divided-difference table*:

x_0	$f[x_0]$		
		$f[x_0, x_1]$	
x_1	$f[x_1]$		$f[x_0, x_1, x_2]$
		$f[x_1, x_2]$	
x_2	$f[x_2]$		

Given more data (x_3, f_3), (x_4, f_4), etc. this table can be extended down-wards in a similar manner. The $f[x_i, x_{i+1}]$ are called *first divided-differences*, the $f[x_i, x_{i+1}, x_{i+2}]$ are *second divided-differences*. We leave until later the extension of these to higher divided-differences and the corresponding horizontal extension of the table.

When the divided-difference table is complete all we need do is read off the coefficients of the interpolating polynomial from the leading diagonal.

EXAMPLE 13.3 To find the interpolating polynomial based on the data given in the following table and using the divided-difference form.

x	0	1	2
$f(x)$	-3	4	15

x_i	$f[x_i]$	$f[x_i, x_{i+1}]$	$f[x_i, x_{i+1}, x_{i+2}]$
0	-3		
		$\dfrac{4-(-3)}{1-0} = 7$	
1	4		$\dfrac{11-7}{2-0} = 2$
		$\dfrac{15-4}{2-1} = 11$	
2	15		

$$\therefore \quad p_2(x) = -3 + 7(x - 0) + 2(x - 0)(x - 1)$$
$$= -3 + 7x + 2x^2 - 2x$$
$$= 2x^2 + 5x - 3.$$

Have a look at Example 13.2 again!

EXERCISE 13.4

In each case below use the data to draw up a table of divided-differences and write down the corresponding interpolating polynomial.

(a)

x	1	3	5
y	2	7	11

(b)

x	-1	1	3
$f(x)$	1	4	2

(c)

t	0	2	4
$f(t)$	-1	1	3

3.5 THE GENERAL LAGRANGIAN FORM OF THE INTERPOLATING POLYNOMIAL

We have seen that given 2 points we get a unique linear interpolating polynomial $p_1(x)$, given 3 points we get a unique quadratic $p_2(x)$, and so we would expect that given $n + 1$ points $(n \in \mathbb{N})$ we would get a unique interpolating polynomial $p_n(x)$ of degree n. $p_2(x)$ is given by

$$p_2(x) = L_0(x)f_0 + L_1(x)f_1 + L_2(x)f_2$$

where

$$L_0(x) = \frac{(x - x_1)(x - x_2)}{(x_0 - x_1)(x_0 - x_2)}$$

$$L_1(x) = \frac{(x - x_0)(x - x_2)}{(x_1 - x_0)(x_1 - x_2)}$$

and

$$L_2(x) = \frac{(x - x_0)(x - x_1)}{(x_2 - x_0)(x_2 - x_1)}.$$

Generalising this result for $p_n(x)$ we would expect that, for the points $(x_0, f_0), (x_1, f_1), \ldots, (x_n, f_n)$,

$$p_n(x) = L_0(x)f_0 + L_1(x)f_1 + \ldots + L_n(x)f_n$$

where, for all i, $L_i(x)$ is a polynomial of degree n with value one for $x = x_i$ and zero for $x = x_j$ $(j \neq i)$; i.e.,

$$L_0(x) = \frac{(x - x_1)(x - x_2) \ldots (x - x_n)}{(x_0 - x_1)(x_0 - x_2) \ldots (x_0 - x_n)}$$

$$L_1(x) = \frac{(x - x_0)(x - x_2) \ldots (x - x_n)}{(x_1 - x_0)(x_1 - x_2) \ldots (x_1 - x_n)}$$

.
.
.

$$L_n(x) = \frac{(x - x_0)(x - x_1) \ldots (x - x_{n-1})}{(x_n - x_0)(x_n - x_1) \ldots (x_n - x_{n-1})}.$$

Notice that in, say, $L_1(x)$ all the factors $(x - x_0), (x - x_1), (x - x_2), \ldots, (x - x_n)$ are present in the numerator *except* $(x - x_1)$ and that the denominator is the same as the numerator except that x_1 replaces x. This fact makes it easy to write down the general formula for the Lagrangian interpolating polynomial of any degree, provided we use standard mathematical notation for sums and products.

Using this notation gives

$$L_0(x) = \frac{\prod\limits_{j=1}^{n} (x - x_j)}{\prod\limits_{j=1}^{n} (x_0 - x_j)} = \prod_{j=1}^{n} \left(\frac{x - x_j}{x_0 - x_j} \right).$$

$L_1(x)$ is more difficult because $x - x_1$ is 'missing' from the numerator and '$x_1 - x_1$' from the denominator. This is done in the following way:

$$L_1(x) = \prod_{\substack{j=0 \\ j \neq 1}}^{n} \left(\frac{x - x_j}{x_1 - x_j} \right).$$

Similarly for $L_i(x)$ we get (remember that i is fixed as far as the product is concerned)

$$L_i(x) = \prod_{\substack{j=0 \\ j \neq i}}^{n} \left(\frac{x - x_j}{x_i - x_j} \right).$$

Hence the Lagrangian form of $p_n(x)$, the interpolating polynomial through the $n + 1$ points (x_i, f_i), $i = 0, 1, \ldots, n$, is

$$p_n(x) = \sum_{i=0}^{n} L_i(x) f_i \qquad \qquad \text{[13-10]}$$

where

$$L_i(x) = \prod_{\substack{j=0 \\ j \neq i}}^{n} \left(\frac{x - x_j}{x_i - x_j} \right)$$

giving a much more compact expression than that originally stated.

One thing we haven't done is to check that $p_n(x)$ has the appropriate properties. This should be checked by the reader, i.e., you should check that $p_n(x_i) = f_i$ for $i = 0(1)n$.

EXAMPLE 13.6   To find the Lagrangian polynomial passing through the points $(-1, -1), (1, 1), (2, 8), (4, 64)$.

$$p_3(x) = -1 \frac{(x-1)(x-2)(x-4)}{(-1-1)(-1-2)(-1-4)} + 1 \frac{(x+1)(x-2)(x-4)}{(1+1)(1-2)(1-4)}$$

$$+ 8 \frac{(x+1)(x-1)(x-4)}{(2+1)(2-1)(2-4)} + 64 \frac{(x+1)(x-1)(x-2)}{(4+1)(4-1)(4-2)}.$$

When this is simplified we get $p_3(x) = x^3$. (Do it for yourself — it's not very nice but doing it is instructive!)

EXERCISE 13.5

Write down and simplify the Lagrange interpolating polynomials for the sets of points:

(a) (0, 0), (1, 1), (2, 8), (3, 27);

(b) $(-2, 4)$, $(-1, 3)$, (1, 7), (2, 12).

_.6 THE GENERAL DIVIDED-DIFFERENCE FORM OF THE INTERPOLATING POLYNOMIAL

We can similarly extend the divided-difference quadratic for (x_0, f_0), (x_1, f_1) and (x_2, f_2):

$$p_2(x) = f[x_0] + (x - x_0)f[x_0, x_1] + (x - x_0)(x - x_1)f[x_0, x_1, x_2]$$

to the nth degree polynomial for (x_0, f_0), (x_1, f_1), . . . , (x_n, f_n):

$$p_n(x) = f[x_0] + (x - x_0)f[x_0, x_1] + (x - x_0)(x - x_1)f[x_0, x_1, x_2]$$
$$+ (x - x_0)(x - x_1)(x - x_2)f[x_0, x_1, x_2, x_3]$$
$$+ \ldots + (x - x_0)(x - x_1) \ldots (x - x_{n-1})f[x_0, x_1, \ldots, x_n] \quad [13\text{-}11]$$

where the third and higher divided-differences are calculated by a fairly obvious extension of the divided-difference table:

x_0	$f[x_0]$				
		$f[x_0, x_1]$			
x_1	$f[x_1]$		$f[x_0, x_1, x_2]$		
		$f[x_1, x_2]$		$f[x_0, x_1, x_2, x_3]$	
x_2	$f[x_2]$		$f[x_1, x_2, x_3]$		$f[x_0, x_1, x_2, x_3, x_4]$
		$f[x_2, x_3]$		$f[x_1, x_2, x_3, x_4]$	
x_3	$f[x_3]$		$f[x_2, x_3, x_4]$		
		$f[x_3, x_4]$			
x_4	$f[x_4]$				

etc., where, for example,

$$f[x_1, x_2, x_3, x_4] = \frac{f[x_2, x_3, x_4] - f[x_1, x_2, x_3]}{x_4 - x_1}$$

and

$$f[x_0, x_1, x_2, x_3, x_4] = \frac{f[x_1, x_2, x_3, x_4] - f[x_0, x_1, x_2, x_3]}{x_4 - x_0}$$

The numbers we need for the interpolating polynomial are those at the head of the columns.

EXAMPLE 13.7   To find the quartic interpolating polynomial for the data $(-1, 1)$, $(-\frac{1}{2}, -\frac{1}{2})$, $(0, 1)$, $(\frac{1}{2}, -\frac{1}{2})$, $(1, 1)$,   using the divided-difference form.

x_i	$f[x_i]$				
-1	1				
		-3			
$-\frac{1}{2}$	$-\frac{1}{2}$		6		
		3		-8	
0	1		-6		8
		-3		8	
$\frac{1}{2}$	$-\frac{1}{2}$		6		
		3			
1	1				

The required quartic is thus

$$p_4(x) = 1 - 3(x + 1) + 6(x + 1)(x + \tfrac{1}{2}) - 8(x + 1)(x + \tfrac{1}{2})x$$
$$+ 8(x + 1)(x + \tfrac{1}{2})x(x - \tfrac{1}{2}).$$

There is no real need to rearrange this; it is in fact

$$p_4(x) = 1 - 8x^2 + 8x^4.$$

Once again you should check that $p_4(x_i) = f[x_i]$ for each x_i.

Note that we have not actually proved anything in this section. If we define the higher divided-differences in such a way that the table can be constructed:

$$f[x_k, x_{k+1}, \ldots, x_m] = \frac{f[x_{k+1}, \ldots, x_m] - f[x_k, \ldots, x_{m-1}]}{x_m - x_k}$$

then it can be shown, but we shall not, that the nth degree polynomial [13-11] does indeed pass through the points (x_i, f_i) for $i = 0(1)n$.

EXERCISE 13.6

Find the divided-difference interpolating polynomials for the data below.

(a)

x	0	1	2	3
$f(x)$	0	1	8	27

(b)

x	-2	-1	1	2
$f(x)$	4	3	7	12

13.7 COMPARISON OF THE LAGRANGE AND DIVIDED-DIFFERENCE FORMS

The first thing to note is that the Lagrange and the divided-difference interpolating polynomials for the same set of data points, are of course, the same polynomial. The polynomial is just found in different ways. You should have noted that the exercises demonstrate this fact.

Lagrange's form is difficult to write down and not at all easy to simplify. It is also not particularly easy to find the value of the interpolating polynomial for a given value of x using Lagrange's form. Moreover, if an extra point is to be added to the set of data already available, the Lagrangian polynomial (or its value for a particular x) must be recalculated from the beginning — a formidable task! However, Lagrange's form is of value in theoretical work.

The divided-difference form on the other hand is easy to write down, if not particularly easy to simplify, and easy to evaluate for a particular value of x. (The evaluation is best done by nesting the polynomial.) The calculation of the coefficients is broken down into easy stages by the table of divided differences and this is a considerable advantage. Best of all, if an extra point is to be added to the set of data points already available, all that is needed is an extra line at the bottom of the divided-difference table and an extra term added to the interpolating polynomial. There is no need to start from scratch.

To sum up then: the Lagrange and divided-difference forms of the interpolating polynomial based on a given set of points are equivalent but the divided-difference form is of much greater practical use. In this lesson the evaluation of the interpolating polynomial for a particular value of x has been mentioned but not stressed. We shall deal more fully with the problems of evaluation and of adding extra data points in Lesson 14.

3.8   UNIQUENESS OF THE INTERPOLATING POLYNOMIAL

Geometrically it is obvious that there is only one straight line through two distinct data points, hence $p_1(x)$ is unique. The uniqueness of higher-degree interpolating polynomials is not quite so obvious, but is a fact.

THEOREM

$n + 1$ data points   with distinct abscissae   (x-coordinates) possess a unique interpolating polynomial of degree at most n.

PROOF

Let p_n and q_n be different interpolating polynomials of degree at most n for the data. Then $p_n - q_n$ is a polynomial of degree at most n which is zero at all $n + 1$ data points.   But a polynomial of degree at most n can have no more than n zeros (Fundamental Theorem of Algebra), unless it is the trivial (zero) polynomial. Hence $p_n - q_n \equiv 0$, i.e., $p_n \equiv q_n$. So there cannot be two different interpolating polynomials for the same data.

NOTE   We have to let the degree be 'at most n' because of the possibility that the points happen to lie on a curve of degree less than n. For example, three points may be collinear.

ANSWERS TO EXERCISES

13.1   (a) $p_1(x) = \dfrac{1(x-2)}{0-2} + \dfrac{3(x-0)}{2-0} = x+1.$

      (b) $p_1(x) = \dfrac{4(x-2)}{3-2} + \dfrac{1(x-3)}{2-3} = 3x-5.$

13.2   (a) $p_1(x) = 1 + 1(x-0) = 1 + x.$

      (b) $p_1(x) = 4 + 3(x-3) = 3x-5.$

      Note — The answers are the same as in Exercise 13.1.

13.3   (a) $p_2(x) = \dfrac{2(x-3)(x-5)}{(1-3)(1-5)} + \dfrac{7(x-1)(x-5)}{(3-1)(3-5)} + \dfrac{11(x-1)(x-3)}{(5-1)(5-3)}$

            $= \tfrac{1}{8}(-x^2 + 24x - 7).$

      (b) $p_2(x) = \dfrac{1(x-1)(x-3)}{(-1-1)(-1-3)} + \dfrac{4(x+1)(x-3)}{(1+1)(1-3)} + \dfrac{2(x+1)(x-1)}{(3+1)(3-1)}$

            $= \tfrac{1}{8}(-5x^2 + 12x + 25).$

      (c) $p_2(x) = \dfrac{-1(x-2)(x-4)}{(0-2)(0-4)} + \dfrac{1(x-0)(x-4)}{(2-0)(2-4)} + \dfrac{3(x-0)(x-2)}{(4-0)(4-2)}$

            $= x - 1$

      i.e., the 3 points are collinear.

13.4   (a)

x_i	$y_i = f[x_i]$	$f[x_i, x_{i+1}]$	$f[x_i, x_{i+1}, x_{i+2}]$
1	2		
		$\dfrac{7-2}{2} = \tfrac{5}{2}$	
3	7		$\dfrac{2-5/2}{5-1} = -\tfrac{1}{8}$
		$\dfrac{11-7}{5-3} = 2$	
5	11		

      $p_2(x) = 2 + \tfrac{5}{2}(x-1) - \tfrac{1}{8}(x-1)(x-3)$

            $= \tfrac{1}{8}(-x^2 + 24x - 7).$

(b)

$$\begin{array}{cc|ccc}
-1 & 1 & & & \\
 & & \tfrac{3}{2} & & \\
1 & 4 & & & -\tfrac{5}{8} \\
 & & -1 & & \\
3 & 2 & & & \\
\end{array}$$

$$\begin{aligned}
p_2(x) &= 1 + \tfrac{3}{2}(x+1) - \tfrac{5}{8}(x+1)(x-1) \\
&= \tfrac{1}{8}(-5x^2 + 12x + 25).
\end{aligned}$$

(c)

$$\begin{array}{cc|cc}
0 & -1 & & \\
 & & 1 & \\
2 & 1 & & 0 \\
 & & 1 & \\
4 & 3 & & \\
\end{array}$$

$$\begin{aligned}
p_2(t) &= -1 + 1(t-0) + 0(t-0)(t-2) \\
&= t - 1.
\end{aligned}$$

Note — The answers are the same as in Exercise 13.3.

13.5 (a) $p_3(x) = \dfrac{0(x-1)(x-2)(x-3)}{(0-1)(0-2)(0-3)} + \dfrac{1(x-0)(x-2)(x-3)}{(1-0)(1-2)(1-3)}$

$\qquad\qquad\quad + \dfrac{8(x-0)(x-1)(x-3)}{(2-0)(2-1)(2-3)} + \dfrac{27(x-0)(x-1)(x-2)}{(3-0)(3-1)(3-2)}$

$\qquad\quad = x^3$ — eventually!

(b) $p_3(x) = \dfrac{4(x+1)(x-1)(x-2)}{(-2+1)(-2-1)(-2-2)}$

$\qquad\qquad\quad + \dfrac{3(x+2)(x-1)(x-2)}{(-1+2)(-1-1)(-1-2)}$

$\qquad\qquad\quad + \dfrac{7(x+2)(x+1)(x-2)}{(1+2)(1+1)(1-2)} + \dfrac{12(x+2)(x+1)(x-1)}{(2+2)(2+1)(2-1)}$

$\qquad\quad = x^2 + 2x + 4$ — eventually!

Note — The 4 points happen to lie on a parabola, so the x^3 term is missing.

13.6 (a)

$$\begin{array}{cc|cccc}
0 & 0 & & & & \\
 & & 1 & & & \\
1 & 1 & & 3 & & \\
 & & 7 & & 1 & \\
2 & 8 & & 6 & & \\
 & & 19 & & & \\
3 & 27 & & & & \\
\end{array}$$

$$p_3(x) = 0 + 1(x-0) + 3(x-0)(x-1) + 1(x-0)(x-1)(x-2)$$
$$= x^3.$$

(b)

-2	4			
		-1		
-1	3		1	
		2		0
1	7		1	
		5		
2	12			

$$p_3(x) = 4 - 1(x+2) + 1(x+2)(x+1) + 0(x+2)(x+1)(x-1)$$
$$= x^2 + 2x + 4.$$

Note — The answers are the same as in Exercise 13.5.

FURTHER READING

For a similar view to ours see Phillips and Taylor Sections 4.1, 4.2 and 4.6. Most books take a more 'old-fashioned' view of interpolation concentrating on finite differences (see Lesson 16). In particular the more elementary ones do this, so we cannot suggest an alternative reference.

LESSON 13 – COMPREHENSION TEST

1. Write down the linear interpolating polynomial which agrees with $f(x) = x^2$ at $x = 1$ and $x = 3$:

(a) in Lagrange form, and

(b) in divided-difference form.

Show that they are identical.

2. What is the quadratic interpolating polynomial which agrees with $f(x) = x^2$, regardless of the choice of the three defining points?

3. Write down in Lagrange form the cubic interpolating polynomial agreeing with $f(x)$ at $x = x_0, x_1, x_2, x_3$.

4. Write down the general cubic divided-difference interpolating polynomial for the function f and the points x_0, x_1, x_2, x_3.

5. Form a divided-difference table from the following table of values and hence write down the quadratic polynomial passing through the points indicated by the table.

x	-1	2	4
y	-3	0	12

6. Write down, without rearrangement, the Lagrange polynomial for the data in Question 5.

7. If an extra data point is added and the associated interpolating polynomial (with degree one higher) is required, which form makes the task easier — Lagrange or divided-difference?

8. Give a geometric description of the distinction between a linear Taylor polynomial and a linear interpolating polynomial.

ANSWERS

1 (a) $(1)\dfrac{(x-3)}{(1-3)} + (9)\dfrac{(x-1)}{(3-1)};$ (b) $1 + \dfrac{(9-1)}{(3-1)}(x-1).$

   Both equal $4x - 3.$

2 $x^2.$

3 $\dfrac{(x-x_1)(x-x_2)(x-x_3)}{(x_0-x_1)(x_0-x_2)(x_0-x_3)} f(x_0) + \dfrac{(x-x_0)(x-x_2)(x-x_3)}{(x_1-x_0)(x_1-x_2)(x_1-x_3)} f(x_1)$

$+ \dfrac{(x-x_0)(x-x_1)(x-x_3)}{(x_2-x_0)(x_2-x_1)(x_2-x_3)} f(x_2) + \dfrac{(x-x_0)(x-x_1)(x-x_2)}{(x_3-x_0)(x_3-x_1)(x_3-x_2)} f(x_3).$

4 $f[x_0] + f[x_0, x_1](x-x_0) + f[x_0, x_1, x_2](x-x_0)(x-x_1)$

   $+ f[x_0, x_1, x_2, x_3](x-x_0)(x-x_1)(x-x_2).$

5
-1	-3		
2	0	$\dfrac{1}{6}$	1
4	12		

$-3 + (1)(x+1) + (1)(x+1)(x-2).$

6 $\dfrac{(x-2)(x-4)}{(-1-2)(-1-4)}(-3) + \dfrac{(x+1)(x-4)}{(2+1)(2-4)}(0) + \dfrac{(x+1)(x-2)}{(4+1)(4-2)}(12).$

7 Divided-difference.

8 The linear Taylor polynomial is a tangent (agrees with function and first derivative at one point). The linear interpolating polynomial is a chord (agrees with the function at two points).

LESSON 13 – SUPPLEMENTARY EXERCISES

13.7. Show by constructing a divided-difference table that the points (x, y) given below lie on the graph $y = p(x)$ of a polynomial function. Construct the polynomial.

x	-2	-1	0	2	4	7
y	-104	-99	-100	-96	-44	229

13.8. Construct the Lagrangian form of the polynomial $p_2(x)$ agreeing with x^3 at $x = 0, 1$ and 2. Use elementary calculus to show that the maximum value of the error modulus $|x^3 - p_2(x)|$ for $x \in [0, 2]$ is 0.385 and occurs at $x = 0.423$ and $x = 1.577$. Draw a sketch illustrating these results.

13.9. Using divided-differences construct the quartic polynomial passing through all these data points.

x	-4	-1	0	2	5
y	3.24	3.37	3.71	4.84	7.64

*13.10. We have defined interpolation to mean agreeing with function values. The idea can be extended to agreement with derivative (and other) values. This is often called Hermite interpolation.

Verify that the cubic polynomials

$$H_0(x) = (2x + 1)(x - 1)^2$$
$$H_1(x) = x^2(3 - 2x)$$
$$H_2(x) = x(x - 1)^2$$
$$H_3(x) = x^2(x - 1)$$

have the following properties:

$$H_0(0) = 1, \quad H_0(1) = H_0^{(1)}(0) = H_0^{(1)}(1) = 0$$
$$H_1(1) = 1, \quad H_1(0) = H_1^{(1)}(0) = H_1^{(1)}(1) = 0$$
$$H_2^{(1)}(0) = 1, \quad H_2(0) = H_2(1) = H_2^{(1)}(1) = 0$$
$$H_3^{(1)}(1) = 1, \quad H_3(0) = H_3(1) = H_3^{(1)}(0) = 0.$$

It follows that the unique cubic polynomial agreeing with $f(x)$ and $f^{(1)}(x)$ at $x = 0$ and $x = 1$ is $p_3(x) = f(0)H_0(x) + f(1)H_1(x) + f^{(1)}(0)H_2(x) + f^{(1)}(1)H_3(x)$.

Hence find the cubic polynomial agreeing with x^4 and its derivatives at $x = 0$ and $x = 1$. Verify that the result actually has this property.

Lesson 14 Interpolation

OBJECTIVES

At the end of this lesson you should

(a) appreciate the problem of interpolation;

(b) be able to do linear interpolation in standard tables;

(c) understand the difficulties of the Lagrange approach to inter-polation;

(d) understand and be able to use the divided-difference approach to interpolation and appreciate its advantages over Lagrange interpolation;

(e) understand and be able to use the Neville-Aitken method of interpolation;

(f) be able to recognise when polynomial interpolation fails.

CONTENTS

14.1 THE INTERPOLATION PROBLEM

In Lesson 13 our attention was concentrated mainly on the mathematical aspects of finding various forms of polynomials whose graphs passed through a given set of data points: so-called interpolating polynomials. However, in practice the problem is to find a reasonable value for the function at a point in its domain which is not tabulated. It will be seen later that such a point must be taken within the range of the data points.

EXAMPLE 14.1   Given the table of values below and $x = -0.2$ the problem is to find a reasonable value for $f(-0.2)$.

x	-0.4	-0.3	-0.1	0.1	0.2
$f(x)$	0.670	0.741	0.905	1.105	1.221

You will notice that the given x values need not be evenly spaced and that $x = -0.2$ lies between the end values $x = -0.4$ and $x = 0.2$.

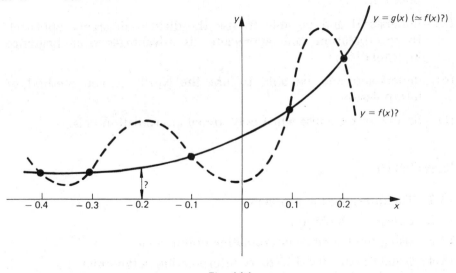

Fig. 14.1

Fig. 14.1 shows that the given points apparently lie on the smooth curve $y = g(x)$. In such a case we can solve this interpolation problem by letting $g(x)$ be a suitable interpolating polynomial for the data and evaluating it at $x = -0.2$. The essential assumption is then that the interpolating polynomial will pass smoothly through the data points. This is not always true, as we shall see later. We are also assuming that the given data are sufficient to represent the function f. (As an extreme example of this *not* being so consider sine being represented by its values at $0, \pi, 2\pi, 3\pi, \ldots$).

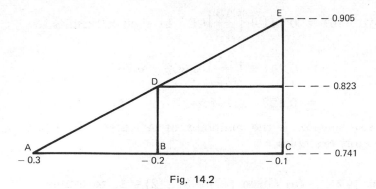

Fig. 14.2

If for example the function $f(x)$ is very similar to the function $g(x)$ in Fig. 14.1 then we shall get a good estimate $g(-0.2)$ of $f(-0.2)$. If on the other hand $f(x)$ looks like the dotted curve in Fig. 14.1 then $g(-0.2)$ will not be a good estimate of $f(-0.2)$. Throughout this lesson we shall assume that the given function values $f(x)$ *do* sufficiently represent the (unknown) function f.

Having decided to use polynomial interpolation one of the most difficult questions to answer is: how much of the data do we need to use? (There might be a lot!) We can follow that question with another: how accurate do we need the answer to be? Consideration of these questions is also deferred until later. For the moment let us consider the simplest approach to the problem — linear interpolation.

4.2 LINEAR INTERPOLATION

To find $f(-0.2)$ in Example 14.1 by linear interpolation we construct a linear interpolating polynomial and evaluate it at $x = -0.2$. The obvious two data points to choose are those at $x = -0.3$ and -0.1. Using the divided-difference form the polynomial is

$$p_1(x) = 0.741 + \left(\frac{0.905 - 0.741}{-0.1 + 0.3} \right)(x + 0.3)$$

and $p_1(-0.2) = 0.823$.

Looking at the data this seems a 'sensible' answer.

There is a geometric interpretation of linear interpolation which is worth considering. Fig. 14.2 shows that we are actually concerned with similar triangles.

A and E are the data points. They are connected by a straight line. The point D is the approximate position of the point $(-0.2, f(-0.2))$ and $f(-0.2)$ is thus given approximately by

$$0.741 + BD = 0.741 + \left(\frac{AB}{AC}\right)CE \quad \text{by similar triangles}$$

$$= 0.741 + \left(\frac{0.1}{0.2}\right)(0.905 - 0.741)$$

$$= 0.823 \quad \text{as before.}$$

In this case since B is the midpoint of AC then BD is just $\frac{1}{2}$CE. Here are some more examples.

EXAMPLE 14.2 (a) Given $f(0) = 5$, $f(2) = 1$, to estimate $f(0.4)$. (See Fig. 14.3(a).)

$$f(0.4) \simeq 5 - \frac{0.4}{2}(4) = 4.2$$

or $\quad f(0.4) \simeq 1 + \dfrac{1.6}{2}(4) = 4.2$.

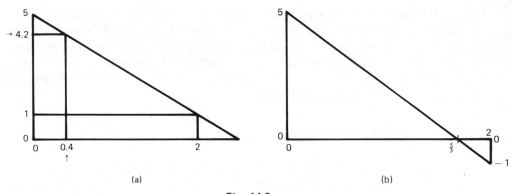

(a)   (b)

Fig. 14.3

(b) Given $f(0) = 5$, $f(2) = -1$, to estimate the point x at which $f(x) = 0$ (assuming f is continuous for $0 \leqslant x \leqslant 2$). Considering Fig. 14.3(b), the point is $\frac{5}{6}(2) = \frac{5}{3}$ by simple proportion, or is given by

$$f(x) = 0 = 5 + \frac{(-1-5)}{2}(x-0)$$

i.e. $x = 5/3$, using 'proper' linear interpolation.

(c) A certain specific heat is 4.218 at 100°C and 4.230 at 110°C; to estimate its value at 107°C. The linear interpolate is

$$4.218 + \tfrac{7}{10}(4.230 - 4.218) = 4.226 \quad \text{to 3D}.$$

(*We dare not quote more than 3D when that is the accuracy of the data.*)

Standard mathematical and other tables use linear interpolation. For example, a typical set of 5-figure tables includes values of $\sin x$ for $x = 0°(6')90°$. The interval of $6'$ is deliberately chosen so that linear interpolation gives five-figure accuracy. The interpolation is almost done for us because we are given 'mean differences' to add or subtract.

EXAMPLE 14.3   We are given that

$$\sin 10°30' = 0.182\,24$$
$$\sin 10°36' = 0.183\,95.$$

By linear interpolation

$$\sin 10°34' = 0.182\,24 + \tfrac{4}{6}(0.183\,95 - 0.182\,24)$$
$$= 0.182\,24 + 0.001\,14$$
$$= 0.183\,38.$$

The given mean difference in the tables is (surprise!) $0.001\,14$.

Apart from tables which are designed for it we shall usually find that linear interpolation is not sufficiently accurate, unless a rough answer is all we need. In the rest of this lesson we shall consider higher-degree interpolation in which we aim to get the maximum justifiable accuracy in the *interpolate* (the estimated function value obtained by interpolation).

EXERCISE 14.1

(a) Estimate $\sqrt{3}$ by linear interpolation using $\sqrt{2} \simeq 1.414$ and $\sqrt{4} = 2$. What is the error in the interpolate?

(b) Estimate $\log_e 1.06$ from $\log_e 1.02 = 0.019\,80$ and $\log_e 1.09 = 0.086\,18$. Comment on the difference between your answer and the value $\log_e 1.06 = 0.058\,27$ correct to 5D.

(c) From the following data estimate by linear interpolation:
(i) $f(0.5)$;   (ii) $f(1.3)$;   (iii) the value of x at which $f(x) = 0$.

x	0	1	2
$f(x)$	0.1071	0.0428	-0.0264

14.3 USING THE LAGRANGE INTERPOLATING POLYNOMIAL

Let us return to the table of values given in Example 14.1

x	-0.4	-0.3	-0.1	0.1	0.2
$f(x)$	0.670	0.741	0.905	1.105	1.221

and use Lagrange's form of the interpolating polynomial to attempt to find a reasonable value for $f(x)$ at $x = -0.2$. As we have seen, if we use linear interpolation based on the x values -0.3 and -0.1 then we obtain the interpolate $p_1(-0.2) = 0.823$.

Now what should we do next? Perhaps a quadratic interpolate would be better (who knows?).

Let us try using the data at $x = -0.4, -0.3, -0.1$ and the Lagrange form of the quadratic:

$$p_2(-0.2) = \frac{0.670(-0.2 + 0.3)(-0.2 + 0.1)}{(-0.4 + 0.3)(-0.4 + 0.1)}$$

$$+ \frac{0.741(-0.2 + 0.4)(-0.2 + 0.1)}{(-0.3 + 0.4)(-0.3 + 0.1)}$$

$$+ \frac{0.905(-0.2 + 0.4)(-0.2 + 0.3)}{(-0.1 + 0.4)(-0.1 + 0.3)}$$

$$= -\frac{0.670}{3} + 0.741 + \frac{0.905}{3}$$

$$= 0.819 \quad \text{to 3D.}$$

(Why to 3D?)

Looking at the graph in Fig. 14.1 we can see that the linear interpolate (0.823) would be high, so $p_2(-0.2)$ looks 'sensible' and is sufficiently close to $p_1(-0.2)$ to give us confidence that we are on the right track.

What next? We could try the quadratic interpolate based on the data at $-0.3, -0.1, 0.1$, or perhaps the cubic interpolate based on the data at $-0.4, -0.3, -0.1, 0.1$. Both of these involve completely new calculations. For this reason the Lagrange form is almost never used for practical interpolation. The divided-difference form on the other hand is highly suitable for this kind of investigation.

14.4 USING THE DIVIDED-DIFFERENCE INTERPOLATING POLYNOMIAL

In complete contrast to the difficulty in using the Lagrange form for practical interpolation, the divided-difference approach provides us with all the answers we need and very economically. Most importantly it usually tells us the most suitable degree of interpolating polynomial (if any) and hence the subset of data to use, so that we obtain the maximum justifiable accuracy in the interpolate.

Consider, for example, the divided-difference table for the data in Example 14.1.

x_i	$f[x_i]$	1st	Divided differences 2nd	3rd	4th
−0.4	0.670				
		0.710			
−0.3	0.741		0.367		
		0.820		0.166	
−0.1	0.905		0.450		0
		1.000		0.166	
0.1	1.105		0.533		
		1.160			
0.2	1.221				

Notice that the divided differences decrease from left to right, with the third divided differences constant and the fourth zero. So, at least for $-0.4 \leqslant x \leqslant 0.2$, f behaves very much like a cubic polynomial. (A set of data made up of values of a cubic polynomial must have zero fourth and higher divided differences, otherwise we could construct an interpolating polynomial for the data of degree greater than three, which would be non-sense. If you look back at the definition of divided differences in Lesson 13 you will soon see that if all the fourth divided differences are zero then the third divided differences must all be equal to each other. Can you generalise this to data which are values of an nth degree polynomial?)

Having decided to use a cubic polynomial (anything less would give less than maximum accuracy) we only have to choose a suitable subset of the data. We need four points and the obvious way to choose them is so that the point at which we wish to interpolate $(x = -0.2)$ is central. So we choose the data at $x = -0.4, -0.3, -0.1, 0.1$, i.e., ignore the last entry in each column. The interpolating polynomial is

$$p_3(x) = 0.670 + 0.710(x + 0.4) + 0.367(x + 0.4)(x + 0.3)$$
$$+ 0.166(x + 0.4)(x + 0.3)(x + 0.1)$$

so that

$$p_3(-0.2) = 0.670 + (0.710)(0.2) + (0.367)(0.2)(0.1)$$
$$+ 0.166(0.2)(0.1)(-0.1)$$
$$= 0.819 \quad \text{to 3D}$$

confirming the quadratic Lagrange interpolate of the last section. In this particular case, because the two third divided differences are equal, the interpolating polynomial based on the data at $x = -0.3, -0.1, 0.1$ and 0.2 will be equivalent to the p_3 above (though looking different) and hence give the same interpolate at $x = -0.2$ (apart from small differences due to rounding errors).

EXERCISE 14.2

Write down the interpolating polynomial based on the last four data points in the table above. Evaluate it at $x = -0.2$.

In the remainder of this section we shall look at some practical problems associated with interpolation in general and the divided-difference approach in particular.

14.4.1 ACCURACY OF THE INTERPOLATE

As a rule of thumb, if there is a clear indication of the degree of inter-polating polynomial to use, then the error bound for the interpolate will roughly equal that for the data. Thus, in the example above, having been given 3D data we quoted the interpolate to 3D. We shall deal more fully with errors in interpolation in the next lesson.

14.4.2 NESTED MULTIPLICATION

The divided-difference form of the polynomial can be rewritten to make its evaluation more efficient. Reconsidering the example above as

$$p_3(x) = 0.670 + 0.710(x + 0.4) + 0.367(x + 0.4)(x + 0.3)$$
$$+ 0.166(x + 0.4)(x + 0.3)(x + 0.1)$$

we note that it involves six multiplications. On the other hand in 'nested' form

$$p_3(x) = 0.670 + (x + 0.4)[0.710 + (x + 0.3)\{0.367 + (x + 0.1)0.166\}]$$

it requires only three multiplications. This can be a considerable economy if many evaluations of p_3 are required, as for example in drawing an accurate graph of p_3 on a computer-driven graph-plotter or visual-display screen.

EXERCISE 14.3

Write the polynomial obtained in Exercise 14.2 in nested form and evaluate it at $x = -0.2$ in that form. Compare the effort involved with the evaluation in Exercise 14.2.

14.4.3 SMALL SETS OF DATA

The data of Example 14.1 turned out to be sufficient to define clearly a suitable interpolating polynomial. Sometimes we have to make do with too few data.

EXAMPLE 14.4   Consider this table of divided differences. Can we deduce a reasonable value for $f(0.5)$?

x_i	$f[x_i]$	1st	Divided differences 2nd	3rd	4th
0.1	1.105				
		1.160			
0.2	1.221		0.650		
		1.355		1.676	
0.4	1.492		1.488		-4.834
		1.950		-2.191	
0.6	1.882		-0.046		
		1.927			
0.9	2.460				

There is no indication in this table as to a suitable degree of interpolating polynomial. In fact everything suggests that no interpolating polynomial based on this data will represent $f(x)$ very well for $0.1 \leqslant x \leqslant 0.9$. However, we can at least investigate the possibilities. Let us write down the quartic polynomial passing through all five points:

$$p_4(x) = 1.105 + 1.160(x - 0.1) + 0.650(x - 0.1)(x - 0.2)$$
$$+ 1.676(x - 0.1)(x - 0.2)(x - 0.4)$$
$$- 4.834(x - 0.1)(x - 0.2)(x - 0.4)(x - 0.6).$$

Using the permanence property of the Newton form, we have

$$p_1(0.5) = 1.105 + 1.160(0.5 - 0.1) = 1.569$$
$$p_2(0.5) = p_1(0.5) + 0.650(0.5 - 0.1)(0.5 - 0.2) = 1.647$$
$$p_3(0.5) = p_2(0.5) + 1.676(0.5 - 0.1)(0.5 - 0.2)(0.5 - 0.4) = 1.667$$
$$p_4(0.5) = p_3(0.5) - 4.834(0.5 - 0.1)(0.5 - 0.2)(0.5 - 0.4)(0.5 - 0.6)$$
$$= 1.673$$

all to 3D.

Although $p_1(0.5)$ is not based on the best choice of points for linear interpolation (those at $x = 0.4$ and 0.6), and similarly $p_2(0.5)$ and $p_3(0.5)$ are not perhaps the best available, nevertheless these results encourage us to assert that $f(0.5)$ is probably 1.67 to 2D. As a check we could evaluate $p_3(0.5)$ based on the *last* four data points in the table (instead of the first four as above).

$$p_3(0.5) = 1.221 + 1.355(0.5 - 0.2) + 1.488(0.5 - 0.2)(0.5 - 0.4)$$
$$- 2.191(0.5 - 0.2)(0.5 - 0.4)(0.5 - 0.6)$$
$$= 1.679 \quad \text{to 3D}$$

rather upsetting the apple-cart, but confirming the indication of the table that we would have trouble finding a 'good answer'. We must clearly be content with '1.67 or 1.68 to 2D' and post a warning that so little data casts doubt on the results. In a later section we shall see that a divided-difference table as inconsistent as that in this example *can* indicate the complete failure of the polynomial interpolation method. So beware! Never calculate and trust a single estimate; always check and confirm by other calculations and estimates.

14.4.4 LARGE SETS OF DATA

Having too many data is a more acceptable problem than having too few! The points to remember are

(a) it is unlikely that you will be able to represent *all* the data with one polynomial, indeed the degree required may vary within the data;

(b) it may not be obvious what degree of polynomial you should choose. If in doubt choose higher rather than lower;

(c) if you require just one interpolate start work on a subset of the data near the interpolation point.

This last piece of advice is possible because the divided-difference method does not require the data to be in any specific order. The difference table can be built up line by line, rather than column by column, until a suitable degree becomes clear.

EXAMPLE 14.5 To estimate $f(0.552)$ from these data.

x	0	0.20	0.40	0.60	0.70	0.80	0.85
$f(x)$	0	0.2013	0.4108	0.6367	0.7586	0.8881	0.9561

x	0.90	0.95	1.00
$f(x)$	1.0265	1.0995	1.1752

We first write down the data in order of closeness to $x = 0.552$. The remainder of the calculation is as before except that the table is constructed line by line. As soon as a suitable polynomial degree emerges we can abandon the unused data.

x_i	f_i	1st	2nd	3rd	4th	5th	6th
0.60	0.6367						
		1.2190					
0.70	0.7586		0.298 50				
		1.1593		0.202 50			
0.40	0.4108		0.339 00		0.034 44		
		1.1932		0.211 11		0.011 03	
0.80	0.8881		0.370 67		0.037 75		− 0.005 925
		1.3600		0.218 66		0.013 40	
0.85	0.9561		0.480 00		0.031 05		
		1.4080		0.212 45			
0.90	1.0265		0.352 53				
		1.1789					
0.20	0.2013						
0.95	1.0995						
1.00	1.1752						
0.00	0						

The sudden reduction from the third divided differences to the fourth, followed by equally small fifth and very small sixth divided differences, together with the evident consistency in each column, suggests that a quartic or at most quintic polynomial will accurately represent $f(x)$ for $0.2 \leqslant x \leqslant 0.9$ at least. At $x = 0.552$ we get

$$f(0.552) \simeq 0.6367 + 1.2190 \ (- 0.048)$$

$$+ \ 0.298 \ 50(- 0.048)(- 0.148)$$

$$+ \ 0.202 \ 50(- 0.048)(- 0.148)(0.152)$$

$$+ \ 0.034 \ 44(- 0.048)(- 0.148)(0.152)(- 0.248)$$

$$+ \ 0.011 \ 03(- 0.048)(- 0.148)(0.152)(- 0.248)(- 0.298)$$

$$= 0.5805 \quad \text{to} \quad 4D.$$

At $x = 0.552$ we can calculate successively higher degree interpolates starting with the linear interpolate

$$p_1(0.552) = 0.578 \ 19$$

$$p_2(0.522) = 0.580 \ 31$$

$$p_3(0.522) = 0.580 \ 53$$

$$p_4(0.522) = 0.580 \ 52$$

$$p_5(0.522) = 0.580 \ 52.$$

Results like these make one pretty sure that $f(0.552) = 0.5805$ to 4D.

Again note that we dare not quote more than 4D (the accuracy of the data) and that we have worked with 5D so that we can be sure that rounding is not affecting the result adversely.

In fact neither the fourth nor fifth degree terms affect the 4D result so that a cubic polynomial would probably be sufficient in $0.2 \leqslant x \leqslant 0.9$. (It may make you more confident of what we are doing to know that $f(x) = \sinh x$ in this problem and that $\sinh 0.552 = 0.5805$ correct to 4D.)

EXERCISE 14.4

Estimate the value of $f(4.4)$ from the following table of data, using the divided-difference method. (*Hint*: don't forget to reorder the data.)

x	0.0	1.0	2.0	2.5	3.0	3.5	4.0
$f(x)$	1.000	1.020	1.081	1.128	1.185	1.255	1.337

x	4.25	4.5	4.75	5.0
$f(x)$	1.384	1.433	1.486	1.543

14.5 THE NEVILLE–AITKEN METHOD

So far, using either the Lagrange or the divided-difference form of the interpolating polynomial, a set of data points has been used to construct the interpolating polynomial and the value of the interpolating polynomial has then been found. This process can be represented by the following scheme.

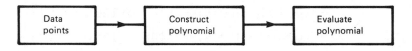

The Lagrange method is cumbersome and the accuracy is difficult to ascertain because the whole calculation must be re-done when changing the subset of data used. The divided-difference method is easier in calculation and is easily extended when an extra point is added to check on the validity of the answer. In either case the construction and evaluation parts can be collapsed into one calculation. That is, we do not actually need to write down the polynomial as a function, but can write down the expression with the value of x already substituted.

In contrast, the Neville–Aitken method (on which there are many variations) constructs from the data points successive *values* for interpolating polynomials directly from the data for a given value of x; the formulae for the interpolating polynomials never appear. It is easily extended by the addition of extra points and the validity of the interpolation is easy to examine. There is very little to choose between the divided-difference and Neville–Aitken methods in terms of efficiency and ease of use. They are superficially similar in operation too, so make sure you can distinguish them.

You may recall from Lesson 13 that the Neville–Aitken form for $p_1(x)$, based on data points (x_0, f_0) and (x_1, f_1), was found to be

$$p_1(x) = \frac{(x - x_0)f_1 - (x - x_1)f_0}{x_1 - x_0}$$

which we will call $p_{0,1}$ as it depends on the points with suffices 0 and 1.

If we have other data, at x_2, x_3, etc., we similarly define $p_{1,2}, p_{2,3}$, etc. Each $p_{i,i+1}$ is the value of a linear interpolating polynomial. For a specific value of x the values will be linear interpolates based on successive pairs of data points. The calculation of these linear interpolates can be tabulated as follows.

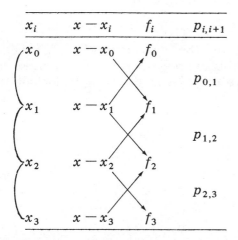

x_i	$x - x_i$	f_i	$p_{i,i+1}$
x_0	$x - x_0$	f_0	
			$p_{0,1}$
x_1	$x - x_1$	f_1	
			$p_{1,2}$
x_2	$x - x_2$	f_2	
			$p_{2,3}$
x_3	$x - x_3$	f_3	

Notice that if $x = x_0$, $p_{0,1} = f_0$ and if $x = x_1$, $p_{0,1} = f_1$ as expected. If we extend the idea and write

$$p_{0,1,2} = \frac{(x - x_0)p_{1,2} - (x - x_2)p_{0,1}}{x_2 - x_0}$$

then $p_{0,1,2}$ is calculated in just the same sort of way as $p_{0,1}, p_{1,2}, p_{2,3}$, etc., and is essentially quadratic in x (why?). What other properties does $p_{0,1,2}$ have? Well, for $x = x_0$

$$p_{0,1,2} = -\frac{(x_0 - x_2)}{x_2 - x_0} p_{0,1} = f_0 \quad \text{since} \quad p_{0,1} = f_0 \text{ when } x = x_0.$$

For $x = x_2$,

$$p_{0,1,2} = \frac{x_2 - x_0}{x_2 - x_0} p_{1,2} = f_2 \quad \text{since} \quad p_{1,2} = f_2 \text{ when } x = x_2.$$

(Check this last statement!)

For $x = x_1$,

$$p_{0,1,2} = \frac{(x_1 - x_0)f_1 - (x_1 - x_2)f_1}{x_2 - x_0}$$

$$= f_1 \frac{(x_1 - x_0 - x_1 + x_2)}{x_2 - x_0} = f_1.$$

(Have you checked? — it isn't all that obvious!)

Thus we have shown that

(a) for $x = x_0$, $\quad p_{0,1,2} = f_0$

(b) for $x = x_1$, $\quad p_{0,1,2} = f_1$ $\quad$ and

(c) for $x = x_2$, $\quad p_{0,1,2} = f_2$

so that $p_{0,1,2}$ is the $p_2(x)$ based on the data points (x_0, f_0), (x_1, f_1) and (x_2, f_2); i.e., it is the interpolating polynomial based on these three points. Defining $p_{1,2,3}$ similarly we can extend the table to include these quadratic interpolates.

x_i	$x - x_i$	f_i	$p_{i,i+1}$	$p_{i,i+1,i+2}$
x_0	$x - x_0$	f_0		
			$p_{0,1}$	
x_1	$x - x_1$	f_1		$p_{0,1,2}$
			$p_{1,2}$	
x_2	$x - x_2$	f_2		$p_{1,2,3}$
			$p_{2,3}$	
x_3	$x - x_3$	f_3		

Thus the calculation of the $p_{i,i+1,i+2}$ column is done using the second column and the fourth column in very similar fashion to the calculation of the $p_{i,i+1}$.

Note that $p_{i,i+1,i+2}$ is a quadratic interpolate and the notation implies that it is based on the data (x_i, f_i), (x_{i+1}, f_{i+1}) and (x_{i+2}, f_{i+2}).

EXERCISE 14.5

Defining $p_{0,1,2,3}$ by

$$p_{0,1,2,3} = \frac{(x - x_0)p_{1,2,3} - (x - x_3)p_{0,1,2}}{x_3 - x_0}$$

show that

(a) $p_{0,1,2,3}$ is a cubic in x

(b) $p_{0,1,2,3} = f_0$ at $x = x_0$

(c) $p_{0,1,2,3} = f_1$ at $x = x_1$

(d) $p_{0,1,2,3} = f_2$ at $x = x_2$ and

(e) $p_{0,1,2,3} = f_3$ at $x = x_3$.

(Answer not given — deliberately!)

Clearly we can continue this mode of calculation, defining successively higher-degree interpolates in terms of those of one degree less. We will just define $p_{i,i+1,...,i+n}$ but not attempt to prove its properties (although this is not difficult, being a generalisation of Exercise 14.5. Try it!)

$$p_{i,i+1,...,i+n-1,i+n} = \frac{(x - x_i)p_{i+1,...,i+n} - (x - x_n)p_{i,i+1,...,i+n-1}}{x_n - x_i}$$

[14-1]

where $i = 0, 1, 2, \ldots$ and $n = 1, 2, 3, \ldots$ extending as far as the data allow. You should notice that we have shown that the $\{p_{i,i+1,...,i+n}\}$ are really the appropriate interpolating polynomials expressed in terms of those of one degree less, but in practice all that would be found would be the *values* of those polynomials as given in the tables. Because of their construction the p values are called *linear cross-means*.

EXAMPLE 14.6

This example uses integer values from a known function $f(x) = x^4 - 2x$ just to see how the calculation proceeds. Let us take the following as the data:

x_i	-2	-1	0	1	3	4
f_i	20	3	0	-1	75	248

and choose $x = 2$. The Neville–Aitken table is as follows.

Interpolates

x_i	$x - x_i$	f_i	Linear	Quadratic	Cubic	Quartic	Quintic
-2	4	20					
			$\dfrac{(4)(3)-(3)(20)}{(-1)-(-2)} = -48$				
-1	3	3		$\dfrac{-24+96}{(0)-(-2)} = 36$			
			$\dfrac{(3)(0)-(2)(3)}{(0)-(-1)} = -6$		-12		
0	2	0		$\dfrac{-6+6}{(1)-(-1)} = 0$		12	
			$\dfrac{(2)(-1)-(1)(0)}{(1)-(0)} = -2$		18		12
1	1	-1		$\dfrac{74-2}{(3)-(0)} = 24$		12	
			$\dfrac{(1)(75)-(-1)(-1)}{(3)-(1)} = 37$		8		
3	-1	75		$\dfrac{-98+74}{(4)-(1)} = -8$			
			$\dfrac{(-1)(248)-(-2)(75)}{(4)-(3)} = -98$				
4	-2	248					

Of course $x^4 - 2x = 12$ when $x = 2$. The table shows exactly what we would expect: that not even the cubic interpolates are near the true value for $f(2)$ (since the data are too widely spaced to represent $x^4 - 2x$ well enough), but that both quartic interpolates are exact (since all the data lie on the same quartic curve). The quintic interpolate is really the same as the quartic ones since the coefficient of x^5 in the interpolating polynomial would be zero. If you look back at the general recurrence for the linear cross-means you will see that if two adjacent nth degree interpolates are equal then the resulting $(n + 1)$th degree interpolate must also have the same value. (Don't just believe us — try for yourself!)

Now to a more realistic example — our old friend, the problem from Example 14.1.

EXAMPLE 14.7

x	-0.4	-0.3	-0.1	0.1	0.2
$f(x)$	0.670	0.741	0.905	1.105	1.221

Given the data above, try to estimate $f(-0.2)$.

x_i	$x - x_i$	f_i	Interpolates			
			Linear	Quadratic	Cubic	Quartic
-0.4	0.2	0.670				
			0.8120			
-0.3	0.1	0.741		0.8193		
			0.8230		0.8190	
-0.1	-0.1	0.905		0.8185		0.8190
			0.8050		0.8190	
0.1	-0.3	1.105		0.8210		
			0.7570			
0.2	-0.4	1.221				

Here we can see that the linear interpolates in the fourth column are fairly near to each other in value. The quadratic interpolates in the fifth column are almost all the same. The sixth column of cubic interpolates clearly gives a consistent value (largely agreeing with the quadratic column), which our rule of thumb suggests is correct to 3D (the accuracy of the data); i.e., we would quote an answer: $f(-0.2) = 0.819$ to 3D.

Apparently we get consistent values for f by using cubics drawn through the points $\{-0.4, -0.3, -0.1, 0.1\}$ and $\{-0.3, -0.1, 0.1, 0.2\}$, as we did with divided differences. Thus we can reasonably suppose that f can be represented by a cubic, at any rate as far as interpolation near $x = -0.2$ is concerned.

In fact we can go further than this. If we drop the point $(0.2, 1.221)$ from the data the interpolation table for $x = -0.2$ is the same as before without the bottom line. Here linear interpolation does not give consistent results, but quadratic interpolation does do so. Adding the extra point confirms the value of 0.819 but requires cubic interpolation to do so. The dividing line in the table indicates the extra numbers to be calculated if the point $(0.2, 1.221)$ is added after the numbers above the line have been found.

Finally, notice that we have worked with one extra decimal place to avoid our round-off errors confusing the accuracy issue.

To conclude this section here again are the problems which are short on data and overgenerous on data respectively. In the second of these we shall use the fact that, like the divided-difference method, we can reorder the data at will. Generally speaking the Neville–Aitken method is best used with the data in order of their closeness to the interpolation point. The table of interpolates then shows as much consistency as is possible, with the best interpolate in each column (in the sense of using the most relevant data) being at the head of the column.

EXAMPLE 14.8   Using the data of Example 14.4

x	0.1	0.2	0.4	0.6	0.9
$f(x)$	1.105	1.221	1.492	1.882	2.460

to estimate $f(0.5)$ by the Neville–Aitken method. In view of the small amount of data we shall not change the order or consider a subset.

x_i	$x - x_i$	f_i	Linear	Quadratic	Cubic	Quartic
0.1	0.4	1.105				
			1.5690			
0.2	0.3	1.221		1.6470		
			1.6275		1.6671	
0.4	0.1	1.492		1.6721		1.6729
			1.6870		1.6787	
0.6	−0.1	1.882		1.6874		
			1.6893			
0.9	−0.4	2.460				

Linear interpolation (4th column) clearly does not give agreement, nor does quadratic (5th column). Quadratic values only tell us that the value of $f(0.5)$ is about 1.6 or 1.7. Cubic values (6th column) tell us that $f = 1.7$

correct to 1D or perhaps that $f = 1.67$ or 1.68 correct to 2D. The values gained by interpolation do not settle down very well and so perhaps the best that can be done is to give $f(0.5)$ as '1.67 or 1.68'. The data for Examples 14.1 and 14.8 are values of e^x except that $e^{0.6}$ is changed from 1.822 to 1.882. This creates a function f with a 'kink' in its graph, which means that the data cannot be represented very well by a polynomial. More data for $f(x)$ near $x = 0.6$ would be needed to give a 'good' result.

EXAMPLE 14.9   Using the data of Example 14.5 here is the Neville–Aitken scheme for estimating $f(0.552)$. The data points are taken in order of closeness to the interpolation point and the scheme is constructed *row by row*, not column by column. (As there is a deal of calculation do not check the following table (unless you are not sure of the method), but concentrate on the interpretation of the table.)

x_i	$x - x_i$	f_i	Linear	Quadratic	Cubic	Quartic
0.60	−0.048	0.6367				
			0.578 19			
0.70	−0.148	0.7586				
				0.580 31		
			0.587 02		0.580 53	
0.40	+0.152	0.4108		0.579 40		0.580 52
			0.592 17		0.580 58	
0.80	−0.248	0.8881		0.578 20		0.580 52
			0.550 82		0.580 66	
0.85	−0.298	0.9561		0.586 28		
			0.536 52			
0.90	−0.348	1.0265				
0.20	+0.352	0.2013				
0.95	−0.398	1.0995				
1.00	−0.448	1.1752				
0.00	+0.552	0				

After including the row corresponding to the point $(0.90, 1.2065)$, we have agreement to 4D among the best cubic and the two best quartic interpolates. Moreover all the other nearby interpolates show a consistency suggesting that all is well. Hence we conclude confidently that

$$f(0.552) = 0.5805 \quad \text{to 4D.}$$

The first exercise below has simple data values so that you may get used to the method. The second is a more typical problem and uses the data of Exercise 14.4.

EXERCISE 14.6

Estimate $f(2)$ from the following data using the Neville–Aitken method. (It is not necessary to reorder the data.)

x	0	1	3	4	5
$f(x)$	5	5	−1	−7	−15

EXERCISE 14.7

Repeat Exercise 14.4 but using the Neville–Aitken method. (*Hint*: the hint of Exercise 14.4 still applies!)

14.6 COMPARING THE DIVIDED-DIFFERENCE AND NEVILLE–AITKEN METHODS

We have seen that it is not always obvious from the divided-difference table which degree of polynomial to use, involving us in calculating interpolates of successively higher degree until we do (or do not) get consistency. You may have noticed a more significant drawback of the method: it is not always clear how accurately we need to calculate the divided differences. On the other hand, the calculation processes involved are essentially very easy.

In the Neville–Aitken method it is much easier to see when the interpolation process is succeeding and there is no problem about the accuracy of the entries in the table. They are all estimates of the function value and should all be quoted to at least one more place than the data. The calculations are only a little more complicated than in the divided-difference method. The following exercise demonstrates the differences between the two methods.

EXERCISE 14.8

Use both the divided-difference and Neville–Aitken method to estimate $f(35)$ from the data given.

x	0	10	30	45	50
$f(x)$	0	0.1736	0.5000	0.7071	0.7660

(The correct answer is 0.5736, and is obtainable — if you perform the calculation properly!)

Example 14.4 showed that polynomial interpolation may not always be entirely successful. In fact it may fail completely.

EXAMPLE 14.10 Consider this data taken from the function $f(x) = x^{1/4}$.

x	0	1	16	81
$f(x)$	0	1	2	3

Now $y = x^{1/4}$ has a very 'smooth' graph and the four points given do represent the shape of the graph quite well. Here is the Neville–Aitken scheme for estimating $f(9) = \sqrt{3} = 1.732\ldots$

(You need not check these figures, just concentrate on the conclusions.)

x_i	$x - x_i$	f_i			
0	9	0			
			9		
1	8	1		4.8000	
			1.5333		4.4410
16	−7	2		1.5692	
			1.8923		
81	−72	3			

These values show no particular consistency. They are not simply inaccurate; they bear almost no relation to the value of $f(9)$.

The reason for this is that $f(x) = x^{1/4}$ cannot be represented well near $x = 0$ by a polynomial, because $y = x^{1/4}$ has an infinite gradient at $x = 0$ and no polynomial can have an infinite gradient. To make things worse the gradient of $y = x^{1/4}$ decreases rapidly between $x = 0$ and $x = 1$. Fig. 14.4 shows the graphs of $y = x^{1/4}$ and $y = p_3(x)$, the cubic interpolating polynomial through all four data points.

It is interesting to note that if we reject the point $(0, 0)$ from the Neville–Aitken table then the interpolates are not as wildly wrong.

The lesson to be learnt from this example is that you must study and interpret the whole Neville–Aitken table. The number in the last column is never 'the answer'.

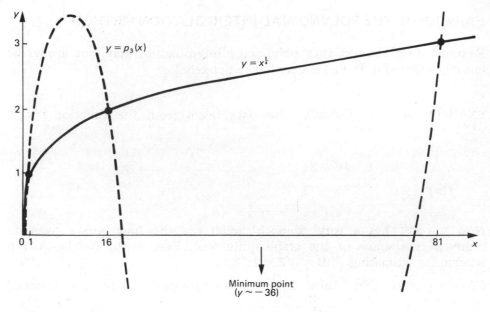

Fig. 14.4

ANSWERS TO EXERCISES

14.1 (a) $p_1(\sqrt{3}) = \sqrt{2} + \dfrac{(2 - \sqrt{2})}{4 - 2}(3 - 2) = \sqrt{2} + \tfrac{1}{2}(2 - \sqrt{2})$

$\qquad = 1 + \tfrac{1}{2}\sqrt{2} = 1 + 0.707 = 1.707.$

$\sqrt{3}$ actually equals 1.732 05 ... implying an error of 0.025, or $1\tfrac{1}{2}\%$ — quite acceptable in many practical situations.

(b) $p_1(1.6) = \ln 1.2 + \left(\dfrac{\ln 1.9 - \ln 1.2}{1.9 - 1.2}\right)(1.6 - 1.2)$

$\qquad = 0.019\,80 + \left(\dfrac{0.086\,18 - 0.019\,80}{0.7}\right)(0.4)$

$\qquad = 0.057\,73.$

Thus we only have 3D accuracy and linear interpolation only gives a roughly correct answer. The two data points are too far apart for accurate linear interpolation. (In 5-figure tables the spacing for accurate linear interpolation is 0.01 for the natural logarithm function.)

(c) (i) $f(0.5) \simeq 0.1071 + \left(\dfrac{0.0428 - 0.1071}{1 - 0}\right)(0.5 - 0)$

$\qquad = 0.0750$ (accuracy unknown)

(ii) $f(1.3) \simeq 0.0428 + \left(\dfrac{-0.0264 - 0.0428}{2-1} \right) (1.3 - 1)$

$= 0.0220$ (accuracy unknown)

(iii) $f(x) = 0$ between $x = 1$ and $x = 2$. Therefore use

$$0 = 0.0428 + \frac{(-0.0264 - 0.0428)}{2-1}(x-1)$$

i.e. $x = 1.62$ approximately.

14.2 $p_3(x) = 0.741 + 0.820(x + 0.3) + 0.450(x + 0.3)(x + 0.1)$

$\qquad + 0.166(x + 0.3)(x + 0.1)(x - 0.1)$

$p_3(-0.2) = 0.741 + 0.820(0.1) + 0.450(0.1)(-0.1)$

$\qquad + 0.166(0.1)(-0.1)(-0.3)$

$\qquad = 0.819$ to 3D.

14.3 $p_3(x) = 0.741 + (x + 0.3)(0.820 + (x + 0.1)(0.450$

$\qquad + (x - 0.1)0.166))$

$p_3(-0.2) = 0.741 + 0.1(0.820 - 0.1(0.450 - 0.3(0.166)))$

$\qquad = 0.819$ to 3D as before.

However, here we do 3 multiplications and 3 additions, whereas in Exercise 14.2 we had 6 multiplications and 3 additions.

14.4

x_i	f_i					
4.5	1.433					
		0.196 00				
4.25	1.384		0.032 00			
		0.204 00		0.021 360		
4.75	1.486		0.021 32		− 0.021 387	
		0.198 67		0.010 667		− 0.034 202
4.0	1.337		0.029 32		0.012 815	
		0.206 00		0.001 056		
5.0	1.543		0.028 00			
		0.192 00				
3.5	1.255					

It is not obvious which degree of polynomial to use, so we must evaluate successive ones.

$p_1(4.4) = 1.433 + 0.196(4.4 - 4.5) = 1.4134$

$p_2(4.4) = 1.4134 + 0.032(4.4 - 4.5)(4.4 - 4.25) = 1.4129$

$p_3(4.4) = 1.4129 + 0.021\,36(4.4 - 4.5)(4.4 - 4.25)(4.4 - 4.75)$

$\qquad = 1.4130.$

So a quadratic was alright and 1.413 to 3D looks a good value to quote.

14.6

x_i	$x - x_i$	f_i	Linear	Quadratic
0	2	5		
			5	
1	1	5		3
			2	
3	−1	−1		3
			5	
4	−2	−7		3
			9	
5	−3	−15		

So $f(2) = 3$ and f appears to be a quadratic polynomial. (In fact $f_i = 5 + x_i - x_i^2$). There is no point in calculating higher-degree interpolates. (What would their values be?)

14.7

x_i	$x - x_i$	f_i	Linear	Quadratic	Cubic	Quartic
4.5	−0.1	1.433				
			1.4134			
4.25	0.15	1.384		1.4129		
			1.4146		1.4130	
4.75	−0.35	1.486		1.4135		1.4129
			1.4165		1.4133	
4.0	0.4	1.337		1.4124		
			1.4194			
5.0	−0.6	1.543				

Clearly $f(4.4) = 1.413$ correct to 3D. We could conclude this even without the data $(5.0, 1.543)$, but the latter confirms the result nicely.

14.8 *Divided-differences*

x_i	f_i				
30	0.5000				
		0.013 806 7			
45	0.7071		−0.000 101 3		
		0.011 780 0		−0.735 × 10⁻⁶	
50	0.7660		−0.000 086 6		0.18 × 10⁻⁸
		0.014 810 0		−0.791 × 10⁻⁶	
10	0.1736		−0.000 051 0		
		0.017 360 0			
0	0				

It is tempting to think that only the first divided differences are non-negligible. This is not so:

$$p_1(35) = 0.5 + 0.013\,806\,7(35 - 30) = 0.569\,03$$

$$p_2(35) = 0.569\,03 - 0.000\,101\,3(35 - 30)(35 - 45) = 0.574\,10$$

$$p_3(35) = 0.574\,10 - 0.735 \times 10^{-6}(35 - 30)(35 - 45)(35 - 50)$$

$$= 0.573\,55$$

$$p_4(35) = 0.573\,55 + 0.18 \times 10^{-8}(35 - 30)(35 - 45)(35 - 50)$$

$$\times (35 - 10) = 0.573\,58$$

indicating that $f(35) = 0.5736$ to 4D. The catch here is that the divided differences are multiplied by numbers of the order of 10, 100, 1000, etc. and must have sufficient accuracy to give the correct fifth decimal place at least.

Neville–Aitken

x_i	$x - x_i$	f_i	Linear	Quadratic	Cubic	Quartic
30	5	0.5000				
			0.569 03			
45	−10	0.7071		0.574 10		
			0.589 30		0.573 55	
50	−15	0.7660		0.576 31		0.573 58
			0.543 85		0.573 35	
10	25	0.1736		0.562 98		
			0.607 60			
0	35	0				

This time we have no problems with accuracy and it is quite clear that we need all the data to give a good answer — which is the same as before of course.

FURTHER READING

As we indicated at the end of Lesson 13 we are taking a fairly new approach to teaching interpolation and this makes it difficult to give good references. Again Phillips and Taylor have a very similar approach to ours and we can recommend Section 4.4, in addition to Sections 4.1, 4.2 and 4.6 previously recommended.

The other problem is the number of variations on the Neville–Aitken theme; unless you are confident of the contents of Section 14.5 it could probably confuse you to read of alternatives. If you are in that happy state of confidence then we recommend Ribbans (Book 1), pp. 32–34 (Sections 3.5.1 and 3.5.2), Williams, P. W., Section 7.2, and Hosking, Joyce and Turner, 'Steps 22 and 23'.

Each of these authors do something we didn't dare do: they show that the divided-difference and Neville–Aitken methods are really the same method in different guise. Are you still confident?

LESSON 14 – COMPREHENSION TEST

1. Given that $f(0) = -1$ and $f(3) = 5$, estimate by linear interpolation: (a) $f(2)$, (b) a root of $f(x) = 0$.

2. Consider this table of divided differences.

x_i	$f(x_i)$			
−4	3.24			
		0.043 333		
−1	3.37		0.074 167	
		0.340 000		0.000 138 83
0	3.71		0.075 000		−0.000 040 129
		0.565 000		−0.000 222 33
2	4.84		0.073 666	
		0.933 333		
5	7.64			

(a) What degree of interpolating polynomial appears to be sufficient to represent this data? What accuracy would you expect to obtain in a value of such a polynomial?

(b) Which of the data points define this quadratic?

$$3.37 + 0.340\,000(x + 1) + 0.075\,000(x + 1)x$$

(c) Write down the quadratic polynomial most suitable for interpolation near $x = 2$.

3. Consider this Neville–Aitken table for $f(1.31)$.

x_i	$x - x_i$	$f(x_i)$		
−4	5.31	3.24		
			3.4701	
−1	2.31	3.37		4.3803
			4.1554	
0	1.31	3.71		4.3824
			4.4502	
2	−0.69	4.84		4.3836
			4.1960	
5	−3.69	7.64		

(a) Why have we stopped at the quadratic interpolates?

(b) Why have we worked to 4D?

(c) What answer would you give for $f(1.31)$?

(d) What is (i) the linear interpolate for $f(1.31)$ using the data at $x = 0$ and $x = 2$; (ii) the quadratic interpolate using the data at $x = 0, 2$ and 5?

ANSWERS

1 (a) 3   (b) $\frac{1}{2}$.

2 (a) 2, 2D   (b) at $x = -1, 0, 2$

    (c) $3.71 + 0.565\,000x + 0.073\,666x\,(x - 2)$

3 (a) The precision of the data is 2D, and all three quadratic interpolates agree to 2D.

    (b) To guard against introducing rounding errors which could affect the 2D accuracy.

    (c) 4.38 (to 2D).

    (d) (i) 4.45(02)   (ii) 4.38(36).

LESSON 14 – SUPPLEMENTARY EXERCISES

14.9. Use the Neville–Aitken method to estimate $f(1.31)$ from these data.

x	-4	-1	0	2	5
$f(x)$	3.24	3.37	3.71	4.84	7.64

What degree of polynomial does the Neville–Aitken tableau suggest will be suitable for interpolation in this data?

Relate your answers to that for Exercise 13.9. In particular check the value of $f(1.31)$.

14.10. Use the Neville–Aitken method to try to estimate x such that $f(x) = 4$, where f is defined by the data in Exercise 14.9. (Treat the $f(x)$ values as values of the independent variable, and the x values as values of the dependent variable; i.e., reverse the role of $f(x)$ and x.)

Draw a graph of $f(x)$ using the data in Exercise 14.9. This should show that you obtained a good result in Exercise 14.9 but a bad result here. Can you see why the method works in one case and not in the other?

14.11. Use the divided-difference and Neville–Aitken methods to estimate $f(0.87)$ from these data.

x	0.8	0.9	1.1	1.2	1.4	1.5	1.6
$f(x)$	0.8187	0.9048	1.1052	1.2214	1.4918	1.6487	1.8221

*14.12. We know that the divided-difference and Neville–Aitken methods must be equivalent by the uniqueness of the interpolating polynomial. In Lesson 13 we saw how the two methods were related for linear interpolation (see Section 13.2). Show that the quadratic interpolates are different arrangements of the same expression; i.e.,

$$p_{0,1,2} \equiv f[x_0] + (x - x_0)f[x_0, x_1] + (x - x_0)(x - x_1)f[x_0, x_1, x_2].$$

Try to describe how the two methods are related for any degree of interpolating polynomial.

Lesson 15 The Interpolation Error Term

OBJECTIVES

At the end of this lesson you should

(a) understand and be able to describe what the interpolation error function E_n represents;

(b) know the expression for $E_n(x)$;

(c) appreciate the usefulness and the limitations of this expression;

(d) be able to find a bound on the error function in simple cases.

CONTENTS

5.1 INTRODUCTION

In Lesson 13 we saw how it is possible to write down the interpolating polynomial of degree n, $p_n(x)$, based on $n + 1$ different points. This polynomial has the property that its graph passes through each of the $n + 1$ given points. The $n + 1$ points may be calculated from some function f of x, or may be data derived by experiment, in which case they can still be regarded as values of some function f.

We will assume that $f(x)$ is properly defined for all values of x in some interval even if we only know its value for a finite number of values of x.

Clearly what we would like to be able to do is to say how good the approximation $p_n(x)$ is to the true value $f(x)$ (for any particular value of x). A simple way of measuring this 'goodness' is to take the *error function*, E_n, to be $f - p_n$ (i.e., the 'data function' minus the approximating function). Likewise we take the *value of the error function*, $E_n(x)$, to be $f(x) - p_n(x)$; i.e., the value of the data function minus the value of the approximating polynomial at x. In other words $E_n(x)$ is the difference between the true value of the function and the value of an interpolate calculated by the methods of Lesson 14.

Fig. 15.1 shows the graphs of a function $f(x)$ and a quadratic polynomial $p_2(x)$ interpolating to f at $x = x_0, x_1, x_2$. The vertical distance between the two graphs is $|E_2(x)| = |f(x) - p_2(x)|$.

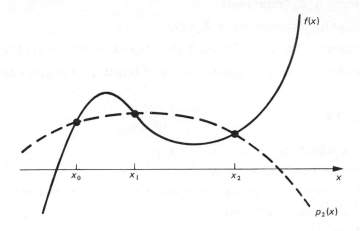

Fig. 15.1

EXERCISE 15.1

Sketch the graph of $E_2(x)$ from Fig. 15.1.

Our inquiry may now be better framed: can we find an estimate for $E_n(x)$ for all values x in the domain of interpolation? In particular, is $E_n(x)$ small for such x? — for some x? — for no x? Well, clearly $E_n(x_i) = 0$ for all $n + 1$ data values x_i, since $p_n(x_i) = f(x_i)$ by design. So we would expect $E_n(x)$ to have $x - x_i$ as a factor for all $i = 0(1)n$. What else does $E_n(x)$ depend on? Look again at Fig. 15.1 and your answer to Exercise 15.1.

Clearly E_n depends on the 'shape' of f as well as the positions of x_0, x_1, x_2. Now the shape of a function is essentially given by its derivatives ($f^{(1)} =$ slope, $f^{(2)}$ is related to curvature, etc.), so the following theorem is no great surprise.

THEOREM

1. Let $[a, b]$ be an interval containing the $n + 1$ points $x_0, x_1, x_2, \ldots, x_n$ and x.

2. Let $f, f^{(1)}, f^{(2)}, \ldots, f^{(n)}$ exist and be continuous on $[a, b]$.

3. Let $f^{(n+1)}$ exist in (a, b).

Then there exists a number ξ (which depends on x) in (a, b) such that

$$E_n(x) = f(x) - p_n(x)$$
$$= (x - x_0)(x - x_1)(x - x_2) \ldots (x - x_n)\frac{f^{(n+1)}(\xi)}{(n+1)!}. \qquad [15\text{-}1]$$

The conditions $1, 2, 3$ mean that f is a nice, smooth, well-behaved function.

Notice that the x_i are all fixed numbers such that $a \leqslant x_i \leqslant b$. As antici-pated, if we let $x = x_i$ $(i = 0, 1, \ldots, n)$ we get $E_n(x_i) = 0$, i.e., $f(x_i) = p_n(x_i)$, since the graph of the interpolating polynomial goes through the points $(x_i, f(x_i))$. We have had to assume a fair amount for the result of this theorem — that f itself is defined in $[a, b]$ together with its first $n + 1$ derivatives, the first n of these being continuous. As with Taylor's Theorem we end up with an expression involving some point ξ in (a, b) which we don't know. We only know the interval in which it lies.

The theorem gives us an expression for $E_n(x)$ but does *not* tell us whether the approximation is good, indifferent or bad. If we want to know more about the smallness (or otherwise) of the value $E_n(x)$ then we need more information or we must examine $f(x)$ and do some further work.

5.2 APPLYING THE ERROR TERM

First let us have a look at the simplest case, that of linear interpolation. Suppose that we are given the points (x_0, f_0) and (x_1, f_1). (Remember that $f(x_r) \equiv f_r$.) We can draw a unique straight line through these two points (with $n = 1$). Assuming that $f, f^{(1)}$ are continuous on $[a, b]$, $f^{(2)}$ exists on (a, b) and $x, x_0, x_1 \in [a, b]$, then

$$E_1(x) = f(x) - p_1(x) = (x - x_0)(x - x_1)\frac{f^{(2)}(\xi)}{2!}$$

where $a < \xi < b$.

If we want to find restrictions on $E_1(x)$ (or perhaps $|E_1(x)|$) clearly we must impose further conditions. A not unreasonable one is to assume that $f^{(2)}$ is bounded in $[a, b]$. We can write this assumption as

$$|f^{(2)}(x)| \leqslant M$$

where M is some positive constant. (For example, if $f(x) = \sin x$ then $f^{(2)}(x) = -\sin x$ and $|f^{(2)}(x)| \leqslant 1$.) With this assumption we have

$$|E_1(x)| \leqslant \tfrac{1}{2}|(x - x_0)(x - x_1)|M.$$

So we must now consider $|(x - x_0)(x - x_1)|$ as x varies.

Well, if $[a, b]$ is a large interval and $[x_0, x_1]$ a small one within it then $|(x - x_0)(x - x_1)|$ can be very large. Perhaps this can best be seen from a picture — Fig. 15.2.

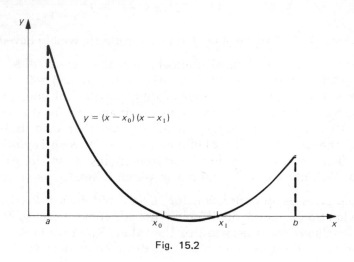

Fig. 15.2

However $|(x - x_0)(x - x_1)|$ looks much smaller near x_0 and x_1 than elsewhere. Since we have enough trouble with $|f^{(2)}(\xi)|$ anyway (with $|f^{(n+1)}(\xi)|$ in the general case) it is usual to restrict linear interpolation to the interior of $[x_0, x_1]$. Use of the interpolation process outside that interval is usually called *extrapolation*, and is not recommended, even though it is sometimes the only way of obtaining a wanted result. In the general case, assuming the data points are ordered, $x_0 < x_1 < x_2 < \ldots < x_n$, we interpolate only in $[x_0, x_n]$, where we expect $|(x - x_0)(x - x_1) \ldots (x - x_n)|$ to be reasonably small.

EXERCISE 15.2

Draw *sketches* of the graphs of
(a) $y = (x - 1)(x - 2)(x - 3)$ in $(-5, 10)$
(b) $y = x(x + 2)(x + 1)(x + 4)$ in $(-10, 4)$

What is the point of doing this? (There is one very relevant to the previous paragraph!)

Now to return to $y = (x - x_0)(x - x_1)$, restricting ourselves from now on to $x \in [x_0, x_1]$! Can you show (a little very elementary calculus?) that the maximum value of $|y|$ in $[x_0, x_1]$ is $\frac{1}{4}(x_1 - x_0)^2$? This means that we finally have

$$|E_1(x)| \leqslant \tfrac{1}{8}(x_1 - x_0)^2 M. \qquad \qquad [15\text{-}2]$$

Well, you may say, a nice little result! But what if M has to be a large number because of the nature of f and hence $f^{(2)}$ on $[x_0, x_1]$? In such a case $|E_1(x)|$ will still be a large number. Just because you have a nice formula does not mean that $|E_n(x)|$ will be small. You have to look at each case individually and carefully.

EXAMPLE 15.1   Let $f(x) = e^x$ on $[0, 1]$ and use linear interpolation at $x = 0$ and 1. $f^{(2)}(x) = e^x$ so

$$E_1(x) = (x - 0)(x - 1)\, e^\xi / 2! \quad \text{where} \quad 0 < \xi < 1.$$

Using inequality [15-2] we have

$$|E_1(x)| \leqslant \tfrac{1}{8}(1 - 0)^2 M$$

where M is a bound for $|f^{(2)}(\xi)| = |e^\xi|$. Since $0 < \xi < 1$, $1 = e^0 < e^\xi < e^1 = 2.718 \dots$, so $M = 3$ will do nicely. Hence

$$|E_1(x)| \leqslant \tfrac{3}{8}.$$

Compare this with the corresponding Taylor result. The Taylor polynomial of degree one constructed at $x = 0$ is

$$p_1(x) = 1 + x$$

with remainder

$$R_1(x) = e^\xi x^2 / 2!$$

where $0 < \xi < 1$ if we restrict x to $[0, 1]$. Hence

$$|R_1(x)| \leqslant \tfrac{3}{2}$$

a rather worse result than that for interpolation. But this is hardly a fair comparison since the Taylor result is biased to one side of the interval. If we consider the Taylor polynomial constructed at $x = \frac{1}{2}$ we get a remainder

$$R_1(x) = e^\xi (x - \tfrac{1}{2})^2 / 2!$$

where $\frac{1}{2} < \xi < x$ or $x < \xi < \frac{1}{2}$. Allowing $x \in [0, 1]$ means in effect $0 < \xi < 1$ again, so that

$$|R_1(x)| \leqslant 3(\tfrac{1}{2})^2 / 2! = \tfrac{3}{8}.$$

Fig. 15.3 shows the various cases considered.

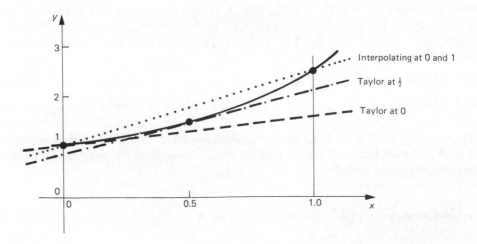

Fig. 15.3

In particular cases we may be able to be more precise than when using inequality [15-2]. For example, if we know constants M_1 and M_2 such that

$$M_1 \leqslant f^{(2)}(x) \leqslant M_2$$

for all $x \in [x_0, x_1]$, then

$$\tfrac{1}{2} M_1 (x - x_0)(x - x_1) \geqslant E_1(x) \geqslant \tfrac{1}{2} M_2 (x - x_0)(x - x_1)$$

since $(x - x_0)(x - x_1)$ is negative in (x_0, x_1).

For the problem of Example 15.1, $f^{(2)}(x) = e^x$ and $[x_0, x_1] = [0, 1]$ means we can choose $M_1 = 1$, $M_2 = 3$, giving

$$\tfrac{1}{2} x(x - 1) \geqslant E_1(x) \geqslant \tfrac{3}{2} x(x - 1).$$

At $x = \tfrac{1}{2}$ we then have

$$-\tfrac{1}{8} \geqslant E_1(\tfrac{1}{2}) \geqslant -\tfrac{3}{8}$$

which is slightly more informative than $|E_1(\tfrac{1}{2})| \leqslant \tfrac{3}{8}$.

EXERCISE 15.3

If $f(x) = \ln x$ estimate upper and lower bounds on $E_1(x)$ for
(a) $x = 3.16$, $x_0 = 3.1$ and $x_1 = 3.2$
(b) $x = 0.16$, $x_0 = 0.1$ and $x_1 = 0.2$.

In which case, (a) or (b), does linear interpolation work better, and why?

You may recall that we discussed briefly the role of linear interpolation in the use and hence construction of mathematical tables. Inequality [15-2] tells us a lot about the problem of making such tables. If we ask for say m-figure tables then we need the spacing $x_1 - x_0$ to be such that

$$\tfrac{1}{8}(x_1 - x_0)^2 M \leqslant \tfrac{1}{2} \times 10^{-m}.$$

If we make a book of 5-figure tables and then a book of 6-figure tables we must expect the latter to be $\sqrt{10}$ times fatter, because $x_1 - x_0$ must be $\sqrt{10}$ times smaller.

So much for linear interpolation. We can do similar analyses for higher degree interpolation but it becomes progressively much harder and is not usually done in practice. Here's a relatively easy problem to give you an idea of what's involved.

EXERCISE 15.4

If $f(x) = \sin x$ estimate a bound on $|E_2(x)|$, the error in quadratic interpolation using $x_0 = 0.6$, $x_1 = 0.7$, $x_2 = 0.8$ and $x \in [0.6, 0.8]$.

15.3 COMPARISON OF INTERPOLATION AND TAYLOR POLYNOMIAL ERROR TERMS

You will remember that the error term gained when approximating $f(x)$ by a Taylor polynomial of degree n constructed at $x = x_0$ was

$$R_n(x) = \frac{(x - x_0)^{n+1}}{(n+1)!} f^{(n+1)}(\xi)$$

where $x_0 < \xi < x$ if $x > x_0$.

For the interpolating polynomial of degree n based on $n + 1$ points $\{x_i : i = 0(1)n\}$ it is

$$E_n(x) = (x - x_0)(x - x_1) \ldots (x - x_n)\frac{f^{(n+1)}(\xi)}{(n+1)!}$$

where $x_0 < \xi < x_n$. The ξ's are not necessarily the same in the two formulae!

To see the relationships between R_n and E_n look at the various cases shown in Fig. 15.4. (We will assume that $f^{(n+1)}(x)$ is roughly constant, which it will be in the cases where the interpolation works well.)

Perhaps it is fairly obvious that the Taylor polynomial approximation approach may well give excellent results *near* the point at which the Taylor polynomial is constructed but very poor results elsewhere. On the other hand, the interpolating polynomial may give less good results than the best Taylor results but can do so over the larger interval $[x_0, x_n]$. Once outside this interval we may expect the error to grow as quickly as the Taylor error does.

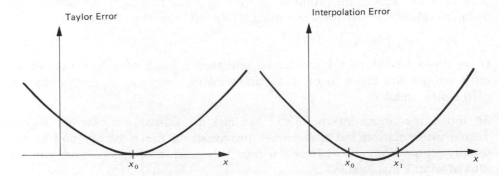

Fig. 15.4(a).   Errors for linear polynomial approximations

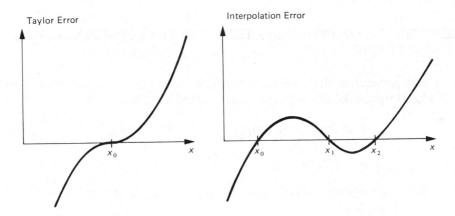

Fig. 15.4(b).   Errors for quadratic polynomial approximations

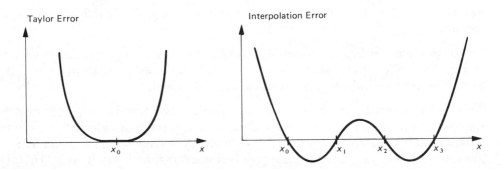

Fig. 15.4(c).   Errors for cubic polynomial approximations

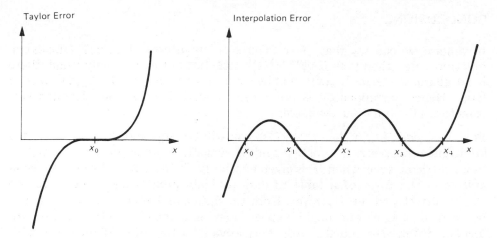

Fig. 15.4(d).   Errors for quartic polynomial approximations

5.4   FAILURE CASES REVISITED

In Lesson 14 we looked at the Neville—Aitken scheme for a set of data
acquired from $y = x^{1/4}$, i.e.

x	0	1	16	81
$f(x)$	0	1	2	3

and tried to estimate $f(9) = \sqrt{3} = 1.732 \ldots$. The Neville—Aitken scheme
showed no particular consistency, indicating the failure of the polynomial-
interpolation method. We can now see why, using our theorem.

The error term is

$$E_3(x) = (x - 0)(x - 1)(x - 16)(x - 81)f^{(4)}(\xi)/4!$$

where $0 < \xi < 81$.

Since $f(x) = x^{1/4}$, $f^{(4)}(x) = (\tfrac{1}{4})(-\tfrac{3}{4})(-\tfrac{7}{4})(-\tfrac{11}{4})x^{-15/4}$. Near $x = 0$, $x^{-15/4}$
is unbounded; i.e., we can make it as large as we like by taking x sufficiently
close to 0. It is now not at all surprising that polynomial interpolation does
not work for these data.

In practice of course we usually don't know the function f. We have seen in
Lesson 14 that the failure of the interpolation process is nevertheless clearly
indicated by the behaviour of the divided-difference or Neville—Aitken table,
whichever is being used. More of this in the next section.

15.5 CONCLUSIONS

In general we can say that, if x is *outside* the interval $[x_0, x_n]$ (the extrapolation case), then even if $|f^{(n+1)}(\xi)| < M$ but not exceedingly small, there is no guarantee that $|E_n(x)|$ will be small. Indeed generally $|E_n(x)|$ will be large. Hence extrapolation is not a recommended technique, although it is sometimes the only one available.

For x *inside* $[x_0, x_n]$, $|f^{(n+1)}(\xi)|$ can still be large and the corresponding approximation poor. In practice finding realistic bounds on $f^{(n+1)}$ may be quite difficult even when f is given explicitly. If the only information available is in the form of a table of data (or data points) we cannot estimate $f^{(n+1)}$ directly at all. However, from the data a divided-difference table can be formed, and as one might expect there is a relation between the nth divided differences and the nth derivative of a function. If the $(n+1)$th divided differences are small we can reasonably infer that $f^{n+1}(x)$ is small, so that the approximation is a good one. In this case the divided differences would show the sort of consistent pattern indicated in Lesson 14. If there is no such consistent pattern it means that $|f^{n+1}(x)|$ is not small and the interpolating approximation is probably a poor one.

Hence in practice we very rarely attempt to estimate bounds on the interpolation error term $E_n(x)$. We use instead our knowledge (backed up by the theory of the error term) of how interpolation processes behave when they are working well (with negligible error) or badly (with non-negligible error). In particular if the Neville–Aitken scheme settles quickly with increasing accuracy from left to right and consistency in each column it is a good bet that the errors are small. The converse is equally true.

ANSWERS TO EXERCISES

15.3 $f(x) = \ln x$, $f^{(1)}(x) = \dfrac{1}{x}$, $f^{(2)}(x) = \dfrac{-1}{x^2}$.

(a) $E_1(x) = (x - 3.1)(x - 3.2)\dfrac{f^{(2)}(\xi)}{2!}$, $3.1 < \xi < 3.2$.

$E_1(3.16) = (0.06)(-0.04)\dfrac{f^{(2)}(\xi)}{2!}$, $3.1 < \xi < 3.2$.

Now

$$0.097\,66 < \frac{1}{\xi^2} < 0.104\,06.$$

$\therefore$ $0.000\,117 < E_1(3.16) < 0.000\,125$.

(b) $E_1(x) = (x - 0.1)(x - 0.2)\dfrac{f^{(2)}(\xi)}{2!}, \quad 0.1 < \xi < 0.2.$

$\qquad E_1(0.16) = (0.06)(-0.04)\dfrac{f^{(2)}(\xi)}{2!}, \quad 0.1 < \xi < 0.2.$

Now

$\qquad 25 < \dfrac{1}{\xi^2} < 100.$

$\therefore \quad 0.03 < E_1(0.16) < 0.12.$

Linear interpolation works better in case (a) because $f^{(2)}(x)$ is smaller than in case (b).

15.4 $f(x) = \sin x.$

$\therefore \quad f^{(3)}(x) = -\cos x \quad$ and $\quad |f^{(3)}(x)| \leqslant 1.$

$\therefore \quad E_2(x) = (x - 0.6)(x - 0.7)(x - 0.8)\dfrac{f^{(3)}(\xi)}{6}, \quad 0.6 < \xi < 0.8$

so $\quad |E_2(x)| \leqslant |(x - 0.6)(x - 0.7)(x - 0.8)|/6.$

Let $\quad y = (x - 0.6)(x - 0.7)(x - 0.8).$

$\qquad y^{(1)} = (x - 0.6)(x - 0.7) + (x - 0.6)(x - 0.8)$

$\qquad\qquad + (x - 0.7)(x - 0.8)$

$\qquad\quad = 3x^2 - 4.2x + 1.46.$

At turning points $y^{(1)} = 0$, i.e.

$\qquad x = \dfrac{4.2 \pm \sqrt{0.12}}{6} = \dfrac{4.2 \pm 0.346}{6}$

$\qquad\quad = 0.758 \quad \text{or} \quad 0.642.$

Now

$\qquad |E_2(0.758)| \leqslant |(0.158)(0.058)(-0.042)|/6$

$\qquad\qquad\qquad = 0.000\,064$

and

$\qquad |E_2(0.642)| \leqslant |(0.042)(-0.058)(-0.158)|/6$

$\qquad\qquad\qquad = 0.000\,064.$

Thus the error bound using quadratic interpolation is approximately 0.000 06 for $x \in [0.6, 0.8]$.

FURTHER READING

If you have been reading the sections in Phillips and Taylor on interpolation you will probably have seen Phillips and Taylor, Section 4.3, on the accuracy of interpolation. It includes a proof of the error formula.

The books by Conte, Conte and deBoor, and Henrici contain similar sections.

LESSON 15 – COMPREHENSION TEST

1. Write down the expression for the error $E_3(x)$ in the interpolating polynomial $p_3(x)$ matching $f(x) = e^{-2x}$ at $x = 0, 1, 3$ and 4. Estimate an upper bound for $|E_3(2)|$.

2. Where, in the range of values of x for which these $f(x)$ are defined, would you expect interpolation to be able to provide good approximation? Bad approximation?

(a) $f(x) = \arcsin x, \quad x \in [-1, 1]$;

(b) $f(x) = e^{-x}, \quad x \geqslant 0$.

ANSWERS

1 $E_3(x) = 2x(x-1)(x-3)(x-4) \, e^{-2\xi}/3, \quad 0 < \xi < 4$.

$|E_3(2)| \leqslant |(2)(2)(1)(-1)(-2) \, e^{-2(0)}/3| = 8/3$ since $e^{-2\xi}$ takes its maximum value at $\xi = 0$.

2 (a) Good near $x = 0$. Bad near $x = \pm 1$ (derivatives infinite there).

(b) Good everywhere. ($|f^{(k)}(x)| = e^{-x} \leqslant 1$ for $x \geqslant 0$, for all $k = 0, 1, 2, \ldots$).

LESSON 15 – SUPPLEMENTARY EXERCISES

15.5. Sketch the graphs of:

(a) $y = 1/x$,

(b) $y = p_1(x)$, the linear interpolating polynomial matching $f(x) = 1/x$ at $x = \frac{1}{2}$ and $x = 2$,

(c) $y = E_1(x) = (1/x) - p_1(x)$.

15.6. Use the formula $\frac{1}{8}(x_1 - x_0)^2 M$ (see Equation [15-2]) to obtain a bound on the $|E_1(x)|$ of Exercise 15.5, where $\frac{1}{2} \leqslant x \leqslant 2$. Relate your answer to the graph of E_1 in Exercise 15.5.

15.7. Use your new-found knowledge of the interpolation-error formula to explain the failure of polynomial interpolation observed in Exercise 14.10.

15.8. Use the interpolation-error formula to obtain a bound on the error $|x^3 - p_2(x)|$ considered in Exercise 13.8. Relate your answer to the true maximum error obtained there.

* 15.9. Consider again the error for linear interpolation

$$E_1(x) = \frac{(x - x_0)(x - x_1)}{2!} f^{(2)}(\xi).$$

Suppose we wish to choose x_0 and x_1 to minimise the maximum value of $|E_1(x)|$. Show that the least possible value of $|(x - x_0)(x - x_1)|$ for $-1 \leqslant x \leqslant 1$ is $\frac{1}{2}$ and is obtained by choosing $x_0 = -1/\sqrt{2}$ and $x_1 = 1/\sqrt{2}$.

Lesson 16 Finite Differences and Interpolation

OBJECTIVES

At the end of this lesson you should

(a)   know what a difference table is and be able to construct one;

(b)   know the forward difference interpolation formula;

(c)   be able to choose a sensible degree for the interpolation formula by examining the difference table;

(d)   be able to calculate an interpolate using forward differences;

(e)   understand backward and central difference notation.

CONTENTS

16.1   INTRODUCTION

So far in all the practical interpolation that we have considered we have not assumed anything about the spacing (in the x direction) of the set of data points that have been used (except that all the x coordinates should be different). In this lesson $x_0, x_1, x_2, \ldots, x_n$ will be assumed to be *equally spaced* and we usually denote the spacing by h, so that

$$h = x_1 - x_0 = x_2 - x_1 = \ldots = x_n - x_{n-1}.$$

Hence

$$x_1 = x_0 + h$$

$$x_2 = x_1 + h = x_0 + 2h$$

$$x_3 = x_2 + h = x_0 + 3h \quad \text{and so on.}$$

Generally we have $x_r = x_0 + rh$, and remember that we write $f(x_r) = f_r$.

6.2 THE DIVIDED-DIFFERENCE INTERPOLATING POLYNOMIAL WITH EQUALLY SPACED DATA

The general divided-difference interpolating polynomial based on the $n + 1$ points $x_0, x_1, x_2, \ldots, x_n$ is

$$p_n(x) = f[x_0] + (x - x_0)f[x_0, x_1] + (x - x_0)(x - x_1)f[x_0, x_1, x_2]$$
$$+ \ldots + (x - x_0)(x - x_1) \ldots (x - x_{n-1})f[x_0, x_1, \ldots, x_n]$$

where $f[x_0] = f_0$,

$$f[x_0, x_1] = \frac{f[x_1] - f[x_0]}{x_1 - x_0},$$

$$f[x_0, x_1, x_2] = \frac{f[x_1, x_2] - f[x_0, x_1]}{x_2 - x_0}$$

and so on. Hence for equally spaced x_i values

$$f[x_0, x_1] = \frac{f_1 - f_0}{h},$$

$$f[x_0, x_1, x_2] = \frac{\dfrac{f_2 - f_1}{h} - \dfrac{f_1 - f_0}{h}}{2h}$$

$$= \frac{f_2 - 2f_1 + f_0}{2h^2}$$

and so on.

EXERCISE 16.1

Show that, for equally spaced points,

(a) $f[x_0, x_1, x_2, x_3] = \dfrac{f_3 - 3f_2 + 3f_1 - f_0}{6h^3}$

(b) $f[x_0, x_1, x_2, x_3, x_4] = \dfrac{f_4 - 4f_3 + 6f_2 - 4f_1 + f_0}{24h^4}$.

*(c) Guess a formula for

$$f[x_j, x_{j+1}, \ldots, x_k], \quad k > j.$$

If we let $x = x_0 + sh$ (so that s is now the variable quantity instead of x), then

$$x - x_0 = sh,$$
$$x - x_1 = x_0 + sh - x_0 - h = (s-1)h$$
$$x - x_2 = x_0 + sh - x_0 - 2h = (s-2)h.$$

(Are you checking?)

In general $x - x_r = (s-r)h$.

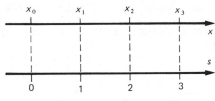

Fig. 16.1

Fig. 16.1 shows the relation between the x scale and the s scale. It is traditional to make the change of variable from x to s because it simplifies the interpolation formulae for equally-spaced data.

Using this substitution the general divided-difference interpolating polynomial becomes

$$p_n(x_0 + sh) = f_0 + sh \frac{(f_1 - f_0)}{h} + sh(s-1)h \frac{(f_2 - 2f_1 + f_0)}{2h^2}$$

$$+ sh(s-1)h(s-2)h \frac{(f_3 - 3f_2 + 3f_1 - f_0)}{6h^3} + \ldots,$$

i.e. $p_n(x_0 + sh) = f_0 + s(f_1 - f_0) + \dfrac{s(s-1)}{2!}(f_2 - 2f_1 + f_0)$

$$+ \frac{s(s-1)(s-2)}{3!}(f_3 - 3f_2 + 3f_1 - f_0) + \ldots.$$

[16-1]

Notice first that the right hand side does *not* contain h. (That was the whole point of changing from x's to s's.) We now have a polynomial in s similar in form to the Newton interpolation form but where the coefficients of the binomial-series functions

$$s, \quad \frac{s(s-1)}{2!}, \quad \frac{s(s-1)(s-2)}{3!} \quad \text{etc.}$$

are not divided-differences but simply combinations of the function values. The question then arises — how can we find these coefficients, i.e. $f_1 - f_0$, $f_2 - 2f_1 + f_0$, $f_3 - 3f_2 + 3f_1 - f_0$, ... in a simple, efficient manner?

The next section answers this question and the following one deals with the evaluation of the right-hand side of Equation [16-1].

16.3 FORWARD DIFFERENCES

To find the coefficients of the binomial-series functions (of s) on the right-hand side of Equation [16-1] we note the following relations.

$$(f_2 - f_1) - (f_1 - f_0) = f_2 - 2f_1 + f_0$$
$$(f_3 - f_2) - (f_2 - f_1) = f_3 - 2f_2 + f_1$$

and

$$(f_3 - 2f_2 + f_1) - (f_2 - 2f_1 + f_0) = f_3 - 3f_2 + 3f_1 - f_0.$$

In fact these are just the numerators of the original divided differences. Hence we can use a table of differences (*with no division*) of the function values, which formally appears as

x_0	f_0			
		$f_1 - f_0$		
x_1	f_1		$f_2 - 2f_1 + f_0$	
		$f_2 - f_1$		$f_3 - 3f_2 + 3f_1 - f_0$
x_2	f_2		$f_3 - 2f_2 + f_1$	
		$f_3 - f_2$		
x_3	f_3			

The third column is gained by differencing the second column (in the sense that we take the upper number from the lower); the fourth column is the third column differenced, and the fifth column is the fourth column differenced in just the same way.

EXAMPLE 16.1   To difference the following data:

x	0	1	2	3	4	5
$f(x)$	7	3	9	11	2	6

We get

x	f(x)				
0	7				
		−4			
1	3		10		
		6		−14	
2	9		−4		7
		2		−7	24
3	11		−11	31	
		−9		24	
4	2		13		
		4			
5	6				

You will observe that the *calculation* of the differences is very easy. The formal difference table (in terms of the *f*'s) is, however, rather clumsy. The whole thing can be simplified by using a notation called the *forward difference* notation. This notation is useful not only in the present context but is used quite widely in numerical mathematics. So here it is!

We define a new function (a rule, remember) Δf by

$$(\Delta f)(x) = f(x + h) - f(x).$$

Δ is called the *forward difference operator* and depends on the value of h. If we evaluate Δf at $x = x_r$ we get

$$(\Delta f)(x_r) = f(x_{r+1}) - f(x_r)$$

which we naturally write as

$$\Delta f_r = f_{r+1} - f_r.$$

In particular

$$\Delta f_0 = f_1 - f_0$$
$$\Delta f_1 = f_2 - f_1$$
$$\Delta f_2 = f_3 - f_2.$$

If we operate on Δf with Δ again we have

$$\Delta(\Delta f)(x) = \Delta f(x + h) - \Delta f(x),$$

i.e. $\Delta^2 f(x) = (f(x + 2h) - f(x + h)) - (f(x + h) - f(x))$
$$= f(x + 2h) - 2f(x + h) + f(x).$$

At $x = x_0$ we get

$$\Delta^2 f_0 = f_2 - 2f_1 + f_0.$$

In general

$$\Delta^n f(x) = \Delta^{n-1} f(x + h) - \Delta^{n-1} f(x)$$

and

$$\Delta^n f_r = \Delta^{n-1} f_{r+1} - \Delta^{n-1} f_r.$$

E.g. $\Delta^3 f_2 = \Delta^2 f_3 - \Delta^2 f_2$

by taking $n = 3$ and $r = 2$.

EXERCISE 16.2

Evaluate in terms of f values

(a) Δf_1 (b) Δf_2 (c) Δf_3 (d) $\Delta^2 f_0$ (e) $\Delta^2 f_1$ (f) $\Delta^2 f_2$
(g) $\Delta^3 f_0$ (h) $\Delta^3 f_1$ (i) $\Delta^4 f_0$.

When you have completed Exercise 16.2 you will see that the difference table for $n = 4$ needed to calculate the coefficients in the expansion of $p_4(x_0 + sh)$ is

x_0	f_0				
		Δf_0			
$x_0 + h$	f_1		$\Delta^2 f_0$		
		Δf_1		$\Delta^3 f_0$	
$x_0 + 2h$	f_2		$\Delta^2 f_1$		$\Delta^4 f_0$
		Δf_2		$\Delta^3 f_1$	
$x_0 + 3h$	f_3		$\Delta^2 f_2$		
		Δf_3			
$x_0 + 4h$	f_4				

Thus from the original table of function values we can easily calculate columns containing the first, second, third and fourth differences. Once this has been done we can read off from the leading (upper) diagonal the coefficients for $p_4(x_0 + sh)$ in the formula

$$p_4(x_0 + sh) = f_0 + s\Delta f_0 + \frac{s(s-1)}{2!}\Delta^2 f_0 + \frac{s(s-1)(s-2)}{3!}\Delta^3 f_0$$

$$+ \frac{s(s-1)(s-2)(s-3)}{4!}\Delta^4 f_0. \qquad [16\text{-}2]$$

Clearly the same sort of procedure for calculating the coefficients in the interpolating polynomial can be followed no matter what the value of n is. We commonly call Equation [16-2] the *forward-difference interpolating polynomial* (of degree 4 in this case). The nth degree forward-difference interpolating polynomial is

$$p_n(x_0 + sh) = f_0 + s\Delta f_0 + \frac{s(s-1)}{2!}\Delta^2 f_0 + \frac{s(s-1)(s-2)}{3!}\Delta^3 f_0 + \ldots$$

$$+ \frac{s(s-1)(s-2)\ldots(s-n+1)}{n!}\Delta^n f_0. \qquad [16\text{-}3]$$

EXAMPLE 16.2   If $x_0 = 0$, $h = 1$ and $f_0 = 2$, $f_1 = 12$, $f_2 = 38$, $f_3 = 86$, $f_4 = 162$, $f_5 = 272$ the difference table is

x_i	f_i	Δf_i	$\Delta^2 f_i$	$\Delta^3 f_i$	$\Delta^4 f_i$	$\Delta^5 f_i$
0	2					
		10				
1	12		16			
		26		6		
2	38		22		0	
		48		6		0
3	86		28		0	
		76		6		
4	162		34			
		110				
5	272					

Here we can see that the third differences are constant and equal 6, and all the higher differences are zero in consequence. In fact the function values (the f_i) are obtained from a polynomial of degree 3 in x and the table illustrates the result of a theorem about the differences of polynomials tabulated at equal intervals. For those who are interested in the details this theorem appears with its proof in Appendix 3. Briefly it says that the nth differences of values of an nth degree polynomial are constant.

EXERCISE 16.3

Tabulate $f(n) = 2n^2 - 4n + 3$ for $n = 1, 2, 3, 4$ and 5 and draw up the corresponding table of differences.

EXERCISE 16.4

Tabulate $f(x) = \sin x + \cos x$ correct to 4D for $x = 5°$, $10°$, $15°$, $20°$, $25°$, $30°$, $35°$, $40°$, $45°$. (An easier way of writing this is $x = 5°(5°)45°$.) Draw up a difference table for $f(x)$ as far as the 8th differences. (Practical hint — write the entries in the table without a decimal point or leading zeros (i.e., multiply everything by 10^4.) It's economical and not misleading provided that you remember that you've done it!)

16.4 PRACTICAL INTERPOLATION USING FUNCTION VALUES TABULATED AT EQUAL INTERVALS

By this time you should understand the way in which we can write down the formula for an interpolating polynomial of a given degree and find a difference table from which to read off the unknown coefficients in the interpolating polynomial. If you look at the difference table in Example 16.2 or Exercise 16.3 you can see that the differences are apparently all zero after a certain stage so that the evaluation of the interpolate should present no problems. However, in Exercise 16.4 we did not start with a polynomial and the differences decrease in size to a sort of minimum at the $\Delta^4 f$ stage and then appear to start to grow in size again. Now what does this mean?

The theorem stated in Appendix 3 says that the nth differences of values of an nth degree polynomial are constant. But in Exercise 16.4 we haven't got a polynomial. Moreover we must remember that the function values are subject to rounding errors. We must have a look at this to see what effect it has on a difference table. Let us construct a table of rounded function values minus the true function values. This gives us a table of rounding errors and the *worst* way in which they can behave in this context is to be alternately plus and minus $\frac{1}{2}$ in the last decimal place. If we assume this to be the case we get the following table.

x	Function errors $\times\ 10^p$	Δ	Δ^2	Δ^3	Δ^4	Δ^5
x_0	$\frac{1}{2}$					
		-1				
x_1	$-\frac{1}{2}$		2			
		1		-4		
x_2	$\frac{1}{2}$		-2		8	
		-1		4		-16
x_3	$-\frac{1}{2}$		2		-8	
		1		-4		
x_4	$\frac{1}{2}$		-2			
		-1				
x_5	$-\frac{1}{2}$					

The original function-data are assumed given to pD.

Thus we can see that the errors may well grow as we construct higher and higher differences. To return to Exericse 16.4, Δf, $\Delta^2 f$ and $\Delta^3 f$ are all too large to be mainly to do with rounding errors but the $\Delta^4 f$ column is smaller than the maximum possible ± 8. Thus we can say that $\Delta^4 f = 0$ (to within rounding error) for the set of data we have taken in Exercise 16.4 and we can treat this set of data as though coming from a polynomial of degree 3 $(n + 1 = 4)$.

EXAMPLE 16.3   Using the table of Exercise 16.4 find a value for $\sin 7° + \cos 7°$.

$$x_0 = 5 \quad \text{and} \quad h = 5, \quad \text{so}$$
$$x = 7 = 5 + s5 \quad \therefore s = \tfrac{2}{5} = 0.4.$$

We have already deduced that $n = 3$, and so (in full)

$$\sin 7° + \cos 7° \simeq p_3(5 + 0.4 \times 5)$$

$$= 1.0834 + 0.4 \times 0.0751 + \frac{0.4(0.4 - 1)}{2!}(-0.0089)$$

$$+ \frac{0.4(0.4 - 1)(0.4 - 2)}{3!}(-0.0003)$$

$$= 1.0834 + 0.030\,04 + 0.001\,07 - 0.000\,02$$

$$= 1.114\,49.$$

Hence we obtain $\sin 7° + \cos 7° = 1.1145$ to 4D.

The correct 4D value is 1.1144. Have we done less well than we might expect from such a good difference table (giving such a clear indication of a suitable degree)? No — the correct 5D value is 1.114 42, differing by only 0.7×10^{-4} from our 5D result. This difference is not much more than the error bound for the data even though we have introduced extra rounding errors in the calculations. There are two lessons here:

(a) giving answers to nD is not always the best way to specify their precision;

(b) interpolation via a difference table which shows a clear indication of polynomial degree is usually extremely accurate.

EXERCISE 16.5

Using the table of Exercise 16.4 find a value for:

(a) $\sin 13° + \cos 13°$,   (b) $\sin 22\frac{1}{2}° + \cos 22\frac{1}{2}°$, and

(c) $\sin 43° + \cos 43°$.

Comment on your results.

16.5 THE ERROR TERM

The error term found in Lesson 15 was given by

$$E_n(x) = (x - x_0)(x - x_1) \ldots (x - x_n)\frac{f^{(n+1)}(\xi)}{(n+1)!}$$

where $x_0 < \xi < x_n$ if the x_i. values are ordered.

Writing $x = x_0 + sh$, $x_r = x_0 + rh$ $(r \in \mathbb{N})$ and $\xi = x_0 + \sigma h$ gives

$$E_n(x_0 + sh) = h^{n+1}\frac{s(s-1)\ldots(s-n)}{(n+1)!}f^{(n+1)}(x_0 + \sigma h)$$

where $0 < \sigma < n$. In practice this expression is rarely used to estimate the accuracy of a result. It is included here for completeness.

16.6 BACKWARD AND CENTRAL DIFFERENCES

In the answer to Exercise 16.5(c) it is stated that *backward difference* and *central difference* interpolating polynomials exist. In this section we will just explain what they are but not cover their use.

The *backward difference operator* ∇ is defined by $\nabla f(x) = f(x) - f(x-h)$ so that with $x = x_1$ for example $\nabla f_1 = f_1 - f_0$. Thus we can see that $\nabla f_1 = \Delta f_0$. ∇^n is defined in a way similar to that for Δ^n, $n > 1$.

The *central difference operator* δ is defined by

$$\delta f(x) = f(x + \tfrac{1}{2}h) - f(x - \tfrac{1}{2}h)$$

so that with $x = x_{1/2} = x_0 + \tfrac{1}{2}h$ for example $\delta f_{1/2} = f_1 - f_0$. Thus we have $\delta f_{1/2} = \Delta f_0 = \nabla f_1$. δ^n is defined in a way similar to that for Δ^n, $n > 1$.

Generalising the result $\delta f_{1/2} = \Delta f_0 = \nabla f_1$ to get $\delta f_{r+1/2} = \Delta f_r = \nabla f_{r+1}$ and with similar results for the higher differences, we can see that the following three difference tables represent the same set of function values and the same differences of these function values. The elements of the tables just have different names for different uses.

x_0	f_0			
		Δf_0		
x_1	f_1		$\Delta^2 f_0$	
		Δf_1		$\Delta^3 f_0$
x_2	f_2		$\Delta^2 f_1$	
		Δf_2		
x_3	f_3			

x_0	f_0			
		∇f_1		
x_1	f_1		$\nabla^2 f_2$	
		∇f_2		$\nabla^3 f_3$
x_2	f_2		$\nabla^2 f_3$	
		∇f_3		
x_3	f_3			

x_0	f_0			
		$\delta f_{1/2}$		
x_1	f_1		$\delta^2 f_1$	
		$\delta f_{3/2}$		$\delta^3 f_{3/2}$
x_2	f_2		$\delta^2 f_2$	
		$\delta f_{5/2}$		
x_3	f_3			

As examples of the different notations for the same element

$$\Delta^2 f_1 = \nabla^2 f_3 = \delta^2 f_2$$
$$\Delta^3 f_0 = \nabla^3 f_3 = \delta^3 f_{3/2}.$$

Notice that in the first table the subscripts go diagonally downwards ('forwards'), in the second diagonally upwards ('backwards'), and in the third horizontally ('centrally').

The chief use for these different notations is in different kinds of interpolation formulae. Here are two simple examples.

Taking $n = 2$ we might use the *backward-difference interpolating polynomial*

$$p_2(x_2 + sh) = f_2 + s\nabla f_2 + \frac{s(s+1)}{2!}\nabla^2 f_2$$

to find an interpolate in the interval (x_1, x_2).

Taking $n = 2$ again we might use the *central-difference interpolating polynomial*

$$p_2(x_1 + sh) = f_1 + \tfrac{1}{2}s(\delta f_{1/2} + \delta f_{3/2}) + \tfrac{1}{2}s^2\,\delta f_1$$

to find an interpolate in the interval (x_0, x_2).

For more information about the origin and use of such formulae see Further Reading.

ANSWERS TO EXERCISES

16.1 (a) $f[x_0, x_1, x_2, x_3] = \dfrac{f[x_1, x_2, x_3] - f[x_0, x_1, x_2]}{x_3 - x_0}$

$$= \frac{1}{3h}\left(\frac{f_3 - 2f_2 + f_1}{2h^2} - \frac{f_2 - 2f_1 + f_0}{2h^2}\right)$$

$$= \frac{1}{6h^3}(f_3 - 3f_2 + 3f_1 - f_0),$$

(b) $f[x_0, x_1, x_2, x_3, x_4] = \dfrac{f[x_1, x_2, x_3, x_4] - f[x_0, x_1, x_2, x_3]}{x_4 - x_0}$

$$= \frac{1}{4h}\left(\frac{f_4 - 3f_3 + 3f_2 - f_1}{6h^3} - \frac{f_3 - 3f_2 + 3f_1 - f_0}{6h^3}\right)$$

$$= \frac{1}{24h^4}(f_4 - 4f_3 + 6f_2 - 4f_1 + f_0).$$

16.2   (a) $\Delta f_1 = f_2 - f_1,$

       (b) $\Delta f_2 = f_3 - f_2,$

       (c) $\Delta f_3 = f_4 - f_3,$

       (d) $\Delta^2 f_0 = \Delta f_1 - \Delta f_0 = (f_2 - f_1) - (f_1 - f_0) = f_2 - 2f_1 + f_0,$

       (e) $\Delta^2 f_1 = \Delta f_2 - \Delta f_1 = (f_3 - f_2) - (f_2 - f_1) = f_3 - 2f_2 + f_1,$

       (f) $\Delta^2 f_2 = \Delta f_3 - \Delta f_2 = (f_4 - f_3) - (f_3 - f_2) = f_4 - 2f_3 + f_2,$

       (g) $\Delta^3 f_0 = \Delta^2 f_1 - \Delta^2 f_0 = (f_3 - 2f_2 + f_1) - (f_2 - 2f_1 + f_0)$
$$= f_3 - 3f_2 + 3f_1 - f_0,$$

       (h) $\Delta^3 f_1 = \Delta^2 f_2 - \Delta^2 f_1 = (f_4 - 2f_3 + f_2) - (f_3 - 2f_2 + f_1)$
$$= f_4 - 3f_3 + 3f_2 - f_1,$$

       (i) $\Delta^4 f_0 = \Delta^3 f_1 - \Delta^3 f_0 = (f_4 - 3f_3 + 3f_2 - f_1) - (f_3 - 3f_2$
$$+ 3f_1 - f_0) = f_4 - 4f_3 + 6f_2 - 4f_1 + f_0.$$

16.3

n	f	Δf	$\Delta^2 f$	$\Delta^3 f$	$\Delta^4 f$
1	1				
		2			
2	3		4		
		6		0	
3	9		4		0
		10		0	
4	19		4		
		14			
5	33				

16.4

x	f	Δf	$\Delta^2 f$	$\Delta^3 f$	$\Delta^4 f$	$\Delta^5 f$	$\Delta^6 f$	$\Delta^7 f$	$\Delta^8 f$
5	1.0834								
		751							
10	1.1585		-89						
		662		-3					
15	1.2247		-92		-3				
		570		-6		6			
20	1.2817		-98		3		-9		
		472		-3		-3		13	
25	1.3289		-101		0		4		-18
		371		-3		1		-5	
30	1.3660		-104		1		-1		
		267		-2		0			
35	1.3927		-106		1				
		161		-1					
40	1.4088		-107						
		54							
45	1.4142								

16.5 (a) For $\sin 13° + \cos 13°$, $x_0 = 10$ and $h = 5$. $\therefore s = 0.6$.

With $n = 3$,

$$\sin 13° + \cos 13° \simeq 1.1585 + 0.6 \times 0.0662 + \frac{0.6(0.6-1)}{2!}(-0.0092)$$

$$+ \frac{0.6(0.6-1)(0.6-2)}{3!}(-0.0006)$$

$$= 1.1585 + 0.039\,72 + 0.001\,10 - 0.000\,03$$

$$= 1.199\,29$$

$$= 1.1993 \quad \text{to 4D.}$$

The correct 4-figure value is 1.993 and the 5-figure value is 1.199 32. The result is even better than that in Example 16.3. It shows that we may lose very little accuracy in a 'good' situation.

(b) For $\sin 22\frac{1}{2}° + \cos 22\frac{1}{2}°$, $x_0 = 20$, and $h = 5$ $\therefore s = 0.5$.

With $n = 3$,

$$\sin 22\frac{1}{2}° + \cos 22\frac{1}{2}° \simeq 1.2817 + 0.5 \times 0.0472$$

$$+ \frac{0.50(0.5-1)}{2!}(-0.0101)$$

$$+ \frac{0.5(0.5-1)(0.5-2)}{3!}(-0.0003)$$

$$= 1.2817 + 0.0236 + 0.001\,26 - 0.000\,02$$

$$= 1.306\,54$$

$$= 1.3065 \quad \text{to 4D.}$$

The correct 4-figure value is 1.3066. The 5-figure value is 1.306 56. Again looking behind the calculation we can see that the error in the interpolate is 0.2×10^{-4} — less than the error in the data, in spite of the rounding in the calculations.

(c) For $\sin 43° + \cos 43°$ we have a problem choosing x_0.

If we choose $x_0 = 40$ (as we would normally) then there aren't sufficient differences to form a cubic interpolating polynomial. In fact to obtain a cubic we have to choose $x_0 = 30$. This is not unreasonable since we obtain the cubic based on the data at $x = 30, 35, 40$ and 45 and this range contains the point ($x = 43$) of interest.

With $x_0 = 30$ and $h = 5$ then, we have $s = 13/5 = 2.6$ (whereas normally $0 < s < 1$) and

$$\sin 43° + \cos 43° \simeq 1.3660 + 2.6(0.0267) + \frac{2.6(2.6 - 1)}{2!}(-0.0106)$$

$$+ \frac{2.6(2.6 - 1)(2.6 - 2)}{3!}(-0.0001)$$

$$= 1.3660 + 0.069\ 42 - 0.022\ 05 - 0.000\ 04$$

$$= 1.413\ 33$$

$$= 1.4133 \quad \text{to 4D.}$$

The correct 4-figure value is 1.4134, but the 5-figure value is 1.413 35, showing again how rounding can make a result look worse than it is.

The dilemma of where to choose x_0 when the normal choice (such that $x_0 < x < x_0 + h$) implies insufficient differences, is usually solved by using interpolation formulae based on *backward* or *central* differences. However, the use of interpolation formulae based on differences is not now as common as it was before high-speed computers were generally available. Hence we shall not consider it further than the brief introduction given in Section 16.6.

FURTHER READING

Just about every book contains material on interpolation using finite differences. There is no single way of developing the formulae. For a readable account of the approach that uses operator algebra see Ribbans, Book 1, Chapters 2 and 3. Ribbans refers to IAT (Interpolation and Allied Tables, HMSO) which was the 'pocket bible' of interpolators in the 'good old days' before high-speed computers.

For a brief summary of the approach used by Ribbans see Williams, Chapter 6, Sections 6.1—6.4.

For a more mathematically satisfying and complete account see Conte, Sections 3.5—3.7 and for a very complete account see Hildebrand, Chapter 4.

More recent books take our approach and do not stress finite-difference interpolation methods, because they are not used as much now as they were. For a similar approach to ours see Conte and deBoor, Section 4.5 in which the backward-difference formula is written using forward differences! Also see Phillips and Taylor, Sections 4.7–4.9, Dixon, Chapters 4 and 5, and Hosking, Joyce and Turner, 'Steps 15 to 20'.

LESSON 16 – COMPREHENSION TEST

1. Construct a difference table for this set of data.

x	1	2	3	4	5
$f(x)$	-13	-6	7	32	75

Write down in forward-difference form the interpolating polynomial for this set of data.

2. What degree of polynomial would be sufficient to provide accurate interpolation among the data giving this difference table.

x	$f(x)$	Δf	$\Delta^2 f$	$\Delta^3 f$	$\Delta^4 f$	$\Delta^5 f$	$\Delta^6 f$
1.0	0.7854						
		476					
1.1	0.8330		-45				
		431		4			
1.2	0.8761		-41		1		
		390		5		-1	
1.3	0.9151		-36		0		-2
		354		5		-3	
1.4	0.9505		-31		-3		
		323		2			
1.5	0.9828		-29				
		294					
1.6	1.0122						

Write down the forward-difference polynomial you would use for estimating $f(1.29)$. Evaluate it at $x = 1.29$.

3. Write down, in terms of function values,

(a) $\delta f_{1/2}$ (b) ∇f_1 (c) $\delta^2 f_1$ (d) $\nabla^2 f_0$.

ANSWERS

1

x	$f(x)$	Δf	$\Delta^2 f$	$\Delta^3 f$	$\Delta^4 f$
1	-13				
		7			
2	-6		6		
		13		6	
3	7		12		0
		25		6	
4	32		18		
		43			
5	75				

$$p_3(x) = -13 + 7s + 3s(s-1) + s(s-1)(s-2)$$

where $s = (x - x_0)/h = (x-1)/1 = x-1$,

or $\qquad -6 + 13s + 6s(s-1) + s(s-1)(s-2)$

where $s = x - 2$.

2 3 (4th and higher differences are effectively zero, i.e., dominated by rounding errors).

$$p_3(x) = 0.8761 + 0.0390s - 0.0018s(s-1) + 0.0005s(s-1)(s-2)/6$$

where $s = (x - 1.2)/0.1$.

$$p_3(1.29) = 0.9114 \quad \text{(to 4D).} \quad (s = 0.9.)$$

3 (a) $f_1 - f_0$, (b) $f_1 - f_0$, (c) $f_2 - 2f_1 + f_0$,
(d) $f_0 - 2f_{-1} + f_{-2}$.

LESSON 16 – SUPPLEMENTARY EXERCISES

16.6. Use forward-difference interpolating polynomials to estimate $f(1.072)$, $f(1.316)$ and $f(1.595)$ from these data.

x	1.0	1.1	1.2	1.3	1.4	1.5	1.6	1.7	1.8
$f(x)$	0.0	0.0953	0.1823	0.2624	0.3365	0.4055	0.4700	0.5306	0.5878

16.7. Show that if $f(x) = \log_e x$ then

$$\Delta^2 f(x) = \log_e \frac{x(x + 2h)}{(x + h)^2}$$

$$\delta^2 f(x) = \log_e \frac{(x-h)(x+h)}{x^2}$$

$$\nabla^2 f(x) = \log_e \frac{(x-2h)x}{(x-h)^2}.$$

16.8.   Show that

$$\int_{x_0}^{x_1} p_1(x) \, dx = \int_0^1 (f_0 + s\Delta f_0)h \, ds = \tfrac{1}{2}h(f_0 + f_1).$$

(This is an approximation to $\int_{x_0}^{x_1} f(x) \, dx$. Such approximations are considered in detail in Lessons 17 to 19.)

16.9.   The data below are believed to satisfy a polynomial relationship $y = p_n(x)$ of some degree n.

Construct a difference table and hence the polynomial.

Evaluate the polynomial at $x = 3.2(0.2)3.6$ by extending the difference table.

x	2.2	2.4	2.6	2.8	3.0
y	14.65	17.31	20.22	23.37	26.77

*16.10. Construct a difference table for these data: $f(x) = 0$ for $x = \pm 1$, $\pm 2, \pm 3, \dots$; $f(0) = \epsilon$.   This will indicate the effect of a mistake in one entry in a difference table.

In the difference table of Exercise 16.9 (for $x = 2.2(0.2)3.6$)   suppose that $y = 23.37$ was   accidentally   given   as 23.77. Reconstruct   the difference table and relate the effect of the mistake to the table constructed above.

*16.11. Use the theory of Exercise 16.10 to find and correct the mistake in this table.

x	-1.0	-0.5	0.0	0.5	1.0	1.5	2.0
$f(x)$	-6.000	-4.625	-4.000	-3.357	-2.000	0.875	6.000

*16.12. Derive the general backward-difference interpolating polynomial for the data (x_i, f_i), $i = 0(1)n$,

$$p_n(x_n + sh) = f_n + s\nabla f_n + \frac{s(s+h)}{2!}\nabla^2 f_n$$

$$+ \dots + \frac{s(s+h)\dots(s+(n-1)h)}{n!}\nabla^n f_n$$

by applying the forward-difference formula to the data taken in reverse order.

PART E   NUMERICAL INTEGRATION

Lesson 17 **Integration Rules**

OBJECTIVES

At the end of this lesson you should

(a)   understand the need for techniques of numerical integration;

(b)   know and be able to apply the rectangular rule, the mid-point rule, the trapezium rule and Simpson's rule;

(c)   be able to use a table of function values to estimate the value of a definite integral.

CONTENTS

7.1   INTRODUCTION

One of the prerequisites of this course is an elementary knowledge of integration and in particular of the definite integral. You should be familiar with the mathematical technique for evaluating $\int_a^b f(x)\,dx$ by first finding the indefinite integral of $f(x)$ (i.e., $\int f(x)\,dx$) as some expression $\phi(x)$ where $d\phi(x)/dx = f(x)$ and then evaluating $\phi(b) - \phi(a)$.

However, in many important applications of the definite integral it is either very difficult or, as far as is known, impossible to find $\phi(x)$. (Example 1.1

(Lesson 1) is a case in point.) In particular this is so when the only information given about f is in the form of a table of values.

To overcome this problem we have to return to the ideas lying behind the definite integral and develop them in a way suitable for numerical computation.

You should be aware that the area between the graph of $y = f(x)$, the x-axis, and the lines $x = a$ and $x = b$ can be expressed as

$$\int_a^b f(x)\, dx.$$

Indeed you may have seen the definite integral defined by this area.

In such an approach the interval $[a, b]$ is divided up into a number of subintervals.

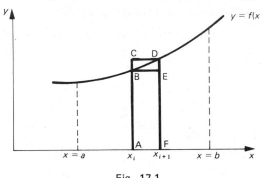

Fig. 17.1

A typical subinterval $[x_i, x_{i+1}]$ is shown in Fig. 17.1, and (as an illustration) the interval $[a, b]$ is divided up into 7 subintervals in Fig. 17.2. In Fig. 17.1 it is easy to see that

$$\text{Area ABEF} < \text{Area under curve BD} < \text{Area ACDF}.$$

Similarly in Fig. 17.2 we can see that the area under $y = f(x)$ between $x = a$ and $x = b$ lies between the area under the upper 'staircase' and the area under the lower 'staircase'. Using mathematical analysis it can be shown that, if $f(x)$ is a reasonably well-behaved function (e.g., a continuous function), then, as we increase the number of subintervals (simultaneously making the lengths of all the subintervals tend to zero), the areas under the two 'staircases' tend towards the same value, which we call the definite integral. We regard this definite integral as the area under the graph. In fact definite integrals can be defined abstractly. Area then becomes just one of their many applications.

In this lesson we will use the idea of area under a curve to derive a number of rules which may be used to obtain approximate values for definite integrals,

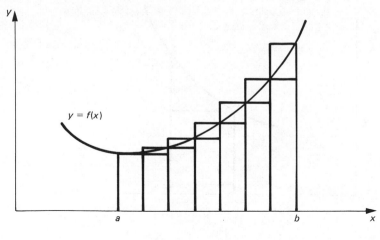

$$y = f(x)$$

Fig. 17.2

when exact values are impossible or too difficult to find. We could develop these rules abstractly, together with expressions for the errors in the resulting approximate integrals, but we leave that for Lesson 18. In Lesson 19 we conclude our look at numerical integration by dealing with interval-halving and extrapolation — methods for obtaining maximum accuracy from the numerical approximations.

7.2 THE GENERAL APPROACH

In pure mathematics we are concerned with letting the number of subintervals tend to infinity while the length of each tends to zero. Practically we cannot do an infinite number of calculations although we may have to have quite a large number of subintervals to obtain some required accuracy. So an integration rule uses a finite number of subintervals on each of which the area under the graph is formed approximately.

The different integration rules are obtained by different ways of approximating the area under the graph. In each case we construct the rule for $\int_a^b f(x)\,dx$ by first considering a single (or in one case double) subinterval, i.e., by approximating an integral of the form $\int_{x_0}^{x_0+h} f(x)\,dx$ (see Fig. 17.3) where x_0 is some fixed value of x inside $[a, b]$. h is the subinterval length and can be thought of as some small number. The rule for $\int_{x_0}^{x_0+h} f(x)\,dx$ is then applied to each subinterval of $[a, b]$ and the results are added to give a general or composite rule for $\int_a^b f(x)\,dx$ (see Fig. 17.4).

7.3 THE RECTANGULAR RULE

In this section the approach outlined above (Section 17.2) is applied using a very simple approximation to the area under the graph. In Fig. 17.3 let

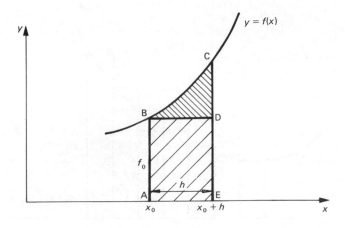

Fig. 17.3

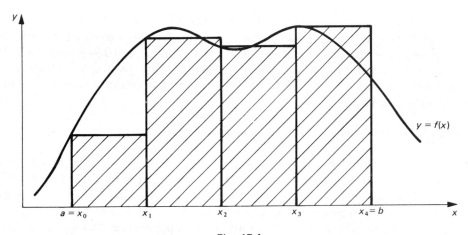

Fig. 17.4

us replace the curve BC by the horizontal line BD and hence the area ABCE by the area ABDE (so that, geometrically, the error is the heavily shaded area). The area ABDE is AE × AB, i.e., $h \times f(x_0)$ or hf_0, using the notation $f(x_r) \equiv f_r$. This gives what is known as the *rectangular rule*

$$\int_{x_0}^{x_0+h} f(x)\,\mathrm{d}x \simeq hf_0.$$

To apply this rule to some 'large' interval $[a, b]$ we can divide the interval $[a, b]$ into n equal subintervals each of length h. Fig. 17.4 illustrates this for $n = 4$.

The area under the curve is approximated by the shaded area, so that

$$\int_a^b f(x)\,dx \simeq h(f_0 + f_1 + f_2 + f_3).$$

Make sure that you understand how this expression was obtained.

EXAMPLE 17.1   To evaluate $\int_1^5 x\,dx$ using the rectangular rule with 8 subintervals (or strips as they are frequently called).

Now $a = 1$, $b = 5$ and $n = 8$. Therefore $8h = 5 - 1$ and $h = \frac{1}{2}$. So

$$\int_1^5 x\,dx \simeq \tfrac{1}{2}(f_0 + f_1 + f_2 + f_3 + f_4 + f_5 + f_6 + f_7)$$

$$= \tfrac{1}{2}(1 + 1.5 + 2 + 2.5 + 3 + 3.5 + 4 + 4.5)$$

$$= 11 \quad \text{(check it!)}$$

Actually

$$\int_1^5 x\,dx = \tfrac{1}{2}(5^2 - 1^2) = 12$$

so it would appear that either h is too large or the rule is not a very good one.

EXERCISE 17.1

Evaluate $\int_1^5 x^2\,dx$ using the rectangular rule with 8 strips. Compare your answer with the exact value.

EXERCISE 17.2

Evaluate $\int_{-1}^1 e^x\,dx$ using the rectangular rule with 4 strips and working with 4D. Compare your answer with the exact value.

If the interval $[a, b]$ is divided into n equal subintervals, each of length h, then $b - a = nh$ and we obtain the *composite rectangular rule*

$$\int_a^b f(x)\,dx \simeq h(f_0 + f_1 + f_2 + \ldots + f_{n-1})$$

where $f_r \equiv f(x_r)$, $x_r = a + rh$. Notice the lopsided form of the rule. All the rectangles use the height of the curve at the left-hand end of each subinterval. This, combined with the rather crude approximation to f (a constant on each subinterval), means that the rectangular rule is not very accurate. The other rules developed in this lesson are symmetric on $[a, b]$ and have better approximations to f.

17.4 THE MID-POINT RULE

For the mid-point rule f is still approximated by a straight line parallel to the x-axis, but instead of using $y = f(x_0)$ as the approximation in $[x_0, x_0 + h]$ we take $y = f(x_0 + \frac{1}{2}h)$, i.e., the height of the curve at the mid-point of the interval.

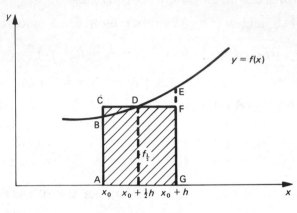

Fig. 17.5

In Fig. 17.5 the curve BE is replaced by the line CF. The area ABDEG is to be approximated by ACDFG. Intuitively this appears to be a better idea than the rectangular rule as the approximation adds on the area BCD and subtracts the area DEF from the area ABDEG so that perhaps the error may be smaller than for the rectangular rule. Using this idea we get the *mid-point rule*

$$\int_{x_0}^{x_0 + h} f(x)\, \mathrm{d}x \simeq hf(x_0 + \tfrac{1}{2}h).$$

Again using the notation $f_r \equiv f(x_r) \equiv f(x_0 + rh)$, and taking $r = \frac{1}{2}$, we have $f_{1/2} \equiv f(x_0 + \frac{1}{2}h)$. Hence we write

$$\int_{x_0}^{x_0 + h} f(x)\, \mathrm{d}x \simeq hf_{1/2}.$$

To approximate $\int_a^b f(x)\, \mathrm{d}x$ we divide up $[a, b]$ as in Section 17.3 and apply the mid-point rule to each subinterval. This gives the general formula (the *composite mid-point rule*)

$$\int_a^b f(x)\, \mathrm{d}x \simeq h(f_{1/2} + f_{3/2} + \ldots + f_{n - 1/2}).$$

EXERCISE 17.3

Draw a diagram similar to Fig. 17.4 demonstrating the geometrical basis of the mid-point rule for $n = 4$. Write down the corresponding approximation to $\int_a^b f(x)\,dx$.

EXAMPLE 17.2   To evaluate $\int_1^5 x\,dx$ using the mid-point rule and $n = 8$. $h = \frac{1}{2}$ as before so that

$$\int_1^5 x\,dx \simeq \tfrac{1}{2}(1.25 + 1.75 + 2.25 + 2.75 + 3.25 + 3.75 + 4.25 + 4.75)$$

$$= 12$$

which is (by chance?) exactly right.

EXERCISE 17.4

Evaluate $\int_1^5 x^2\,dx$ using the mid-point rule and $n = 8$. Comment on your result, comparing it with that obtained in Exercise 17.1 and with the exact value.

EXERCISE 17.5

Evaluate $\int_{-1}^1 e^x\,dx$ using the mid-point rule with 4 strips and using 4D. Again compare your result with that of Exercise 17.2 and with the exact value.

17.5   THE TRAPEZIUM RULE

In Sections 17.3 and 17.4 the approximations used for f over $[x_0, x_0 + h]$ were fairly crude, being just constants. Geometrically the graph of $y = f(x)$ on $[x_0, x_0 + h]$ was replaced by a straight line parallel to the x-axis in both cases. In this section the graph of the approximating function is taken to be the chord joining the end-points of the graph in $[x_0, x_0 + h]$ (i.e., a linear interpolating polynomial!) The situation is shown in Fig. 17.6. Again, intuitively, we would expect the resulting approximate area to be closer to the exact value than that given by the rectangular and mid-point rules.

The area under $y = f(x)$ between $x = x_0$ and $x = x_0 + h$ is approximated by the area of the trapezium ABCD and we have the *trapezium rule*

$$\int_{x_0}^{x_0 + h} f(x)\,dx \simeq \text{area of trapezium ABCD}$$

$$= \tfrac{1}{2}(\text{sum of parallel sides}) \times (\text{distance between})$$

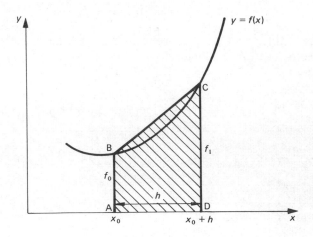

Fig. 17.6

$$= \tfrac{1}{2}(f_0 + f_1)h$$
$$= \tfrac{1}{2}h(f_0 + f_1).$$

To approximate $\int_a^b f(x)\,dx$ we divide up $[a, b]$ as in Section 17.3 and apply the trapezium rule to each subinterval. This gives the *composite trapezium rule*

$$\int_a^b f(x)\,dx \simeq \tfrac{1}{2}h(f_0 + f_1) + \tfrac{1}{2}h(f_1 + f_2) + \ldots + \tfrac{1}{2}h(f_{n-2} + f_{n-1})$$
$$+ \tfrac{1}{2}h(f_{n-1} + f_n)$$
$$= \tfrac{1}{2}h(f_0 + 2f_1 + 2f_2 + \ldots + 2f_{n-2} + 2f_{n-1} + f_n)$$
$$= h(\tfrac{1}{2}f_0 + f_1 + f_2 + \ldots + f_{n-2} + f_{n-1} + \tfrac{1}{2}f_n).$$

EXERCISE 17.6

Draw a diagram similar to Fig. 17.4 demonstrating the geometrical basis of the trapezium rule for $n = 4$. Write down the corresponding approximation to

$$\int_a^b f(x)\,dx.$$

EXAMPLE 17.3

To evaluate $\int_{-1}^{1} e^x\,dx$ using the trapezium rule with 4 strips and working with 4D. $h = \tfrac{1}{2}$ as before, so

$$\int_{-1}^{1} e^x\,dx \simeq h(\tfrac{1}{2}e^{-1} + e^{-1/2} + e^0 + e^{1/2} + \tfrac{1}{2}e^1)$$

$$= 0.5(\tfrac{1}{2}(0.3679) + 0.6065 + 1.0000 + 1.6487 + \tfrac{1}{2}(2.7183))$$
$$= 0.5(\tfrac{1}{2}(3.0862) + 3.2552)$$
$$= 2.3992.$$

Comparing this with your previous results we have

rectangular rule 1.8116
mid-point rule   2.3261 } each using 4 strips.
trapezium rule   2.3992

exact value   2.3504

It is interesting to note that *in this case* the mid-point rule is better than the trapezium rule. If you draw the graph of e^x on $[-1, 1]$ and superimpose the corresponding approximations you will see why.

EXERCISE 17.7

Evaluate $\int_1^5 x^2 \, dx$ using the trapezium rule and $n = 8$.

Compare your result with those of Exercises 17.1 and 17.4 and with the exact value.

Why is the result of applying the trapezium rule to $\int_1^5 x \, dx$ obvious?

EXERCISE 17.8

Write down the Lagrangian interpolating polynomial of degree 1 passing through the points (x_0, f_0) and $(x_0 + h, f_1)$. Integrate this expression formally between $x = x_0$ and $x = x_0 + h$ to obtain the trapezium rule. (Note that $y = f(x)$ is being approximated by the Lagrangian polynomial in question. We'll be using this sort of approach again!)

17.6 SIMPSON'S RULE

One might have hoped that the trapezium rule would have had marked advantages over the rectangular and mid-point rules. In terms of accuracy it is not always so. Nevertheless it is symmetric and so more attractive than the rectangular rule and it does not require function values at mid-points as in the mid-point rule. It also has the apparent advantage of being based on a linear (rather than constant) polynomial approximation to f, as we have seen in Exercise 17.8. Perhaps we can pursue this approach and obtain a 'better' formula for numerical integration by trying a higher degree approximation to f.

Exercise 17.8 used a linear interpolating polynomial — so why not use a quadratic interpolating polynomial? (Little by little — don't look for an nth degree polynomial over $[x_0, x_0 + h]$ yet — it may not be worth it!) To do so we will have to have three data points, instead of two as for the trapezium rule. This suggests the points (x_0, f_0), (x_1, f_1) and (x_2, f_2), or perhaps $(x_{-1}, f_{-1})(x_0, f_0)$ and (x_1, f_1) if we are looking for a symmetrical formula. (Here $x_1 = x_0 + h$, $x_2 = x_0 + 2h$, $x_{-1} = x_0 - h$, so that we are continuing to use subintervals of equal length.)

The Lagrangian interpolating polynomial of degree 2 through the points $(x_0 - h, f_{-1})$, (x_0, f_0) and $(x_0 + h, f_1)$ is

$$p_2(x) = \frac{(x - x_0)(x - x_1)f_{-1}}{(x_{-1} - x_0)(x_{-1} - x_1)} + \frac{(x - x_{-1})(x - x_1)f_0}{(x_0 - x_{-1})(x_0 - x_1)} + \frac{(x - x_{-1})(x - x_0)f_1}{(x_1 - x_{-1})(x_1 - x_0)}$$

$$= \frac{(x - x_0)(x - x_0 - h)f_{-1}}{(-h)(-2h)} + \frac{(x - x_0 + h)(x - x_0 - h)f_0}{(h)(-h)}$$

$$+ \frac{(x - x_0 + h)(x - x_0)f_1}{(2h)(h)}.$$

Having obtained p_2 as our approximation to f on $[x_0 - h, x_0 + h]$ we need to integrate it, but over what interval? Well it's not obvious but we shall choose $[x_0 - h, x_0 + h]$, obtaining an approximation to $\int_{x_0 - h}^{x_0 + h} f(x)\,dx$ indicated by the shaded area in Fig. 17.7, i.e., *our basic formula requires two strips in this case.*

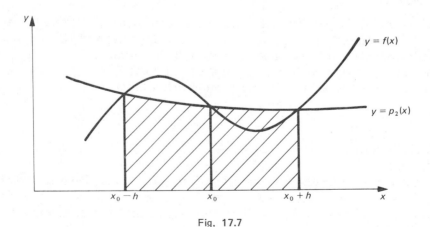

Fig. 17.7

In the integration of p_2 we will use the substitution $u = x - x_0$ (and hence $du = dx$) as a cunning ploy to make the algebraic manipulation a lot easier. Hence

$$\int_{x_0-h}^{x_0+h} p_2(x)\,dx$$

$$= \int_{u=-h}^{u=h} \left(\frac{u(u-h)f_{-1}}{2h^2} + \frac{(u+h)(u-h)f_0}{-h^2} + \frac{(u+h)uf_1}{2h^2} \right) du$$

$$= \left[\frac{f_{-1}}{2h^2}\left(\frac{u^3}{3} - \frac{hu^2}{2}\right) - \frac{f_0}{h^2}\left(\frac{u^3}{3} - h^2u\right) + \frac{f_1}{2h^2}\left(\frac{u^3}{3} + \frac{hu^2}{2}\right) \right]_{u=-h}^{u=h}$$

$$= \frac{f_{-1}}{2h^2}\left(\frac{h^3}{3} - \frac{h^3}{2} + \frac{h^3}{3} + \frac{h^3}{2}\right) - \frac{f_0}{h^2}\left(\frac{h^3}{3} - h^3 + \frac{h^3}{3} - h^3\right)$$

$$+ \frac{f_1}{2h^2}\left(\frac{h^3}{3} + \frac{h^3}{2} + \frac{h^3}{3} - \frac{h^3}{2}\right)$$

$$= \frac{1}{2h^2}\left(\frac{2h^3}{3}f_{-1} + \frac{8h^3}{3}f_0 + \frac{2h^3}{3}f_1\right)$$

$$= \frac{h}{3}(f_{-1} + 4f_0 + f_1).$$

Hence we have what is best known as *Simpson's rule*

$$\int_{x_0-h}^{x_0+h} f(x)\,dx \simeq \frac{h}{3}(f_{-1} + 4f_0 + f_1).$$

EXERCISE 17.9

Write down the Lagrangian interpolating polynomial $p_2(x)$ of degree 2 based on the data points (x_0, f_0), $(x_0 + h, f_1)$ and $(x_0 + 2h, f_2)$. Integrate $p_2(x)$ to gain the formula

$$\int_{x_0}^{x_0+2h} f(x)\,dx \simeq \frac{h}{3}(f_0 + 4f_1 + f_2)$$

which is, of course, Simpson's rule again, in a slightly different guise.

To apply Simpson's rule to $\int_a^b f(x)\,dx$ we must divide $[a, b]$ into $2n$ (i.e., an even number of) equal subintervals each of length h, so that $b - a = 2nh$.

Using the basic formula on each of the n non-overlapping 'double strips' of width $2h$ we get

$$\int_a^b f(x)\,dx \simeq \frac{h}{3}(f_0 + 4f_1 + f_2) + \frac{h}{3}(f_2 + 4f_3 + f_4)$$

$$+ \frac{h}{3}(f_4 + 4f_5 + f_6) + \ldots + \frac{h}{3}(f_{2n-4} + 4f_{2n-3} + f_{2n-2})$$

$$+ \frac{h}{3}(f_{2n-2} + 4f_{2n-1} + f_{2n})$$

$$= \frac{h}{3}(f_0 + 4f_1 + 2f_2 + 4f_3 + 2f_4 + \ldots + 2f_{2n-2} + 4f_{2n-1} + f_{2n}).$$

Notice the pattern of coefficients.

For 2 strips	1	4	1					
for 4 strips	1	4	2	4	1			
for 6 strips	1	4	2	4	2	4	1	
and in general	1	4	2	4	2	... 4	2	4	1.

Here we are approximating f by a series of quadratics, a so-called *piecewise-quadratic approximation*. The trapezium rule is based on a *piecewise-linear approximation*.

EXAMPLE 17.4   Evaluate $\int_{-1}^{1} e^x \, dx$ using $2n = 4$ strips (i.e., $h = \frac{1}{2}$ as before) and 4D working (also as before).

$$\int_{-1}^{1} e^x \, dx \simeq \frac{\frac{1}{2}}{3}(e^{-1} + 4e^{-1/2} + 2e^0 + 4e^{1/2} + e^1)$$

$$= \tfrac{1}{6}(0.3679 + 4 \times 0.6065 + 2 \times 1 + 4 \times 1.6487 + 2.7183)$$

$$= \tfrac{1}{6}(3.0862 + 4 \times 2.2552 + 2 \times 1)$$

$$= 2.3512.$$

Now let's compare again the results we have for $\int_{-1}^{1} e^x \, dx$.

rectangular rule	1.8116	
mid-point rule	2.3261	each with 4 subintervals.
trapezium rule	2.3992	
Simpson's rule	2.3512	
exact value	2.3504	

Clearly Simpson's rule does much the best in this case. Simpson's rule is a good one to use because

(a) it gives good results for fairly large h, and yet

(b) it is not too complicated to use.

Higher-accuracy formulae can clearly be found in the same way, but they will be more cumbersome. Simpson's rule is a good compromise between the requirements of accuracy and simplicity.

EXERCISE 17.10

Evaluate $\int_0^1 1/(1+x)\, dx$ by Simpson's rule using

(a) $(2n) = 2$ strips,   (b) $(2n) = 4$ strips,   (c) $(2n) = 8$ strips.

Work with 4D and comment on your results.

Why have we not used $\int_1^5 x^2\, dx$ as an exercise on Simpson's rule?

EXERCISE 17.11

Draw a diagram similar to Fig. 17.4 demonstrating the geometrical basis of Simpson's rule for $2n = 4$. Write down the corresponding approximation to $\int_a^b f(x)\, dx$.

17.7   INTEGRATING FUNCTIONS GIVEN BY TABLES OF VALUES

So far we have assumed that the value of the function f can be found for any value of x in the domain on which f is defined. For example, in Exercise 17.10 $f(x) = 1/(1+x)$ on $[0, 1]$. However it is quite possible that all we have available is a table of values of $f(x)$. In such a case any estimation of $\int_a^b f(x)\, dx$ will depend on the nature of the table of data.

In particular, as for interpolation, we have to assume that the data sufficiently represent the function. We can usually apply the numerical integration rules, the obvious limitation being that the choice of the number of subintervals will be restricted. In particular if there is an even number of data points (and hence an odd number of subintervals) then Simpson's rule will be inapplicable.

Another point to note is that we shall not have unlimited accuracy in the data. We shall prove in Lesson 18 that the maximum accumulated round-off error in an approximate value for an integral equals the maximum round-off error in each function value (assuming the maximum errors are all the same), times the range of integration. For example, if the data each have error bounds 0.5×10^{-n} then an integration rule estimate for $\int_a^b f(x)\, dx$ will have (round-off) error bound $0.5(b-a) \times 10^{-n}$.

EXAMPLE 17.5   Estimate $\int_{1.0}^{1.4} f(x)\, dx$ given the following data.

x	1.0	1.1	1.2	1.3	1.4
$f(x)$	0.010	0.252	0.586	1.024	1.578

Here the error bound for each $f(x)$ is 0.0005 and the range of integration is 0.4. So the accumulated round-off error in any integration rule will be, at most, 0.2×10^{-3} so we must not give more than 3D in any answer. Note that we are lucky that the x values are equally spaced.

(a) First let us try the rectangular rule.

Taking $h = 0.4$ (seems a little silly but let's try) gives

$$\int_{1.0}^{1.4} f(x)\, dx \simeq 0.4 \times 0.010 = 0.004 \quad \text{(to 3D)}.$$

If $h = 0.2$

$$\int_{1.0}^{1.4} f(x)\, dx \simeq 0.2(0.010 + 0.586)$$

$$= 0.119 \quad \text{(to 3D)}.$$

If $h = 0.1$

$$\int_{1.0}^{1.4} f(x)\, dx \simeq 0.1(0.010 + 0.252 + 0.586 + 1.024)$$

$$= 0.187 \quad \text{(to 3D)}.$$

These three results don't give us much clue to the answer!

(b) What about the mid-point rule?

With $h = 0.4$

$$\int_{1.0}^{1.4} f(x)\, dx \simeq 0.4 \times 0.586 = 0.234 \quad \text{(to 3D)}.$$

(The mid-point rule using one subinterval is a very handy method of obtaining a quick, rough estimate for an integral.)

With $h = 0.2$

$$\int_{1.0}^{1.4} f(x)\, dx \simeq 0.2 \times (0.252 + 1.024)$$

$$= 0.255 \quad \text{(to 3D)}.$$

With $h = 0.1$ we are given no mid-point function values.

Well, these values aren't too far apart but do not enable us to give even one decimal place for the integral.

(c) Now for the trapezium rule.

With $h = 0.4$

$$\int_{1.0}^{1.4} f(x)\, dx \simeq \tfrac{1}{2} \times 0.4(0.010 + 1.578) = 0.318 \quad \text{(to 3D)}.$$

With $h = 0.2$

$$\int_{1.0}^{1.4} f(x)\, dx \simeq 0.2(\tfrac{1}{2}(0.010) + 0.586 + \tfrac{1}{2}(1.578))$$

$$= 0.276 \quad \text{(to 3D)}.$$

With $h = 0.1$

$$\int_{1.0}^{1.4} f(x)\, dx \simeq 0.1(\tfrac{1}{2}(0.010) + 0.252 + 0.586 + 1.024 + \tfrac{1}{2}(1.578))$$

$$= 0.266 \quad \text{(to 3D)}.$$

These last two values are fairly close together and one would expect the second to be the better of the two. We might guess at 0.27 to 2D but can we improve on that? Perhaps Simpson's rule can help. It should be more accurate than any used so far.

(d) Simpson's rule.

With $h = 0.2$ (why not $h = 0.4$?)

$$\int_{1.0}^{1.4} f(x)\, dx \simeq \tfrac{1}{3} \times 0.2(0.010 + 4 \times 0.586 + 1.578)$$

$$= 0.262 \quad \text{(to 3D)}.$$

With $h = 0.1$

$$\int_{1.0}^{1.4} f(x)\, dx \simeq \tfrac{1}{3} \times 0.1(0.010 + 4 \times 0.252 + 2 \times 0.586$$

$$+ 4 \times 1.024 + 1.578)$$

$$= 0.262 \quad \text{(to 3D)}.$$

We can't take any other h value so these are the only Simpson's rule values available. However, they do agree to 3D so that we can be sure that the integral is 0.26 (to 2D) and fairly confident that it is 0.262 (to 3D).

One might object to doing the work in (a), (b) and (c) but it does get the most out of the data, and in this case the results in (b) and (c) give us confidence in the excellent answer of (d).

Both Exercise 17.10 and Example 17.5 show how we try to obtain good estimates of an integral by using more and more subintervals and looking for consistency. In theory, as the number of subintervals tends to infinity and the subinterval width to zero, the approximate values for the integral approach the exact value. We shall return to this idea in Lesson 19. Here's another one for you to try.

EXERCISE 17.12

Using the rules developed in this lesson find as many estimates of $\int_{0.0}^{0.6} f(x)\, dx$ as you can from the following data. Give the best value obtainable for the integral.

x	0.0	0.1	0.2	0.3	0.4	0.5	0.6
$f(x)$	1.000	1.005	1.020	1.045	1.081	1.128	1.185

Prove that $(2M_n + T_n)/3 = S_{2n}$ where M_k, T_k, S_k are respectively the mid-point, trapezium and Simpson's rule results for k subintervals. (*Hint*: If you find it difficult look at the cases $n = 1$, $n = 2$, first.)

ANSWERS TO EXERCISES

17.1 $\quad a = 1, \ b = 5, \ n = 8, \ h = \frac{1}{2}$.

$$\therefore \quad \int_1^5 x^2 \, dx \simeq \tfrac{1}{2}(1^2 + 1.5^2 + 2^2 + 2.5^2 + 3^2 + 3.5^2 + 4^2 + 4.5^2)$$

$$= \underline{35.5}.$$

Actually

$$\int_1^5 x^2 \, dx = \tfrac{1}{3}(5^3 - 1^3) = 41\tfrac{1}{3}.$$

17.2 $\quad a = -1, \ b = 1, \ n = 4$. Therefore $h = (1 - (-1))/4 = \frac{1}{2}$.

$$\int_{-1}^1 e^x \, dx \simeq \tfrac{1}{2}(e^{-1} + e^{-1/2} + e^0 + e^{1/2})$$

$$= \tfrac{1}{2}(0.3679 + 0.6065 + 1.0000 + 1.6487)$$

$$= \underline{1.8116}.$$

Actually

$$\int_{-1}^1 e^x \, dx = e - e^{-1} = 2.3504 \quad \text{to 4D}.$$

17.3 $\quad \int_a^b f(x) \, dx \simeq h(f_{1/2} + f_{3/2} + f_{5/2} + f_{7/2})$.

17.4 $\quad \int_1^5 x^2 \, dx \simeq \tfrac{1}{2}(1.25^2 + 1.75^2 + 2.25^2 + 2.75^2 + 3.25^2$

$$+ 3.75^2 + 4.25^2 + 4.75^2) = \underline{41.25}.$$

This result is much nearer the true result (which we happen to know!) than the estimate obtained by the rectangular rule (Exercise 17.1).

17.5 $\quad \int_{-1}^1 e^x \, dx \simeq \tfrac{1}{2}(e^{-3/4} + e^{-1/4} + e^{1/4} + e^{3/4})$

$$= \tfrac{1}{2}(0.4724 + 0.7788 + 1.2840 + 2.1170)$$

$$= \underline{2.3261}.$$

Again the result is much nearer the true result than in Exercise 17.2.

17.6 $\int_a^b f(x)\,dx \simeq h(\frac{1}{2}f_0 + f_1 + f_2 + f_3 + \frac{1}{2}f_4).$

17.7 $\int_1^5 x^2\,dx \simeq \frac{1}{2}(\frac{1}{2}(1^2) + 1.5^2 + 2^2 + 2.5^2 + 3^2$

$$+ 3.5^2 + 4^2 + 4.5^2 + \frac{1}{2}(5^2))$$

$$= \underline{41.5}$$

compared with $41\frac{1}{3}$ (exact)

                         35.5   (rectangular)

                         41.25   (mid-point).

The approximating curve for $y = x$ is in fact $y = x$ if the trapezium rule is used. So the trapezium rule result for $\int_1^5 x\,dx$ would be exact. (Try it!)

17.8 $p_1(x) = \dfrac{x - x_0}{x_1 - x_0} f_1 + \dfrac{x - x_1}{x_0 - x_1} f_0$ where $x_1 = x_0 + h$.

$$\int_{x_0}^{x_0 + h} p_1(x)\,dx = \int_{x_0}^{x_0 + h} \left(\frac{x - x_0}{x_1 - x_0} f_1 + \frac{x - x_1}{x_0 - x_1} f_0 \right) dx$$

$$= \left[\frac{(x - x_0)^2}{2} \frac{f_1}{h} + \frac{(x - x_0 - h)^2}{2} \frac{f_0}{-h} \right]_{x_0}^{x_0 + h}$$

$$= \frac{h^2}{2} \frac{f_1}{h} - 0 + 0 - \frac{(-h)^2}{2} \frac{f_0}{-h}$$

$$= \frac{1}{2}h(f_0 + f_1).$$

17.10   (a) $h = \frac{1}{2}$

$$\int_0^1 \frac{1}{1 + x}\,dx = \frac{\frac{1}{2}}{3}(1 + \frac{8}{3} + \frac{1}{2}) = \underline{0.6944} \quad \text{(to 4D)},$$

(b) $h = \frac{1}{4}$

$$\int_0^1 \frac{1}{1 + x}\,dx \simeq \frac{\frac{1}{4}}{3}\left(\frac{1}{1} + 4\frac{1}{1 + \frac{1}{4}} + 2\frac{1}{1 + \frac{1}{2}} + 4\frac{1}{1 + \frac{3}{4}} + \frac{1}{1 + 1} \right)$$

$$= \tfrac{1}{12}(1 + 4 \times \tfrac{4}{5} + 2 \times \tfrac{2}{3} + 4 \times \tfrac{4}{7} + \tfrac{1}{2})$$

$$= \tfrac{1}{12}(1.5 + 4 \times 1.3714 + 2 \times 0.6667)$$

$$= \underline{0.6932} \quad \text{(to 4D)}.$$

(c) $h = \frac{1}{8}$

$$\int_0^1 \frac{1}{1+x}\,dx \simeq \frac{\frac{1}{8}}{3}\left(\frac{1}{1} + 4\frac{1}{1+\frac{1}{8}} + 2\frac{1}{1+\frac{1}{4}} + 4\frac{1}{1+\frac{3}{8}} + 2\frac{1}{1+\frac{1}{2}}\right.$$

$$\left. + 4\frac{1}{1+\frac{5}{8}} + 2\frac{1}{1+\frac{3}{4}} + 4\frac{1}{1+\frac{7}{8}} + \frac{1}{1+1}\right)$$

$$= \tfrac{1}{24}\left(1.5 + 4\left(\tfrac{8}{9} + \tfrac{8}{11} + \tfrac{8}{13} + \tfrac{8}{15}\right) + 2\left(\tfrac{4}{5} + \tfrac{2}{3} + \tfrac{4}{7}\right)\right)$$

$$= \tfrac{1}{24}\left(1.5 + 4 \times 2.7649 + 2 \times 2.0381\right)$$

$$= \underline{0.6932} \quad \text{(to 4D)}.$$

(b) and (c) agree to 4D and so we can certainly say that

$$\int_0^1 \frac{1}{1+x}\,dx \simeq 0.693 \quad \text{to 3D}$$

(and 0.6932 is probably correct to 4D).

In fact

$$\int_0^1 \frac{dx}{(1+x)} = \ln 2 \quad \text{(check it!)} = 0.693\,15 \quad \text{(to 5D)}.$$

17.11 $\displaystyle\int_a^b f(x)\,dx \simeq \frac{h}{3}(f_0 + 4f_1 + 2f_2 + 4f_3 + f_4).$

17.12

Rule	No. of strips	Estimate
Rectangular	1	0.600
	2	0.614
	3	0.620
	6	0.628
Mid-point	1	0.627
	3	0.636
Trapezium	1	0.656
	2	0.641
	3	0.639
	6	0.637
Simpson's	2	0.636
	6	0.637

Best value is 0.637 correct to 3D. (The accumulated round-off error is at most 0.3×10^{-3}.

17.13 $T_n = h(\frac{1}{2}f_0 + f_1 + f_2 + \ldots + f_{n-1} + \frac{1}{2}f_n)$

   $\underline{2M_n = 2h(f_{1/2} + f_{3/2} + f_{5/2} + \ldots + f_{n-3/2} + f_{n-1/2})}$

$$(2M_n + T_n)/3 = \frac{h}{6}(f_0 + 4f_{1/2} + 2f_1 + 4f_{3/2} + 2f_2 + \ldots + 2f_{n-1}$$

$$+ 4f_{n-1/2} + f_n)$$

$$= S_{2n}$$

because the h appearing here (for n subintervals) is twice the h required for $2n$ subintervals. (If you can't see the last step replace h by $2h$ and double all the subscripts.)

FURTHER READING

We have tried to make this a very simple introduction to numerical integration. We can find no book which does it as simply and in such detail. When we have covered errors in Lesson 18 we shall be able to recommend some further reading.

LESSON 17 – COMPREHENSION TEST

1. State the rectangular, trapezium, mid-point and Simpson rules in both basic and composite forms.

2. For which of these integrands is the trapezium rule exact:

(a) constant functions,

(b) linear functions,

(c) quadratic functions?

3. What is the maximum round-off error in a Simpson's rule approximation to $\int_{-1}^{1} f(x)\,\mathrm{d}x$ if the values of $f(x)$ are available correct to 4D?

4. If $f(x)$ is given at $x = 0(1)12$ how many different approximations to $\int_0^{12} f(x)\,\mathrm{d}x$ can be obtained using

(a) the trapezium rule,   (b) Simpson's rule?

ANSWERS

1   *Rectangular*

$$\int_{x_0}^{x_0+h} f(x)\,\mathrm{d}x \simeq hf_0$$

$$\int_{x_0}^{x_0 + nh} f(x)\, dx \simeq h(f_0 + f_1 + f_2 + \ldots + f_{n-1}).$$

Trapezium

$$\int_{x_0}^{x_0 + h} f(x)\, dx \simeq \tfrac{1}{2}h(f_0 + f_1)$$

$$\int_{x_0}^{x_0 + nh} f(x)\, dx \simeq h(\tfrac{1}{2}f_0 + f_1 + f_2 + \ldots + f_{n-1} + \tfrac{1}{2}f_n).$$

Mid-point

$$\int_{x_0}^{x_0 + h} f(x)\, dx \simeq hf_{1/2}$$

$$\int_{x_0}^{x_0 + nh} f(x)\, dx \simeq h(f_{1/2} + f_{3/2} + f_{5/2} + \ldots + f_{n-1/2}).$$

Simpson

$$\int_{x_0}^{x_0 + 2h} f(x)\, dx \simeq \tfrac{1}{3}h(f_0 + 4f_1 + f_2)$$

$$\int_{x_0}^{x_0 + 2nh} f(x)\, dx \simeq \tfrac{1}{3}h(f_0 + 4f_1 + 2f_2 + 4f_3 + \ldots + 2f_{2n-2}$$
$$+ 4f_{2n-1} + f_{2n}).$$

2   (a) and (b).

3   $(1 - (-1))\tfrac{1}{2} \times 10^{-4} = 10^{-4}$.

4   (a) 6, using 1, 2, 4, 3, 6, 12 subintervals.
    (b) 4, using 2, 4, 6, 12 subintervals.

LESSON 17 – SUPPLEMENTARY EXERCISES

17.14. Use the rectangular, trapezium, mid-point and Simpson rules with 2, 4 and 8 subintervals to find approximate values for

$$\int_0^1 \frac{dx}{1 + x^2}.$$

Compare your answers with each other and with the true value $\tfrac{1}{4}\pi = 0.785\,398$ correct to 6D. In particular relate your answers to the shape of the graph of $y = 1/(1 + x^2)$. Work throughout to 6D.

17.15. Obtain all possible estimates of $\int_1^3 f(x)\, dx$ using the rectangular, trapezium, mid-point and Simpson rules, given only the following data.

x	1	1.25	1.5	1.75	2.0
$f(x)$	0	0.096 91	0.176 09	0.243 04	0.301 03

x	2.25	2.5	2.75	3.0
$f(x)$	0.352 18	0.397 94	0.439 33	0.477 12

Hence state the most accurate value for the integral which is justified by your results.

*17.16. Obtain the so-called $\frac{3}{8}$ths rule

$$\int_{x_0}^{x_0+3h} f(x)\, dx \simeq \tfrac{3}{8}h(f_0 + 3f_1 + 3f_2 + f_3)$$

by replacing $f(x)$ in the integral by the cubic interpolating polynomial agreeing with $f(x)$ at $x = x_0 + kh$, $k = 0(1)3$. Derive a composite $\frac{3}{8}$ths rule for $\int_{x_0}^{x_0 + 3nh} f(x)\, dx$ and apply it with $n = 1$ and 2 (i.e., with 3 and 6 subintervals respectively) to estimating the integral in Exercise 17.14.

Lesson 18 Errors in Numerical Integration

OBJECTIVES

At the end of this lesson you should

(a) appreciate that error terms for approximate integration formulae can be obtained;

(b) know the orders of the error terms for each of the rectangular, mid-point, trapezium and Simpson rules;

(c) appreciate the effect that halving the strip-width has on the error.

CONTENTS

Answers to exercises. Further reading.
Comprehension test and answers. Supplementary exercises.

18.1 INTRODUCTION

In the last lesson we saw how, using geometrical ideas, various approximate integration rules could be obtained. Of course at this elementary stage only some of the simplest, if very useful, rules were found. In this lesson we shall look at a rather more mathematical way of obtaining integration rules. It is by no means the only way but it does enable us to get some idea of the sort of errors involved in using the rules.

You will find references to appendices containing certain mathematical theorems in this lesson. These theorems will be of interest to those more mathematically inclined. If you do not wish to bother too much about them you may ignore them and concentrate on the results obtained by using the theorems.

The results that we shall obtain (apart from reconstructing the rules we already know from Lesson 17) give expressions for what are called the *truncation errors* for the approximate integration rules. These are the errors due to replacing f in $\int_a^b f(x)\,dx$ by a simpler function — an interpolating polynomial — and should be carefully distinguished from the accumulated round-off error due to round-off errors in values of $f(x)$, which we shall also consider.

The basic strategy for this lesson can be summarised as follows. Let

$$f(x) = p_n(x) + e_n(x)$$

where $p_n(x)$ is an interpolating polynomial for $f(x)$ and $e_n(x)$ is the corresponding error term (see Lesson 15) then

$$\int_a^b f(x)\,dx = \int_a^b p_n(x)\,dx + \int_a^b e_n(x)\,dx$$

(assuming the integrals exist). The first term on the right gives the integration rule (remember Exercises 17.8 and 17.9) and the second gives the truncation error. In Lesson 17 we used Lagrangian forms for $p_n(x)$. Since we always choose equal-length subintervals for the kind of integration rules we are discussing, it is slightly simpler to use the Newtonian forward difference form, and that is what we shall do here. To remind you, here is the general version, for a polynomial of degree m.

$$f(x_0 + sh) = f_0 + s\Delta f_0 + \frac{s(s-1)}{2!}\Delta^2 f_0 + \frac{s(s-1)(s-2)}{3!}\Delta^3 f_0 + \dots$$

$$+ \frac{s(s-1)(s-2)\dots(s-m+1)}{m!}\Delta^m f_0$$

$$+ \frac{s(s-1)(s-2)\dots(s-m)}{(m+1)!}h^{m+1}f^{(m+1)}(\xi_s)$$

where $\xi_s \in (x_0, x_m)$ and depends on s. $\Delta f_0, \Delta^2 f_0, \dots, \Delta^m f_0$ are forward differences based on the data $(x_0, f_0), (x_1, f_1), \dots, (x_m, f_m)$, and $x_r = x_0 + rh$ for all r. (It is worth recalling that the equality only holds if $f^{(m+1)}$ is continuous on (x_0, x_m). We shall assume that this is so.)

18.2 THE RECTANGULAR RULE

As in Lesson 17 we first look for a rule to approximate $\int_{x_0}^{x_0+h} f(x)\,dx$.

Let us take $m = 0$ in the forward difference form, so that

$$f(x_0 + sh) = f_0 + hsf^{(1)}(\xi_s)$$

where we shall impose $0 \leqslant s \leqslant 1$ and ξ_s must therefore satisfy $x_0 < \xi_s < x_0 + h$.

(This is actually the first mean value theorem again (see Appendix 2)). The number ξ_s has a suffix s to indicate that ξ_s depends on s (i.e., it varies as s varies).

Writing $x = x_0 + sh$ (so that $dx = h\,ds$) we get

$$\int_{x_0}^{x_0+h} f(x)\,dx = \int_{s=0}^{s=1} f(x_0 + sh)h\,ds$$

$$= \int_0^1 (f_0 + hsf^{(1)}(\xi_s))h\,ds$$

$$= hf_0 + h^2 \int_0^1 sf^{(1)}(\xi_s)\,ds. \qquad\qquad [18\text{-}1]$$

To make progress with this expression we must do something about $\int_0^1 sf^{(1)}(\xi_s)\,ds$. The theorem that will now be applied is the *mean value theorem for integrals* which is stated (but not proved) in Appendix 4. You should have a look at this appendix but not bother too much about the details at a first reading.

To apply this theorem we observe that s does not change sign in $[0, 1]$.

Assuming that $f^{(1)}(\xi_s)$ is a continuous function of s for $0 < s < 1$, the theorem gives

$$\int_0^1 sf^{(1)}(\xi_s)\,ds = f^{(1)}(\xi) \int_0^1 s\,ds = \tfrac{1}{2}f^{(1)}(\xi)$$

where ξ is some number satisfying $x_0 < \xi < x_0 + h$ and no longer depending on s. As in previous work we don't know the value of ξ, only that it lies between x_0 and $x_0 + h$. Substituting back into equation [18-1] we obtain

$$\int_{x_0}^{x_0+h} f(x)\,dx = hf_0 + \tfrac{1}{2}h^2 f^{(1)}(\xi). \qquad\qquad [18\text{-}2]$$

hf_0 is to be the approximation to $\int_{x_0}^{x_0+h} f(x)\,dx$ (the rectangular rule of the last lesson) and the truncation error is $\tfrac{1}{2}h^2 f^{(1)}(\xi)$.

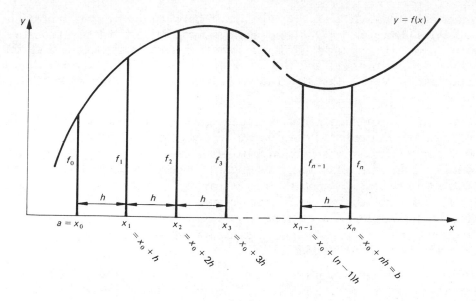

Fig. 18.1

To obtain a formula for $\int_a^b f(x)\,dx$ the interval $[a, b]$ is divided up into n equal sub-intervals as usual, so that $b - a = nh$ (see Fig. 18.1). The formula [18-2] is then applied to each sub-interval in turn and the results are added to give

$$\int_a^b f(x)\,dx = hf_0 + hf_1 + \ldots + hf_{n-1}$$
$$+ \tfrac{1}{2}h^2 f^{(1)}(\xi_1) + \tfrac{1}{2}h^2 f^{(1)}(\xi_2) + \ldots + \tfrac{1}{2}h^2 f^{(1)}(\xi_n)$$

where $x_0 < \xi_1 < x_0 + h,\quad x_0 + h < \xi_2 < x_0 + 2h,\ \ldots,\quad x_0 + (n-1)h < \xi_n < x_0 + nh$.

As before the approximation to the integral is $h(f_0 + f_1 + \ldots + f_{n-1})$ (the composite rectangular rule) and the error term is $\tfrac{1}{2}h^2(f^{(1)}(\xi_1) + f^{(1)}(\xi_2) + \ldots + f^{(1)}(\xi_n))$. This error term is rather messy but because $f^{(1)}$ is continuous on $[x_0, x_0 + nh]$ it can be written as

$$\tfrac{1}{2}h^2 n f^{(1)}(\xi)$$

where $x_0 < \xi < x_0 + nh$, i.e., $a < \xi < b$, so that $f^{(1)}(\xi)$ is a kind of representative value for all the $f^{(1)}(\xi_i)$. An explanation for this step can be found in Appendix 5.

Now n and h are connected by $nh = b - a$, so the normal way of expressing the above result is

$$\int_a^b f(x)\,dx = h(f_0 + f_1 + \ldots + f_{n-1}) + \tfrac{1}{2}(b - a)hf^{(1)}(\xi) \qquad [18\text{-}3]$$

where $a < \xi < b$. It is usual to distinguish between the truncation errors in Equations [18-2] and [18-3] by calling $\frac{1}{2}h^2 f^{(1)}(\xi)$ the *local* truncation error, and $\frac{1}{2}(b-a)hf^{(1)}(\xi)$ the *global* truncation error. Note that the former depends on h^2 and the latter on h. It is the global error that is important, so we say that the rectangular rule has truncation error of $O(h)$ (order h).

What can we conclude about the rectangular rule? Well, if $f^{(1)}(x)$ has magnitude of the order of unity in the interval (a, b), then we need h to be about the size of the maximum tolerable error in the result, and that could be quite small of course. This might involve calculating a great number of function values. Put another way, we can deduce that if h is halved (and $f^{(1)}(\xi)$ does not change too much) we may expect the error to be halved. Thus for quite a lot of work in halving h and hence in calculating the extra function values, we cannot expect a great increase in the accuracy of the new approximation to the integral.

Notice that our conclusions are mostly qualitative. Just as with the errors in Taylor-polynomial approximation and in interpolation, we rarely use the truncation error expression to attempt to *calculate* the error.

EXAMPLE 18.1 In Example 17.5 we found these rectangular-rule estimates of an integral which eventually was evaluated at 0.262 to $3D$.

h	Estimate	Error (0.262-estimate)
0.4	0.004	0.258
0.2	0.119	0.143
0.1	0.187	0.075

We can clearly see that the error halves approximately as h is halved. Work it out for yourself!

18.3 THE TRAPEZIUM RULE

If we now take $m = 1$ then

$$f(x_0 + sh) = f_0 + s\Delta f_0 + \frac{s(s-1)}{2!}h^2 f^{(2)}(\xi_s)$$

where $x_0 < \xi_s < x_0 + h$ and $0 \leqslant s \leqslant 1$.

As in the previous section we formally integrate both sides (i.e., we assume the integrals exist).

$$\int_{x_0}^{x_0+h} f(x)\, dx = \int_0^1 (f_0 + s\Delta f_0 + \frac{s(s-1)}{2!}h^2 f^{(2)}(\xi_s))h\, ds$$

$$= hf_0 + \tfrac{1}{2}h\Delta f_0 + \frac{h^3}{2!}\int_0^1 s(s-1)f^{(2)}(\xi_s)\,\mathrm{d}s.$$

Before applying the mean value theorem for integrals again we must check that $s(s-1)$ does not change sign in $[0,1]$. (We are assuming the other requirement — that $f^{(2)}(\xi_s)$ is continuous for $0 \leqslant s \leqslant 1$.)

In fact $s(s-1)$ is negative throughout $(0,1)$ and zero at $s = 0$ and 1, so we can apply the theorem, obtaining

$$\int_{x_0}^{x_0+h} f(x)\,\mathrm{d}x = hf_0 + \tfrac{1}{2}h\Delta f_0 + \frac{h^3}{2!}f^{(2)}(\xi)\int_0^1 s(s-1)\,\mathrm{d}s$$

$$= hf_0 + \tfrac{1}{2}h\Delta f_0 + \frac{h^3}{2!}f^{(2)}(\xi)(\tfrac{1}{3} - \tfrac{1}{2})$$

$$= hf_0 + \tfrac{1}{2}h(f_1 - f_0) - \frac{h^3}{12}f^{(2)}(\xi)$$

$$= \tfrac{1}{2}h(f_0 + f_1) - \tfrac{1}{12}h^3 f^{(2)}(\xi) \qquad\qquad [18\text{-}4]$$

where $x_0 < \xi < x_0 + h$.

Hence the approximation is the trapezium rule $\tfrac{1}{2}h(f_0 + f_1)$ and the local truncation error is $-(h^3/12)f^{(2)}(\xi)$.

Applying this result to $\int_a^b f(x)\,\mathrm{d}x$ we get

$$\int_a^b f(x)\,\mathrm{d}x = \tfrac{1}{2}h(f_0 + 2f_1 + 2f_2 + \ldots + 2f_{n-2} + 2f_{n-1} + f_n)$$

$$- \frac{h^3}{12}(f^{(2)}(\xi_1) + f^{(2)}(\xi_2) + \ldots + f^{(2)}(\xi_{n-1}) + f^{(2)}(\xi_n))$$

where $x_0 + (k-1)h < \xi_k < x_0 + kh$, for each k.

As in Section 18.2 we can replace

$$f^{(2)}(\xi_1) + f^{(2)}(\xi_2) + \ldots + f^{(2)}(\xi_n)$$

by $nf^{(2)}(\xi)$, where $x_0 < \xi < x_0 + nh$, i.e., $a < \xi < b$, and nh by $b - a$, so that

$$\int_a^b f(x)\,\mathrm{d}x = h(\tfrac{1}{2}f_0 + f_1 + \ldots + f_{n-1} + \tfrac{1}{2}f_n) - \tfrac{1}{12}(b-a)h^2 f^{(2)}(\xi).$$

$$[18\text{-}5]$$

Now we can see that the global truncation error is $O(h^2)$ for the composite trapezium rule. Hence if h is halved and the number of subintervals doubled then the global error will be divided by 4 (assuming that $f^{(2)}(\xi)$ does not change much!). We can now see that the trapezium rule really is an improvement on the rectangular rule. Perhaps we can show that Simpson's rule must be better than the trapezium rule. But first let's illustrate Equation [18-5].

EXAMPLE 18.2   Also found in Example 17.5 were these trapezium rule results.

h	Estimate	Error (0.262 − estimate)
0.4	0.318	− 0.056
0.2	0.276	− 0.014
0.1	0.266	− 0.004

Again the approximate division by four occasioned by the halving of h is clearly seen. (Do you agree?)

18.4   SIMPSON'S RULE

If we now take $m = 2$,

$$f(x_0 + sh) = f_0 + s\Delta f_0 + \frac{s(s-1)}{2!}\Delta^2 f_0 + \frac{s(s-1)(s-2)}{3!}h^3 f^{(3)}(\xi_s)$$

where $x_0 < \xi_s < x_0 + 2h$ and $0 \leqslant s \leqslant 2$. Following the previous line of attack we get

$$\int_{x_0}^{x_0 + 2h} f(x)\, dx = \int_0^2 \left(f_0 + s\Delta f_0 + \frac{s(s-1)}{2!}\Delta^2 f_0 + \frac{s(s-1)(s-2)}{3!}h^3 f^{(3)}(\xi_s) \right) h\, ds$$

$$= 2hf_0 + \tfrac{4}{2}h\Delta f_0 + \frac{h}{2!}\left(\frac{2^3}{3} - \frac{2^2}{2}\right)\Delta^2 f_0 + \frac{h^4}{3!}\int_0^2 s(s-1)(s-2)f^{(3)}(\xi_s)\, ds.$$

Now unfortunately $s(s-1)(s-2)$ does change sign in $[0, 2]$ so that the theorem of Appendix 4 cannot be applied, and anyway

$$\int_0^2 s(s-1)(s-2)\, ds = 0 .$$

In fact (and we shall not prove this result!) it can be shown that the local error term involves not the third but the fourth derivative of f and can be written in the form

$$h^5 f^{(4)}(\xi) \int_0^2 \frac{s(s-1)(s-2)(s-3)}{4!}\, ds$$

where $x_0 < \xi < x_0 + 2h$. You should show that this simplifies to $-h^5 f^{(4)}(\xi)/90$.

The approximation rule turns out to be (surprise!) $\tfrac{1}{3}h(f_0 + 4f_1 + f_2)$ which, again, you can check for yourself! (Remember that working through these details makes sure that you really get to grips with the material — and it will help you to remember it too!) Hence we have the local result

$$\int_{x_0}^{x_0+2h} f(x)\,dx = \frac{h}{3}(f_0 + 4f_1 + f_2) - \frac{1}{90}h^5 f^{(4)}(\xi).$$ [18-6]

Applying this to $\int_a^b f(x)\,dx$, with $[a, b]$ divided into $2n$ subintervals of length h, gives

$$\int_a^b f(x)\,dx = \frac{h}{3}(f_0 + 4f_1 + 2f_2 + 4f_3 + \ldots + 4f_{2n-1} + f_{2n})$$

$$-\frac{h^5}{90}(f^{(4)}(\xi_1) + f^{(4)}(\xi_2) + \ldots + f^{(4)}(\xi_n))$$

where

$$x_0 < \xi_1 < x_0 + 2h, \; x_0 + 2h < \xi_2 < x_0 + 4h, \ldots, x_0 + 2(n-1)h < \xi_n < x_0 + 2nh.$$

As before the sum of these fourth derivatives of f can be replaced by $nf^{(4)}(\xi)$ where $x_0 < \xi < x_0 + 2nh$, i.e., $a < \xi < b$.

The global truncation error therefore simplifies to $-(h^5/90)nf^{(4)}(\xi)$ and hence to $-\frac{1}{180}(b-a)h^4 f^{(4)}(\xi)$. (Why?) The result is

$$\int_a^b f(x)\,dx = \frac{h}{3}(f_0 + 4f_1 + 2f_2 + \ldots + 2f_{2n-2} + 4f_{2n-1} + f_{2n})$$

$$-\frac{1}{180}(b-a)h^4 f^{(4)}(\xi).$$ [18-7]

So the global truncation error for Simpson's rule is $O(h^4)$. Again we don't know the value of ξ and therefore of $f^{(4)}(\xi)$. However, if we bravely assume that $f^{(4)}(x)$ does not vary much in (a, b), then the global error is roughly proportional to h^4, so that halving h (doubling the number of subintervals) will reduce the error by a factor of 16 — rather better than the trapezium rule.

EXAMPLE 18.3   Consider these Simpson's rule estimates for

$$\int_0^1 \frac{1}{1+x}\,dx = \ln 2 = 0.693\,147 \quad \text{to 6D.}$$

h	Estimate
1/2	0.694 444
1/4	0.693 254
1/8	0.693 155

Can we infer from these results that the answer is 0.6932 to 4D? Well, compared with $0.694\,444$ and $0.693\,254$ we would expect $0.693\,155$ to be quite close to the answer (because the errors decrease so quickly with Simpson's rule). So the errors in the first two estimates are, approximately,

$$h = 1/2: \quad 0.693\,155 - 0.694\,444 = -0.001\,289$$

$$h = 1/4: \quad 0.693\,155 - 0.693\,254 = -0.000\,099.$$

Now $-0.001\,289/16 \simeq -0.000\,081$, so it appears that $f^{(4)}(x)$ is not changing very much. We can therefore assume reasonably confidently that the error in $0.693\,155$ is approximately $-0.000\,099/16 \simeq -0.6 \times 10^{-5}$. This means that it is just possible for the answer to be less than $0.693\,15$ and hence to equal 0.6931 correct to 4D. This is an occasion to be flexible and quote an answer such as '0.6931 or 0.6932 to 4D' or perhaps '$0.693\,15 \pm 0.000\,05$' (playing safe) or '0.693\,15 or 0.693\,16 to 5D' (more daring).

In fact we know that the error in $0.693\,155$ is $0.693\,147 - 0.693\,155 = -0.8 \times 10^{-5}$, showing how close this kind of analysis can be (provided $f^{(4)}(\xi)$ doesn't vary too much!)

In this particular example we could, of course, use the fact that $f^{(4)}(x) = 24/(1+x)^5$ and hence that $\frac{24}{32} < f^4(\xi) < 24$ (since $\xi \in (0,1)$ and $f^{(4)}(x)$ has no turning points in $(0,1)$). This leads to

$$-\tfrac{1}{180}(b-a)h^4 f^{(4)}(\xi) = -\tfrac{24}{180}h^4 \quad \text{(at most)}$$
$$= -\tfrac{2}{15}h^4.$$

For $h = \tfrac{1}{8}$ we obtain for the maximum global error $-0.000\,033$, a much worse result than that obtained by interval-halving and assuming that $f^{(4)}(\xi)$ remains roughly constant. We must, however, accept that $f^4(\xi)$ may not always vary slowly, and recognise this when the reduction factor is not close to 16.

18.5 SUMMARY OF TRUNCATION ERRORS

We have seen in this lesson how approximate integration rules and their truncation error formulae can be found from interpolating polynomials and their errors.

It is important to realise that, apart from investigations like that in Example 18.3, the error expressions are of little practical interest. They are of considerable theoretical importance however, establishing a firm logical basis for numerical integration and, as we shall see in Lesson 19, allowing us to develop even better methods.

You should remember the main results of this lesson as the orders of the truncation errors:

Rectangular rule	$O(h)$
Trapezium rule	$O(h^2)$
Simpson's rule	$O(h^4)$

You are probably asking — what about the mid-point rule?

We leave as an optional exercise the derivation of the results

$$\int_{x_0-h}^{x_0+h} f(x)\,dx \; = \; 2hf_0 + \tfrac{1}{3}h^3 f^{(2)}(\xi), \quad x_0-h<\xi<x_0+h \qquad \text{[18-8]}$$

and

$$\int_a^b f(x)\,dx \; = \; h(f_{1/2}+f_{3/2}+\ldots+f_{n-1/2}) + \tfrac{1}{3}(b-a)h^2 f^{(2)}(\xi), \quad a<\xi<b.$$
$$\text{[18-9]}$$

(See Exercise 18.2 below.) So the order of the global truncation error is

  Mid-point rule $O(h^2)$

the same as for the trapezium rule.

EXERCISE 18.1

Estimate $\int_0^1 1/(1+x)\,dx$ by the mid-point and trapezium rules, using $h = 1, \tfrac{1}{2}, \tfrac{1}{4}, \tfrac{1}{8}$ in turn and working with 6D.

Assuming that $f^{(2)}(x)$ is approximately constant in $(0,1)$ estimate the errors in the $h=\tfrac{1}{8}$ results and hence give the best value for the integral obtainable from your analysis.

*EXERCISE 18.2

Derive Equations [18-8] and [18-9] by first evaluating the integral on the right of

$$\int_{x_0-h}^{x_0+h} f(x)\,dx \; = \; \int_{-1}^{1} \left(f_0 + s\Delta f_0 + \frac{s(s-1)}{2!}h^2 f^{(2)}(\xi_s)\right)h\,ds$$

Assume that

$$\int_{-1}^{1} s(s-1)f^{(2)}(\xi_s)\,ds \; = \; f^{(2)}(\xi)\int_{-1}^{1} s(s-1)\,ds$$

for some $\xi \in (x_0-h, x_0+h)$. (Notice that the approximation used here is not strictly as described in Lesson 17. We are not using the constant f_0 at the mid-point of $[x_0-h, x_0+h]$ to approximate f but rather the linear function $f_0 + s\Delta f_0$. Why is the end-result the same?)

8.6 ERROR ANALYSIS FOR INTEGRATION FROM DATA

If the integrand is given by a table of values then we have to consider the effect of rounding errors in the data as well as the truncation error.

18.6.1 ROUNDING ERRORS

Suppose that each function value used in an integration rule is rounded correctly to pD, i.e., has an error bound $\frac{1}{2} \times 10^{-p}$. Using the rule for the error bound of a sum of values, we simply add the error bounds for each function value appearing in the integration rule.

EXAMPLE 18.4 For the rectangular rule the accumulated rounding error in

$$h(f_0 + f_1 + \ldots + f_{n-1})$$

is at most

$$h(n \times \tfrac{1}{2} \times 10^{-p}) \quad \text{i.e.} \quad (b-a)(\tfrac{1}{2} \times 10^{-p}).$$

EXAMPLE 18.5 For the trapezium rule the maximum accumulated error in

$$h(\tfrac{1}{2}f_0 + f_1 + f_2 + \ldots + f_{n-1} + \tfrac{1}{2}f_n)$$

is

$$h(2 \times \tfrac{1}{2} + n - 1)\tfrac{1}{2} \times 10^{-p} = (b-a)(\tfrac{1}{2} \times 10^{-p})$$

as before.

EXERCISE 18.3

Show that the maximum accumulated error for the mid-point and Simpson's rules is also $(b-a)(\tfrac{1}{2} \times 10^{-p})$.

The conclusions to draw from these interesting results are that, provided $b-a$ is not large, the accuracy of the estimated integral can be of the same order as that of the data (the function values). We are assuming of course that the calculations themselves do not contribute any significant round-off errors (i.e., that extra decimal places are used if necessary). This result is only of interest if the subinterval length is sufficiently small for the truncation error to be less than the accumulated round-off error; otherwise the truncation error will dominate. On the other hand there is no point in making h smaller once the truncation error has fallen to the level of the round-off error.

18.6.2 TRUNCATION ERRORS

We try to avoid estimating values of truncation errors, because we have so much difficulty with the $f^{(m)}(\xi)$ part of the expression. When f is given by values only the problem appears to be even more difficult. It can be overcome to some extent because we can show that, if $x_0 < x_1 < \ldots < x_m$, then there is a point $\zeta \in (x_0, x_m)$ such that the divided difference

$$f[x_0, x_1, \ldots, x_m] = \frac{f^{(m)}(\zeta)}{m!}.$$

So if we again assume that $f^{(m)}$ does not vary radically, we can assert, for example, that

$$-\tfrac{1}{12}(b-a)h^2 f^{(2)}(\xi) \simeq -\tfrac{1}{12}(b-a)h^2 2! \, f[x_0, x_1, x_2]$$

where $x_0 < x_1 < x_2$ are three 'suitable' points in $[a, b]$. Not only is this the only way to estimate truncation errors in the data case, it is probably the best approach even when f is given explicitly. However, we shall not pursue the idea in this course.

ANSWERS TO EXERCISES

18.1 *Mid-point Rule*

$$h = 1; \quad \int_0^1 \frac{1}{1+x} \, dx \simeq 1 \left(\frac{1}{1 + \frac{1}{2}} \right) = \tfrac{2}{3} = 0.666\,667$$

$$h = \tfrac{1}{2}; \quad \int_0^1 \frac{1}{1+x} \, dx \simeq \frac{1}{2} \left(\frac{1}{1 + \frac{1}{4}} + \frac{1}{1 + \frac{3}{4}} \right) = 0.685\,714$$

$$h = \tfrac{1}{4}; \quad \int_0^1 \frac{1}{1+x} \, dx \simeq \frac{1}{4} \left(\frac{1}{1 + \frac{1}{8}} + \frac{1}{1 + \frac{3}{8}} + \frac{1}{1 + \frac{5}{8}} + \frac{1}{1 + \frac{7}{8}} \right)$$

$$= 0.691\,220$$

$$h = \tfrac{1}{8}; \quad \int_0^1 \frac{1}{1+x} \, dx \simeq \frac{1}{8} \left(\frac{1}{1 + \frac{1}{16}} + \frac{1}{1 + \frac{3}{16}} + \frac{1}{1 + \frac{5}{16}} + \frac{1}{1 + \frac{7}{16}} \right.$$

$$\left. + \frac{1}{1 + \frac{9}{16}} + \frac{1}{1 + \frac{11}{16}} + \frac{1}{1 + \frac{13}{16}} + \frac{1}{1 + \frac{15}{16}} \right) = 0.692\,661.$$

h	Estimate
1	0.666 667
$\tfrac{1}{2}$	0.685 714
$\tfrac{1}{4}$	0.691 220
$\tfrac{1}{8}$	0.692 661

Remember that the truncation error is $O(h^2)$ and take $0.692\,661$ as the 'true' answer. Then the errors are

$h = 1; \quad 0.692\,661 - 0.666\,667 = 0.025\,994$

$h = \tfrac{1}{2}; \quad 0.692\,661 - 0.685\,714 = 0.006\,947$

$h = \tfrac{1}{4}; \quad 0.692\,661 - 0.691\,220 = 0.001\,441$

Ratio $\simeq 3.7$

Ratio $\simeq 4.8.$

Thus we can see that halving the interval divides the error by 4 (approximately) which means that $f^{(2)}(\xi)$ is not changing much. Thus we can assume fairly safely that the error in $0.692\,661$ is $\frac{1}{4} \times 0.001\,442 \simeq 0.0004$.

Thus, almost certainly, we can say that $\int_0^1 1/(1+x)\,dx$ is 0.693 to 3D.

Trapezium Rule

$$h = 1; \quad \int_0^1 \frac{1}{1+x}\,dx \simeq 1\left(\frac{1}{2} \times 1 + \frac{1}{2} \times \frac{1}{1+1}\right) = \frac{3}{4} = 0.750\,000$$

$$h = \frac{1}{2}; \quad \int_0^1 \frac{1}{1+x}\,dx \simeq \frac{1}{2}\left(\frac{1}{2} \times \frac{1}{1} + \frac{1}{1+\frac{1}{2}} + \frac{1}{2} \times \frac{1}{1+1}\right)$$

$$= 0.708\,333$$

$$h = \frac{1}{4}; \quad \int_0^1 \frac{1}{1+x}\,dx \simeq \frac{1}{4}\left(\frac{1}{2} \times 1 + \frac{1}{1+\frac{1}{4}} + \frac{1}{1+\frac{1}{2}} + \frac{1}{1+\frac{3}{4}}\right.$$

$$\left. + \frac{1}{2} \times \frac{1}{1+1}\right) = 0.697\,024$$

$$h = \frac{1}{8}; \quad \int_0^1 \frac{1}{1+x}\,dx \simeq \frac{1}{8}\left(\frac{1}{2} \times 1 + \frac{1}{1+\frac{1}{8}} + \frac{1}{1+\frac{1}{4}} + \frac{1}{1+\frac{3}{8}}\right.$$

$$+ \frac{1}{1+\frac{1}{2}} + \frac{1}{1+\frac{5}{8}} + \frac{1}{1+\frac{3}{4}} + \frac{1}{1+\frac{7}{8}}$$

$$\left. + \frac{1}{2} \times \frac{1}{1+1}\right) = 0.694\,122 \; .$$

h	Estimate
1	0.750\,000
$\frac{1}{2}$	0.708\,333
$\frac{1}{4}$	0.697\,024
$\frac{1}{8}$	0.694\,122

Again the truncation error is $O(h^2)$. Take $0.694\,122$ as the 'true' answer. Then the errors are

$h = 1; \quad 0.694\,122 - 0.750\,000 \; = \; -0.055\,878$

$\qquad\qquad\qquad\qquad\qquad\qquad\qquad$ Ratio $\simeq 3.9$

$h = \frac{1}{2}; \quad 0.694\,122 - 0.708\,333 \; = \; -0.014\,211$

$\qquad\qquad\qquad\qquad\qquad\qquad\qquad$ Ratio $\simeq 4.8.$

$h = \frac{1}{4}; \quad 0.694\,122 - 0.697\,024 \; = \; -0.002\,902$

Here again $f^{(2)}(\xi)$ appears to vary little and so the error in $0.694\,122$ will be about $-0.002\,902/4 \simeq -0.0007$, implying that the integral is approximately 0.6934 and that we can almost certainly quote 0.693 as correct to 3D.

18.2   The end result is the same because the area under $f_0 + s\Delta f_0$ between $x_0 - h$ and $x_0 + h$ is the trapezoidal area

$$2h \left\{ \frac{(f_0 - \Delta f_0) + (f_0 + \Delta f_0)}{2} \right\} = 2hf_0.$$

FURTHER READING

Most books deal with the integration rules and their errors simultaneously. See, for example, Phillips and Taylor, Section 6.3 up to Example 6.4.

A nicely written approach is that of Stark, Chapter 6 up to Section 6.8 inclusive.

Stark's account differs from ours in two ways. Firstly he is primarily concerned with computing the value of a definite integral for an explicit function and pays a lot of attention to the effects of the finite arithmetic of the digital computer. Secondly he does his truncation error analysis using Taylor series.

Another good reference, at a similar level of difficulty to our own, is Dixon, Chapter 6 up to Section 6.3 inclusive. This gives a detailed description of the trapezium and Simpson rules and their errors, but without actually deriving the truncation-error formulae.

LESSON 18 – COMPREHENSION TEST

1. State the orders of the errors (in the form $O(h^p)$) for the composite rectangular, trapezium, mid-point and Simpson rules.

2. True or false?

(a) Truncation-error formulae are often used to calculate error bounds for approximate integrals.

(b) Truncation-error formulae tell us the relative accuracy of different integration rules.

3. Given these trapezium-rule results:

2 subintervals	1.2214
4 subintervals	1.2170
8 subintervals	1.2160
16 subintervals	1.2156

and

$$1.2156 - 1.2214 = -0.0058$$
$$1.2156 - 1.2170 = -0.0014$$
$$1.2156 - 1.2160 = -0.0004$$

explain how these numbers relate to the truncation-error formula for the trapezium rule, and estimate the error in the 1.2156.

4. Why is interval-halving efficient with the trapezium rule but inefficient with the mid-point rule?

ANSWERS

1   Rectangular $- O(h)$

     Trapezium $- O(h^2)$

     Mid-point $- O(h^2)$

     Simpson $- O(h^4)$.

2   (a) False.   (b) True, provided we assume that the derivatives of the integrand do not vary considerably.

3   -0.0058, -0.0014 and -0.0004 are approximate errors for the 2, 4 and 8-subinterval estimates of the integral. They reduce successively by a factor of 4 approximately, agreeing with the $O(h^2)$ error for the trapezium rule when the second-derivative part of the error is approximately constant over the range of integration. Hence the error in 1.2516 is approximately $-0.0004/4 = -0.0001$.

4   With the trapezium rule all the function values used in previous stages are reused as the number of subintervals is doubled. With the mid-point rule a completely new set of function values is required at each stage.

LESSON 18 – SUPPLEMENTARY EXERCISES

18.4. Consider the trapezium-rule results T_1, T_2, T_4 and T_8 of Exercise 17.15. Show that the ratios $T_8 - T_1 : T_8 - T_2$ and $T_8 - T_2 : T_8 - T_4$ are approximately 4:1. Hence deduce an approximate value for the error in T_8.

Now consider the Simpson's-rule estimates S_2, S_4 and S_8 of the same exercise. Consider the ratio $S_8 - S_2 : S_8 - S_4$, using it to deduce, as far as possible, the error in S_8. Compare the corrected values of T_8 and S_8.

18.5. Obtain T_n for $n = 1, 2, 4$ and 8, for the integral

$$\int_0^8 \frac{dx}{\sqrt{(1 + \sqrt{x})}} .$$

By considering the ratios $T_8 - T_n : T_8 - T_{2n}$, $n = 1, 2$, estimate, if possible, the error in T_8.

Repeat the calculations for the integral

$$\int_1^9 \frac{dx}{\sqrt{(1 + \sqrt{x})}} .$$

Why is this second problem more tractable?

*18.6. As further evidence of the difference between the two integrals in Exercise 18.5 show that

$$\max_{0 \leqslant x \leqslant 8} |f^{(2)}(x)| \simeq 0.44$$

$$\max_{1 \leqslant x \leqslant 9} |f^{(2)}(x)| \simeq 0.035$$

where $f(x) = 1/\sqrt{(1 + \sqrt{x})}$, so that truncation error bounds for the trapezium rule are $8(0.44h^2)/12$ and $8(0.035h^2)/12$ respectively.

Compare these bounds with the actual errors. [*Hint*: the integrals can be evaluated exactly using the substitution $x = \tan^4 \theta$.]

*18.7. Form a difference table for the function values in Exercise 17.15. Hence obtain the approximate error bounds $\frac{1}{12}(3 - 1) \max \Delta^2 f$ and $\frac{1}{180}(3 - 1) \max \Delta^4 f$ for the T_8 and S_8 estimates of the integral. Compare these with the approximate errors obtained in Exercise 18.4. (You should find that the error bounds are about three times larger.)

Lesson 19 **Romberg Integration**

OBJECTIVES

At the end of this lesson you should

(a) understand the need for interval-halving in integration;

(b) be able to interpret a sequence of approximations to the value of an integral;

(c) be able to derive and use Romberg's method.

CONTENTS

19.1   INTERVAL-HALVING IN INTEGRATION

We have seen in Lesson 18 how expressions can be found for the truncation errors associated with integration rules. You should have noticed that they all took the form

$$kh^p f^{(q)}(\xi)$$

where k is some constant and h is the strip-width. As with the errors in Taylor approximation and in interpolation we observed that estimating the error is not a practical proposition in general, because of the problem of estimating the $f^{(q)}(\xi)$ component. What we can see however is that, provided $f^{(q)}(\xi)$ is bounded (i.e., doesn't become infinite), then the error tends to zero as h does. In particular we saw that, if $f^{(q)}(\xi)$ does not vary quickly, then the error decreases by a factor of 2^p as h is successively halved and the number of strips doubled.

This will be the strategy of this lesson. We shall successively halve h and study the resulting sequence of approximations to the integral, looking for a specific accuracy in the answer. Later we shall see how to use the results of interval-halving with the trapezium rule to obtain even better estimates of the integral.

The reason for halving h, as opposed to dividing by 3 or 5 or whatever, is that, with the rectangular, trapezium and Simpson's rules at any rate, we can re-use function values previously calculated.

EXAMPLE 19.1   Suppose the trapezium rule estimate

$$T_2 = h(\tfrac{1}{2}f_0 + f_1 + \tfrac{1}{2}f_2)$$

has been made (Fig. 19.1).

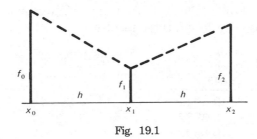

Fig. 19.1

If we halve h and renumber subscripts accordingly (Fig. 19.2) then

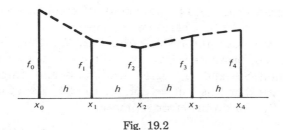

Fig. 19.2

$$T_4 = h(\tfrac{1}{2}f_0 + f_1 + f_2 + f_3 + \tfrac{1}{2}f_4)$$
$$= \tfrac{1}{2}T_2 + h(f_1 + f_3)$$

since

new f_0 = old f_0

new f_2 = old f_1

new f_4 = old f_2

new h = $\tfrac{1}{2} \times$ old h.

You might like to convince yourself that, in general, if T_n is the trapezium rule result for n subintervals of width $2h$, and T_{2n} likewise for $2n$ subintervals of width h, then

$$T_{2n} = \tfrac{1}{2}T_n + h(f_1 + f_3 + f_5 + \ldots + f_{2n-1}).$$

If we require a particular accuracy in the estimated integral then we need to know when to stop the interval-halving process. The technique described in Lesson 18, where the error of each result is estimated by assuming that the best result is exact, is useful but restrictive. It requires us to reassess the errors at each interval-halving. In general we are content to look for consistency in a sequence of estimates, and then make a final check on the answer we deduce.

EXAMPLE 19.2   For $\int_0^1 dx/(1+x)$ using   Simpson's   rule   and 4D working we have already found these results.

n	h	S_n
2	1/2	0.6944
4	1/4	0.6932
8	1/8	0.6932

where S_n means the Simpson's rule result for n subintervals. Since the domain of integration is $[0, 1]$ the round-off error in each S_n is at most $(1-0)\tfrac{1}{2} \times 10^{-4} = \tfrac{1}{2} \times 10^{-4}$. Therefore we can concentrate on the truncation error. If we want the result correct to 2D then S_2 and S_4 agree to 2D, and S_8 confirms the answer 0.69 correct to 2D. If we want 3D accuracy then S_4 and S_8 agree on 0.693 and imply by agreeing in the fourth place that this will be correct to 3D. As a final check we can either calculate S_{16} or do the error estimation, i.e.,

represent error in S_2 by $0.6932 - 0.6944 = -0.0012$

and error in S_4 by $0.6932 - 0.6932 = 0.$

Now we expect errors to diminish by a factor of 16 approximately, and $-0.0012/16 \simeq -0.0001$ which is near enough to zero in 4D working. Hence the error in 0.6932 is likely to be of the order of 0.0001/16 in magnitude. This makes 0.693 correct to 3D certain and 0.6932 correct to 4D very likely. Even so, calculating S_{16} is advisable since it is possible to construct examples where convergence seems to occur, but in fact the sequence is merely temporarily stationary. In this case S_{16} is in fact 0.6932 to 4D.

EXERCISE 19.1

Find T_n for $n = 2, 4, 8, 16$ and 32, for $\int_{0.6}^{1.4} \ln x \, dx$, working with 8D. Comment on the accuracy quotable at each stage of the interval-halving.

EXERCISE 19.2

Find T_3, T_6, T_{12} and T_{24} for $\int_0^{\pi/2} \cos x \, dx$, working with 6D. Again state the available accuracy at each stage (if any).

.2 ROMBERG'S METHOD

From the results of Exercise 19.1 and 19.2 it can easily be seen that, in spite of halving the strip-width (and doubling the number of subintervals), in neither case did the sequence of approximations converge at all quickly. Moreover a considerable amount of work has been done for these not too impressive results. In Exercise 19.1, 8 decimal places have been used and with $n = 32$ the best we can say in terms of decimal places is that $\int_{0.6}^{1.4} \ln x \, dx = -0.02$ to 2D and probably -0.022 to 3D. In Exercise 19.2, 6 decimal places have been used, and with $n = 24$ we can only say that $\int_0^{\pi/2} \cos x \, dx = 1.00$ to 2D and probably 1.000 to 3D. However, the trapezium-rule approximation to the integral has a certain property that can be used to improve these estimates of the integral, and with a minimal amount of extra work.

In fact it can be shown that, if $x_n = x_0 + nh$,

$$\int_{x_0}^{x_n} f(x) \, dx - T_n = E_2 h^2 + E_4 h^4 + E_6 h^6 + \ldots \qquad [19\text{-}1]$$

where the numbers $E_2, E_4, E_6, \ldots$ do *not* depend on h.

Here the error is, as we say, of order h^2, i.e. $O(h^2)$. The idea we shall consider is the obtaining of an expression which is an approximation to the integral in Equation [19-1] with an error of $O(h^4)$, by eliminating the $E_2 h^2$ term.

Let I stand for the integral in Equation [19-1]; then

$$I - T_n = E_2 h^2 + E_4 h^4 + E_6 h^6 + \ldots \qquad [19\text{-}2]$$

and also

$$I - T_{2n} = E_2 (\tfrac{1}{2}h)^2 + E_4 (\tfrac{1}{2}h)^4 + E_6 (\tfrac{1}{2}h)^6 + \ldots \qquad [19\text{-}3]$$

when the number of subintervals is doubled (and hence the strip-width halved). Eliminating E_2 by subtracting Equation [19-2] from four times Equation [19-3], we obtain

$$3I - 4T_{2n} + T_n = -\tfrac{3}{4}E_4h^4 - \tfrac{15}{16}E_6h^6 - \ldots$$

i.e.

$$I - \frac{4T_{2n} - T_n}{3} = -\tfrac{1}{4}E_4h^4 - \tfrac{5}{16}E_6h^6 - \ldots \qquad [19\text{-}4]$$

Hence $(4T_{2n} - T_n)/3$ has the required properties — it differs from I by an error of $O(h^4)$. Moreover it requires only two trapezium-rule estimates to generate it.

Call this new approximation $T_n^{(1)}$, and let $-\tfrac{1}{4}E_4 = E_4^{(1)}$, $-\tfrac{5}{16}E_6 = E_6^{(1)}$, and so on. Note that these are not derivatives! The superscript indicates that we have completed one stage in reducing the order of the error. Thus Equation [19-4] is

$$I - T_n^{(1)} = E_4^{(1)}h^4 + E_6^{(1)}h^6 + E_8^{(1)}h^8 + \ldots \qquad [19\text{-}5]$$

Comparing this with Equation [19-2] suggests that we could now eliminate $E_4^{(1)}$ just as we eliminated E_2, obtaining an approximation to I with error of $O(h^6)$. Doubling the number of subintervals again, we have

$$I - T_{2n}^{(1)} = E_4^{(1)}(\tfrac{1}{2}h)^4 + E_6^{(1)}(\tfrac{1}{2}h)^6 + E_8^{(1)}(\tfrac{1}{2}h)^8 + \ldots \qquad [19\text{-}6]$$

where

$$T_{2n}^{(1)} = (4T_{4n} - T_{2n})/3.$$

From Equations [19-4] and [19-6], eliminating $E_4^{(1)}$ gives

$$15I - 16T_{2n}^{(1)} + T_n^{(1)} = -\tfrac{3}{4}E_6^{(1)}h^6 - \tfrac{15}{16}E_8^{(1)}h^8 - \ldots,$$

i.e.

$$I - \frac{16T_{2n}^{(1)} - T_n^{(1)}}{15} = -\tfrac{1}{20}E_6^{(1)}h^6 - \tfrac{1}{16}E_8^{(1)}h^8 - \ldots \qquad [19\text{-}7]$$

Hence

$$T_n^{(2)} = (16T_{2n}^{(1)} - T_n^{(1)})/15$$

is an approximation to I with error of $O(h^6)$, requiring only three trapezium-rule estimates (T_n, T_{2n}, T_{4n}). We can tabulate these results thus:

$$T_n$$
$$\qquad T_n^{(1)}$$
$$T_{2n} \qquad\qquad T_n^{(2)}$$
$$\qquad T_{2n}^{(1)}$$
$$T_{4n}$$

EXERCISE 19.3

Rewriting Equation [19-7] as

$$I - T_n^{(2)} = E_6^{(2)}h^6 + E_8^{(2)}h^8 + E_{10}^{(2)}h^{10} + \ldots$$

and eliminating $E_6^{(2)}$, derive the approximation

$$T_n^{(3)} = (64T_{2n}^{(2)} - T_n^{(2)})/63 \qquad\qquad\qquad\qquad [19\text{-}8]$$

to I, where

$$I - T_n^{(3)} = E_8^{(3)}h^8 + E_{10}^{(3)}h^{10} + \ldots \qquad\qquad\qquad [19\text{-}9]$$

and $E_8^{(3)}$ and $E_{10}^{(3)}$ are constant multiples of $E_8^{(2)}$ and $E_{10}^{(2)}$ respectively.

*EXERCISE 19.4

Defining $T_n^{(0)} \equiv T_n$ show by mathematical induction that

$$I - T_n^{(m)} = E_{2m+2}^{(m)}h^{2m+2} + E_{2m+4}^{(m)}h^{2m+4} + \ldots \quad m = 0, 1, 2, \ldots$$

where

$$T_n^{(m)} = (4^m T_{2n}^{(m-1)} - T_n^{(m-1)})/(4^m - 1) \qquad\qquad [19\text{-}10]$$

and $E_k^{(m)}$ is a constant multiple of $E_k^{(m-1)}$ for each k.

Applying Equation [19-8] we have this table of approximations to I:

$$
\begin{array}{cccc}
T_n & & & \\
& T_n^{(1)} & & \\
T_{2n} & & T_n^{(2)} & \\
& T_{2n}^{(1)} & & T_n^{(3)} \\
T_{4n} & & T_{2n}^{(2)} & \\
& T_{4n}^{(1)} & & \\
T_{8n} & & &
\end{array}
$$

That is, four trapezium-rule estimates give an approximation $T_n^{(3)}$ with error of $O(h^8)$. If we need to go further then Equation [19-10] can be applied with $m > 3$, but it is unusual to need to go further. Equation [19-10] can be used in a computer program to generate all the improved approximations, once $T_n, T_{2n}, T_{4n}, \ldots$ are available.

The general process of reduction of the error order illustrated above is called *extrapolation*. When applied to the trapezium rule it is called *Romberg's method*. Let us apply this method to the results of Exercises 19.1 and 19.2 to see if the error reduction is as good as the theory predicts.

EXAMPLE 19.3   From Exercise 19.2 we have (using the extrapolation technique)

$$T_3 = 0.977\,049$$
$$T_3^{(1)} = 1.000\,026$$
$$T_6 = 0.994\,282 \qquad\qquad T_3^{(2)} = 1.000\,000$$
$$T_6^{(1)} = 1.000\,002$$
$$T_{12} = 0.998\,572 \qquad\qquad T_6^{(2)} = 1.000\,000.$$
$$T_{12}^{(1)} = 1.000\,000$$
$$T_{24} = 0.999\,643$$

We have already observed that the T_n column tells us at best that $I = 1.00$ to 2D. As we look down that first column the values calculated do agree more and more nearly (but rather slowly!) From the second column we can clearly say that $I = 1.0000$ to 4D and comparing $T_6^{(1)}$ and $T_{12}^{(1)}$, $I = 1.000\,00$ to 5D! Of course the third column finally gives $I = 1.000\,000$ to 6D and there is no need to calculate $T_3^{(3)}$. (Why not?)

Note that we should look at each column to see that the numbers in that column agree more and more. Moreover, as we look at the results taken from the columns going from left to right, we should look for increasing agreement between answers. Warning — it is not sufficient to look at the bottom line of the calculations even though they are in theory the best. If the method works we should see a process of consistently increasing accuracy as we proceed through the table in the appropriate way. If this does not happen then it suggests that something unusual is going on which deserves closer attention.

In Example 19.3 we already had T_3 to T_{24}. If we start from scratch then the order of the calculation is quite important. To avoid unnecessary trapezium-rule calculations a sensible order of evaluation is T_n, T_{2n}, $T_n^{(1)}$, T_{4n}, $T_{2n}^{(1)}$, $T_n^{(2)}$, T_{8n}, $T_{4n}^{(1)}$, $T_{2n}^{(2)}$, $T_n^{(3)}$, etc.

EXERCISE 19.5

Write out the calculation table (as shown after Exercise 19.4) and indicate on it with arrows the order of the calculations, continuing the sequence beyond that given above.

EXERCISE 19.6

Use the Trapezium rule estimates of $\int_{0.6}^{1.4} \ln x \, dx$ that you found in Exercise 19.1 to form a Romberg extrapolation table. Give the most accurate value *that you can justify* for the value of the integral.

THE CONNECTION BETWEEN THE TRAPEZIUM AND SIMPSON RULES

You may be thinking that, if we can obtain such spectacular success with the trapezium rule, then doing the same with Simpson's rule might be even better. We already are! In fact $T_n^{(1)} = S_{2n}$, $T_{2n}^{(1)} = S_{4n}$, etc., so that if we delete the first column of a Romberg table we are left with an extrapolated Simpson's rule.

EXERCISE 19.7

Prove that $T_n^{(1)} = S_{2n}$.

ANSWERS TO EXERCISES

19.1

n	T_n
2	$-0.034\,870\,68$
4	$-0.025\,599\,74$
8	$-0.023\,235\,97$
16	$-0.022\,641\,83$
32	$-0.022\,493\,09$

Comparison of T_2 and T_4 tells us nothing.

Comparison of T_4 and T_8 tells us almost nothing as

$$T_4 = -0.026 \quad \text{to 3D} \quad (\text{or} -0.03 \text{ to 2D})$$

$$T_8 = -0.023 \quad \text{to 3D} \quad (\text{or} -0.02 \text{ to 2D}).$$

Comparison of T_8 and T_{16} suggests $I = -0.023$ to 3D.

Comparison of T_{16} and T_{32} and the progression of values gives $I \neq -0.023$ to 3D, but suggests $-0.0224 < I < -0.0225$.

This shows that we can be misled by simply comparing two successive estimates. We have also done a lot of work for very little reward.

19.2

n	T_n
3	$0.977\,049$
6	$0.994\,282$
12	$0.998\,572$
24	$0.999\,643$

T_3 and T_6 give 1.0 to 1D.

T_6 and T_{12} imply 1.00 to 2D even though $T_6 = 0.99$ to 2D.

T_{12} and T_{24} imply 1.000 to 3D even though $T_{12} = 0.999$ to 3D.

19.3 $\qquad I - T_{2n}^{(2)} = E_6^{(2)}(\tfrac{1}{2}h)^6 + E_8^{(2)}(\tfrac{1}{2}h)^8 + E_{10}^{(2)}(\tfrac{1}{2}h)^{10} + \dots$

$\therefore \quad 63I - 64T_{2n}^{(2)} + T_n^{(2)} = -\tfrac{3}{4}E_8^{(2)}h^8 - \tfrac{15}{16}E_{10}^{(2)}h^{10} - \dots$

i.e. $\quad I - \dfrac{64T_{2n}^{(2)} - T_n^{(2)}}{63} = -\tfrac{1}{84}E_8^{(2)}h^8 - \tfrac{5}{336}E_{10}^{(2)}h^{10} - \dots$

as required.

19.4 The inductive hypothesis is this statement $P(m)$:

$$I - T_n^{(m)} = E_{2m+2}^{(m)}h^{2m+2} + E_{2m+4}^{(m)}h^{2m+4} + \dots \qquad \text{[19-11]}$$

where $T_n^{(m)}$ is given by Equation [19-10] and $E_k^{(m)}$ is a constant multiple of $E_k^{(m-1)}$ for each k.

Now $P(1)$ has already been established, i.e.,

$$I - T_n^{(1)} = E_4^{(1)}h^4 + E_6^{(1)}h^6 + \dots$$

where $T_n^{(1)} = (4T_{2n}^{(0)} - T_n^{(0)})/3$ and $E_k^{(1)}$ is a constant multiple of $E_k^{(0)} \equiv E_k$ of Equation [19-2], for each k.

Assume $P(j)$ for some $j \in \mathbb{N}$. Doubling the number of subintervals in Equation [19-11] with $m = j$ gives

$$I - T_{2n}^{(j)} = E_{2j+2}^{(j)}(\tfrac{1}{2}h)^{2j+2} + E_{2j+4}^{(j)}(\tfrac{1}{2}h)^{2j+4} + \dots.$$

Eliminating $E_{2j+2}^{(j)}$ between this and Equation [19-11] with $m = j$ gives

$$(4^{j+1} - 1)I - 4^{j+1}T_{2n}^{(j)} + T_n^{(j)} = -\tfrac{3}{4}E_{2j+4}^{(j)}h^{2j+4} - \dots.$$

$$\therefore \quad I - \frac{4^{j+1}T_{2n}^{(j)} - T_n^{(j)}}{4^{j+1} - 1} = -\frac{3}{4}\left(\frac{1}{4^{j+1} - 1}\right)E_{2j+4}^{(j)}h^{2j+4} - \dots$$

i.e. $\quad I - T_n^{(j+1)} = E_{2j+4}^{(j+1)}h^{2j+4} + \dots$

where $E_{2j+4}^{(j+1)}$ is a constant multiple of $E_{2j+4}^{(j)}$ (and similarly higher coefficients). Hence $P(j) \Rightarrow P(j + 1)$.

We have shown that $P(1)$ is true, and that $P(j) \Rightarrow P(j + 1)(j \in \mathbb{N})$. Therefore, by the Principle of Mathematical Induction $P(m)$ is true for all $m \in \mathbb{N}$.

19.5 T_n

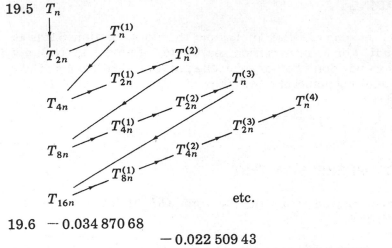

etc.

19.6 $-0.034\,870\,68$

$-0.022\,509\,43$

$-0.025\,599\,74$

$-0.022\,443\,96$

$-0.022\,448\,05$

$-0.023\,235\,97$

$-0.022\,443\,49$

$-0.022\,443\,50$

$-0.022\,443\,78$

$-0.022\,641\,83$

$-0.022\,443\,49$

$-0.022\,443\,49$

$-0.022\,443\,51$

$-0.022\,493\,09$

As already discussed, the first column is difficult to interpret. However, the second column gives $I = -0.022\,444$ to 6D. The third column gives $I = -0.022\,443\,5$ to 7D. The fourth column gives $I = -0.022\,443\,49$ to 8D.

The table shows a pattern of *consistently* improving results. However, we have to be a little careful with the Romberg method with respect to rounding errors since these can be magnified. Since we have worked to 8D we should not be greedy but accept

$$\int_{0.4}^{1.6} \ln x \; dx \;=\; -0.022\,443\,5 \quad \text{correct to } 7D\,.$$

19.7 Let h be the subinterval length associated with $2n$ subintervals then

$$3T_n^{(1)} \;=\; 4T_{2n} - T_n$$
$$= 4h(\tfrac{1}{2}f_0 + f_1 + f_2 + \ldots + f_{2n-1} + \tfrac{1}{2}f_{2n})$$
$$- (2h)(\tfrac{1}{2}f_0 + f_2 + \ldots + f_{2n-2} + \tfrac{1}{2}f_{2n})$$
$$= h(f_0 + 4f_1 + 2f_2 + 4f_3 + \ldots + 2f_{2n-2} + 4f_{2n-1} + f_{2n})$$

as required.

FURTHER READING

Following up the recommendation of Lesson 18, Stark, Section 6.9 is an interesting account. For a fuller version see Conte, Section 4.4, (N.B. *not* Conte and deBoor) but don't bother about the proof of Theorem 4.4. There are some very interesting points also in Phillips and Taylor, Section 6.4.

LESSON 19 – COMPREHENSION TEST

1. State the orders of the errors (in the form $O(h^P)$) for the Romberg extrapolates $T_n^{(1)}, T_n^{(2)}, T_n^{(3)}$.

2. State the formulae for $T_n^{(1)}$ and $T_{2n}^{(1)}$ in terms of T_n, T_{2n}, T_{4n}, and for $T_n^{(2)}$ in terms of $T_n^{(1)}$ and $T_{2n}^{(1)}$.

3. Apply Romberg's method to these trapezium-rule results:

$$T_2 = 1.2214 \qquad T_4 = 1.2170 \qquad T_8 = 1.2160.$$

4. Which of the following is identically equal to S_{2n}:

(a) T_{2n} (b) $T_n^{(1)}$ (c) $T_{2n}^{(1)}$ (d) $T_n^{(2)}$?

ANSWERS

1   $T_n^{(1)}$ is $O(h^4)$; $T_n^{(2)}$ is $O(h^6)$; $T_n^{(3)}$ is $O(h^8)$.

2   $T_n^{(1)} = (4T_{2n} - T_n)/3$
    $T_{2n}^{(1)} = (4T_{4n} - T_{2n})/3$
    $T_n^{(2)} = (16T_{2n}^{(1)} - T_n^{(1)})/15.$

3   1.2214
          1.2155
   1.2170   1.2157.
          1.2157
   1.2160

(Note the confirmation of the answer to Question 3 in the comprehension test of Lesson 18.)

4   (b).

LESSON 19 — SUPPLEMENTARY EXERCISES

19.8.   Apply Romberg's method to the values of T_2, T_4 and T_8 obtained in Exercise 17.14. Compare the extrapolates with the true value for the integral: $\frac{1}{4}\pi = 0.785\,398$ correct to 6D.

19.9.   Apply Romberg's method to the values of T_1, T_2, T_4 and T_8 obtained in Exercise 17.15. Decide on the best possible value that the method provides for the integral and compare it with the corrected value of S_8 found in Exercise 18.4.

19.10. Apply Romberg's method to the trapezium-rule estimates obtained in Exercise 18.5. You should again find that the second integral allows better results than the first. (See also Exercise 18.6 where the true values of the integrals are asked for.)

*19.11. Draw a flow chart for Romberg's method as applied to an explicit integrand; i.e., assume that $f(x)$ in $\int_a^b f(x)\,dx$ can be evaluated for any $x \in [a, b]$. The process should include the generation of the necessary trapezium-rule estimates (note that the efficient way to do this is given at the end of Example 19.1) and should produce these and the extrapolates in the order given by the answer to Exercise 19.5. The input data should be a, b, n (the initial number of subintervals) and m (the number of extrapolations). Assume that $f(x)$ is to be defined in a subroutine.

19.12. Use Romberg's method to evaluate the integral in Example 1.1 of Lesson 1, correct to the nearest minute.

(The integral gives the rise time in minutes.)

PART F   **NUMERICAL SOLUTION OF ORDINARY DIFFERENTIAL EQUATIONS**

Lesson 20 Taylor-Polynomial Methods I

OBJECTIVES

At the end of this lesson you should

(a) be able to recognise an ordinary differential equation (ODE);

(b) know what the order of an ODE is;

(c) be able to distinguish between linear and non-linear ODE's;

(d) understand the concept of a first-order initial-value problem;

(e) realise that a solution may not exist for an ODE;

(f) understand the idea of global Taylor-polynomial solutions to first-order initial-value problems and be able to calculate them in simple cases, together with estimates of their errors.

CONTENTS

20.1 ORDINARY DIFFERENTIAL EQUATIONS

Differential equations arise in physics, economics, business studies, chemistry, engineering, biology and many other subjects. We shall confine our attention to what are called *ordinary* differential equations (ODE's for short) as compared with what are called *partial* differential equations.

An ODE describes mathematically the way in which one variable (called the dependent variable) is related to another (called the independent variable), where the relationship is a 'dynamic' one involving *rate of change*.

EXAMPLE 20.1   In a culture of bacteria the number of bacteria N (the dependent variable) is supposed to increase at a rate proportional to the number of bacteria present at a particular time, t (the independent variable). In symbols:

$$\frac{dN}{dt} = kN \qquad\qquad [20\text{-}1]$$

where k is a positive constant and does not depend on N or t.

Equation [20-1] is a *first-order* ODE of *first degree*. A general equation of that kind has the form

$$\frac{dy}{dx} = f(x, y) \qquad\qquad [20\text{-}2]$$

where x is the independent variable, y the dependent variable, and f is some function of the two variables x and y.

As implied by the above definition ODE's may have second and higher derivatives in them, and may be other than first-degree. Here are some more complicated examples.

EXAMPLE 20.2

$$\frac{d^2y}{dx^2} = -p^2y \quad (p \text{ constant})$$

$$2\frac{d^3y}{dx^3} + 3\frac{d^2y}{dx^2} + 4\frac{dy}{dx} - 10y = e^{-2x}\cos x$$

$$\left[1 + \left(\frac{dy}{dx}\right)^2\right]^{3/2} = 10\frac{d^2y}{dx^2}$$

In this course we shall be concerned only with ODE's of the type shown in Equation [20-2], although the methods we consider can often be extended to deal with equations like those in Example 20.2 and also with simultaneous ODE's in which more than one dependent variable (but again only one independent variable) appears.

The solution of an ODE consists of the values of the dependent variable at all relevant values of the independent variable. For example, in Equation [20-1], we might want the values of N for all $t > 0$. In other words, the solution is the function $N(t)$, $t > 0$. We shall see later that we may have to be content with the solution values at so-called discrete values of the independent variable, e.g., $N(1), N(2), N(3), \ldots$ at $t = 1, 2, 3, \ldots$.

20.2   THE ORDER OF AN ODE

The order of a derivative of y, say $d^n y / dx^n$, is defined to be n, the number of times that y has been differentiated. Similarly the *order* of an ODE is the largest of the orders of the derivatives of y contained in the ODE.

EXAMPLE 20.3

$$\frac{d^2 y}{dx^2} + 2\frac{dy}{dx} + 2y = x^3$$

is an ODE of the second order.

$$\frac{d^4 y}{dx^4} + EI\frac{d^2 y}{dx^2} = 0, \quad E \text{ and } I \text{ constant,}$$

is an ODE of the fourth order.

20.3   LINEAR AND NON-LINEAR ODE'S

A *linear* nth-order ODE is one which can be expressed in the form

$$a_n(x)\frac{d^n y}{dx^n} + a_{n-1}(x)\frac{d^{n-1} y}{dx^{n-1}} + \ldots + a_1(x)\frac{dy}{dx} + a_0(x)y = f(x) \qquad [20\text{-}3]$$

where $a_0(x), a_1(x), \ldots, a_n(x)$ and $f(x)$ are functions of x alone. Any other equation is *non-linear*. Since we are only considering first order equations in this course the most general linear ODE we shall see has the form

$$a_1(x)\frac{dy}{dx} + a_0(x)y = f(x) \qquad [20\text{-}4]$$

which we shall usually rewrite as

$$\frac{dy}{dx} = \left\{ f(x) - a_0(x)y \right\} / a_1(x)$$

to fit the form of Equation [20-2]. Note that this means that $a_1(x)$ must be non-zero for all relevant values of x, i.e., throughout the domain of the function $y(x)$. Otherwise any numerical method for solving the equation may be upset by having 'infinity' in the calculation.

EXAMPLE 20.4

$$x^2 \frac{d^2y}{dx^2} + x \frac{dy}{dx} + (\tfrac{1}{4} - x^2)y = 5, \quad x > 0$$

$$\frac{d^2y}{dx^2} + p^2y \qquad\qquad = 0, \quad p \text{ constant}$$

$$2\frac{d^3y}{dx^3} + 3\frac{d^2y}{dx^2} + 4\frac{dy}{dx} - 10y = e^{-2x} \cos x$$

are all of the form of Equation [20-3] and hence linear ODE's.

$$\frac{dy}{dx} \bigg/ (1 - y) = x^2$$

is also linear because it can be rearranged to give

$$\frac{dy}{dx} + x^2 y = x^2 .$$

On the other hand

$$\frac{dy}{dx} = \frac{1}{y^2},$$

i.e. $y^2 \dfrac{dy}{dx} = 1$

is non-linear, as is

$$\left\{ 1 + \left(\frac{dy}{dx}\right)^2 \right\}^{3/2} = 10 \frac{d^2y}{dx^2} .$$

20.4   FIRST-ORDER INITIAL-VALUE PROBLEMS

Let us consider the first-order ODE

$$\frac{dy}{dx} = 2x$$

with the obvious solution (by simple integration)

$$y = x^2 + c$$

where c is any real number.

The solution of this ODE must contain an (as yet) unknown constant c as we have performed one integration. In fact every member of the family of parabolas $y = x^2 + c$ satisfies the ODE and so is a solution. Three of the family of solutions are shown in Fig. 20.1.

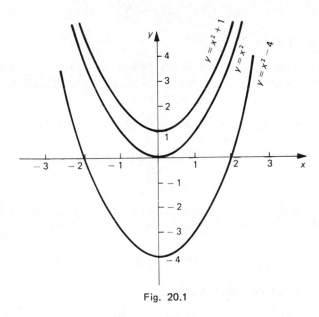

Fig. 20.1

If a solution of the ODE is to be defined uniquely we must have enough information to be able to pick out one member of the family, and this means that a further condition must be imposed. For example, if we impose the condition that $y = 0$ when $x = 2$, then $c = -4$ and the particular member of the above family that we have by implication chosen is

$$y = x^2 - 4$$

Note that the solution is now a single, particular function.

In general we can now guess that if the ODE

$$\frac{dy}{dx} = f(x, y)$$

is to have a unique solution function $y(x)$ then we must impose a condition on the value of y at some value of x. When this situation occurs we call the combination of the ODE and the condition an *initial-value* problem, viz.

$$\frac{dy}{dx} = f(x, y) \quad \text{and} \quad y = y_0 \quad \text{when} \quad x = x_0. \qquad [20\text{-}5]$$

Here y_0 and x_0 are understood to be given, fixed numbers. The condition '$y = y_0$ when $x = x_0$' is often called an *initial condition*.

EXAMPLE 20.5   If we know the number N_0 of bacteria (in the culture of Example 20.1) at time $t = 0$ we have the first-order initial-value problem

$$\frac{dN}{dt} = kN \quad \text{for} \quad t > 0 \quad \text{and} \quad N = N_0 \quad \text{when} \quad t = 0$$

with particular solution $N = N_0 e^{kt}$.

A general initial-value problem is so called because information is given at one (initial) point $x = x_0$ which, together with the ODE, is sufficient to define the solution y as a single function.

In contrast 'boundary-value problems' have the appropriate conditions given at more than one point.

EXAMPLE 20.6

$$\frac{d^2y}{dx^2} - 2\frac{dy}{dx} + 5y = 0, \quad x \geqslant 0$$

$$y = 1 \quad \text{and} \quad \frac{dy}{dx} = 0 \quad \text{when} \quad x = 0$$

is a second-order initial-value problem.

EXAMPLE 20.7

$$\frac{d^2y}{dx^2} - 2\frac{dy}{dx} + 5y = 0, \quad x \in [0, 1]$$

$$y = 1 \quad \text{at} \quad x = 0 \quad \text{and} \quad y = 2 \quad \text{at} \quad x = 1$$

is a second-order boundary-value problem.

In the following work we shall confine our attention to first-order initial-value problems.

20.5   THE EXISTENCE OF SOLUTIONS TO ODE'S

It may be thought that ODE's can always be solved by suitable, entirely mathematical methods, especially after one has studied an introductory, pure mathematics course in ODE's. This is not so; indeed the large majority of ODE's cannot be solved in this way, and in such cases we must use a numerical technique. Even when an analytical solution (i.e. an explicit function) is obtainable it may be so complicated that in practice we still prefer to use a numerical method. For example, the ODE

$$\frac{dy}{dx} = 2y/(1 - x^4)$$

has the family of solutions

$$y = c\left(\frac{1 + x}{1 - x}\right)^{1/2} e^{\arctan x} \quad c \in \mathbb{R}$$

The evaluation of such a function by computer, involving as it does three standard function routines (square root, inverse tangent and exponential functions), would probably take longer than a suitable numerical approach.

Not only may a formula-type solution not exist, a solution may not exist at all! — at least at some points of the domain of interest. For example,

$$\frac{dy}{dx} = y^2 \quad x \in [0, 2]$$

and

$$y = 1 \quad \text{when} \quad x = 0$$

has the apparent theoretical solution $y = 1/(1 - x)$ (check this!) which is not defined at $x = 1$. If, in all innocence, we attempt a numerical solution of such a problem we will either obtain a very poor answer or the method will fail in some way due to the 'singularity' at $x = 1$.

20.6 APPROXIMATE SOLUTIONS USING TAYLOR POLYNOMIALS

Consider the first-order initial-value problem

$$\frac{dy}{dx} = f(x, y), \quad y = y_0 \quad \text{when} \quad x = x_0$$

One approach to finding an approximate solution $y(x)$ is to use a Taylor polynomial for y, constructed at $x = x_0$. To simplify the solution it is usual to replace $x - x_0$ by h (i.e., write $x - x_0 = h$) as in Lessons 11 and 12.

The Taylor polynomial solution of degree n constructed at $x = x_0$ is obtained from

$$y(x_0 + h) = y(x_0) + hy^{(1)}(x_0) + \frac{h^2}{2!} y^{(2)}(x_0) + \ldots + \frac{h^n}{n!} y^{(n)}(x_0)$$

$$+ \frac{h^{n+1}}{(n+1)!} y^{(n+1)}(x_0 + \theta h), \quad 0 < \theta < 1. \qquad \text{[20-6]}$$

by ignoring the remainder term (i.e., the last term in Equation [20-6]). (See Lessons 11 and 12 for the details of Taylor polynomials and their remainder terms.)

Writing $y_0^{(p)}$ for $y^{(p)}(x_0)$, $p = 0, 1, 2, \ldots$ we get the commonly used form

$$y(x_0 + h) = y_0 + hy_0^{(1)} + \frac{h^2}{2!} y_0^{(2)} + \ldots + \frac{h^n}{n!} y_0^{(n)}$$

$$+ \frac{h^{n+1}}{(n+1)!} y^{(n+1)}(x_0 + \theta h), \quad 0 < \theta < 1. \qquad \text{[20-7]}$$

When the remainder term is ignored we get an approximation to $y(x_0 + h)$.

We shall again use the notation $x_r = x_0 + rh$ so that $x_1 = x_0 + h$, and *shall denote by y_r a numerical approximation to $y(x_r)$*. So we have

$$y_1 = y_0 + hy_0^{(1)} + \frac{h^2}{2!} y_0^{(2)} + \ldots + \frac{h^n}{n!} y_0^{(n)} \qquad \text{[20-8]}$$

as the approximation and

$$y(x_1) - y_1 = \frac{h^{n+1}}{(n+1)!} y^{(n+1)}(x_\theta) \qquad \text{[20-9]}$$

as its error, where $x_\theta = x_0 + \theta h$ and $0 < \theta < 1$. (Remember that the error is the true value minus the approximate value.)

So far we have only written down the mathematical expressions for the approximation y_1 to $y(x_1)$, together with the error, $y(x_1) - y_1$. Now we must ask how we find $y_0^{(1)}, y_0^{(2)}$, etc. from the initial condition $y = y_0$ at $x = x_0$. Well $y_0^{(1)} = f(x_0, y_0)$ can be calculated directly (using the equation $dy/dx = f(x, y)$). The higher derivative values $y_0^{(2)}, y_0^{(3)}, \ldots$ can be found by differentiating $dy/dx = f(x, y)$ *implicitly* and then substituting $x = x_0$, $y = y_0$, $dy/dx = y_0^{(1)}$, etc.

When we have done all this we have found a mathematical expression for y_1 *in terms of h* to approximate $y(x_0 + h)$ *for any value of h* (or if you like any value of $x = x_0 + h$). In other words we have found an approximation for $y(x)$ for *any* x (or if you like h) and not just a fixed x (or h). This is sometimes called a *global* solution to contrast it with the so-called *step-by-step* solutions considered in the next lesson.

EXAMPLE 20.8   Given   that $y^{(1)} = y - 2x/y$ and $y = 2$ when $x = 0$, we shall find a cubic Taylor-approximation for y at $x = h$. (Note that $x_0 = 0$ and $y_0 = 2$ in this case.)

Since $y^{(1)} = y - 2x/y$,

$$y_0^{(1)} = y_0 - 2x_0/y_0$$
$$= 2 - 2(0)/2 = \underline{2}.$$

Differentiating gives

$$y^{(2)} = y^{(1)} - 2/y + 2xy^{(1)}/y^2$$
$$\therefore \quad y_0^{(2)} = 2 - 2/2 + 0 = \underline{1}.$$

Similarly,

$$y^{(3)} = y^{(2)} + 4y^{(1)}/y^2 + 2xy^{(2)}/y^2 - 4x(y^{(1)})^2/y^3$$
$$\therefore \quad y_0^{(3)} = 1 + 4(2)/2^2 + 0 - 0 = \underline{3}.$$
$$\therefore \quad y_1 = 2 + 2h + 1h^2/2! + 3h^3/3!$$
$$= 2 + 2h + \tfrac{1}{2}h^2 + \tfrac{1}{2}h^3.$$

To take a particular case let $h = 0.5$, then

$$y(0.5) \simeq 2 + 2(0.5) + \tfrac{1}{2}(0.5)^2 + \tfrac{1}{2}(0.5)^3 = 3.1875.$$

This example shows that, even for relatively low-degree polynomials, the differentiations required may involve more work than we would like.

EXERCISE 20.1

Find the quadratic Taylor-approximation to $y(h)$ when $y^{(1)} = 1/(1 + y)$ and $y = 1$ when $x = 0$. Evaluate the approximation to $y(0.2)$.

EXERCISE 20.2

If $y^{(1)} = 1/(1 + y^2)$ and $y = y_n$ when $x = x_n$ show that the cubic Taylor-polynomial approximation to $y(x_n + h)$ constructed at $x = x_n$ is given by

$$y_{n+1} = y_n + \frac{h}{1 + y_n^2} + \frac{h^2}{2!} \frac{(-2y_n)}{(1 + y_n^2)^3} + \frac{h^3}{3!} \frac{2(5y_n^2 - 1)}{(1 + y_n^2)^5}.$$

Evaluate y_1 to 4D given that $y_0 = 1$, $x_0 = 0$ and $h = 0.2$; then evaluate y_2 to 4D using the y_1 obtained, $x_1 = 0.2$ and $h = 0.2$. What is y_2 approximating?

THE RELATIONS BETWEEN THE DEGREE, THE INTERVAL LENGTH AND THE ACCURACY

Following Lesson 12, if it can be shown that $y^{(n+1)}(x_\theta)$ is bounded, i.e., $|y^{(n+1)}(x_\theta)| < M$ (some fixed positive number) for all θ $(0 < \theta < 1)$, then we are able to consider these three problems.

(a) Given the degree n and the interval length h find an upper bound for the error (remainder term).

(b) Given the interval length h and a maximum acceptable value for the error (presumably a 'fairly small' number to give a 'good' result, e.g., $\frac{1}{2} \times 10^{-4}$ for 4D accuracy), find the degree of polynomial necessary to achieve this accuracy over the interval $[x_0, x_0 + h]$ or $[x_0 + h, x_0]$, depending on the sign of h.

(c) Given the degree n and a maximum acceptable value for the error find the interval containing x_0 over which the problem is solved, to the accuracy required, by the nth-degree Taylor-polynomial for y constructed at $x = x_0$.

No doubt all this looks very complicated. The best thing to do is to have a look at a very simple example.

EXAMPLE 20.9   Consider the first-order initial-value problem

$$\frac{dy}{dx} = 4x^3 + 3x^2 + 2x + 1, \quad y = 1 \quad \text{when} \quad x = 0.$$

(The true solution is, of course, $y = x^4 + x^3 + x^2 + x + 1$.)

Then

$$y_0 = 1 \quad \text{(given)}$$

and $y_0^{(1)} = 1$ (by direct calculation).

Differentiating, $y^{(2)}(x) = 12x^2 + 6x + 2$ so $y_0^{(2)} = 2$.

Similarly,

$$y^{(3)}(x) = 24x + 6 \qquad y_0^{(3)} = 6$$

$$y^{(4)}(x) = 24 \qquad y_0^{(4)} = 24$$

$$y^{(n)}(x) = 0, \quad n > 4.$$

Case (a)

Let $n = 2$, $h = 0.2$, say. Then from

$$y(x_0 + h) = y(h) = 1 + h + \frac{2h^2}{2!} + \frac{h^3}{3!}(24\xi + 6), \quad 0 < \xi < h$$

we get $y(0.2) \simeq y_1 = 1 + 0.2 + 0.04 = 1.24$.

Since $24\xi + 6 < 24(0.2) + 6 = 10.8$ we have

$$0 \leqslant \text{remainder} < \frac{0.008}{6}(10.8) = 0.0144.$$

So a quadratic approximation for $y(0.2)$ could have an error as large as 0.0144. In fact $y(0.2) = 1.2496$ so that the error in $y_1 = 1.24$ is 0.0096 and the quadratic solution is nearly correct to 2D for $x = 0.2$ (and so for all $x \in [0, 0.2]$ in this case).

Case (b)

Let $h = 0.2$ and impose the condition $|R_n| \leqslant 10^{-4}$.

For $h = 0.2$,

$$R_n = \frac{0.2^{n+1}}{(n+1)!} y^{(n+1)}(\xi).$$

We have seen that R_n could be as much as 0.0144 for $n = 2$, so try $n = 3$.

$$R_3 = \frac{0.0016}{24}(24) = 0.0016$$

since $y^{(4)}(x) = 24$ for all x. So $n = 3$ is not good enough either, and only $n = 4$ will give the required accuracy (when the error will be precisely zero in this example).

Case (c)

Now let $n = 2$, and impose $|R_2| \leqslant 10^{-4}$ again; i.e.,

$$\left| \frac{h^3}{3!}(24\xi + 6) \right| \leqslant 10^{-4}.$$

But

$$\left| \frac{h^3}{3!}(24\xi + 6) \right| = |h^3(4\xi + 1)| < |h^3(4h + 1)|$$

since $0 < \xi < h$.

We know that we cannot achieve an error as small as 10^{-4} for $|h| = 0.2$ (Case (b) above) so $|4h + 1| < 1.8$ for certain. So we will certainly achieve our aim if we impose $1.8|h^3| \leqslant 10^{-4}$

i.e. $|h^3| \leqslant 0.5 \times 10^{-4}$ (approximately) $= 50 \times 10^{-6}$

i.e. $|h| \leqslant 3 \times 10^{-2}$ will certainly suffice.

So the quadratic Taylor solution will be correct to within 10^{-4} for $x \in [-0.03, 0.03]$, a rather small interval. We have, of course, been pessimistic at various points in this analysis so that the true interval for the given degree and accuracy will be a little larger, but not significantly so.

Example 20.9 was untypically simple. The example which follows shows what can happen in a slightly more realistic problem.

EXAMPLE 20.10   Consider the problem $y^{(1)} = x - y$ with $y = y_0$ when $x = 0$.

Substituting $x = 0$ gives $y_0^{(1)} = -y_0$. Also

$$y^{(2)} = 1 - y^{(1)}, \quad \text{so } y_0^{(2)} = 1 + y_0$$

$$y^{(3)} = -y^{(2)}, \quad \text{so } y_0^{(3)} = -(1 + y_0)$$

$$y^{(4)} = -y^{(3)}, \quad \text{so } y_0^{(4)} = 1 + y_0$$

and in general

$$y^{(n)} = -y^{(n-1)}, \quad \text{so } y_0^{(n)} = (-1)^n (1 + y_0), \quad n \geqslant 2.$$

So $y(h) \simeq y_1 = y_0 + h y_0^{(1)} + \dfrac{h^2}{2!} y_0^{(2)} + \ldots + \dfrac{h^p}{p!} y_0^{(p)}$

$$= y_0 - h y_0 + (1 + y_0) \left(\frac{h^2}{2!} - \frac{h^3}{3!} + \ldots + \frac{(-1)^p h^p}{p!} \right)$$

with an error

$$R_p(h) = \frac{(-1)^{p+1} h^{p+1}}{(p+1)!} (1 - \theta h + y_\theta)$$

where $y_\theta = y(\theta h)$ and $0 < \theta < 1$, since $y^{(p+1)}(x) = (-1)^{p+1} (1 - x + y)$.

Case (a)

Let $n = 3$, $h = 0.2$, say, then $|\text{remainder term}| = \dfrac{(0.2)^4}{4!} |1 - 0.2\theta + y_\theta|$ where $y_\theta = y(0.2\theta)$, $0 < \theta < 1$.

Suppose $|y_\theta - 0.2\theta| < 1$ (an assumption — see a later comment) then

$$|R_3| < \frac{10^{-4}}{24} \times 2^5 = \frac{4}{3} \times 10^{-4}$$

and a cubic gives 'good' accuracy over $[0, 0.2]$.

Case (b)

Let $h = 0.4$ and require $|R_n| \leqslant 10^{-4}$, say. On looking at Case (a) we can see that n must be more than 3. Try $n = 4$, supposing $|y_\theta - 0.2\theta| < 1$ again.

$$|R_4| < \frac{(0.4)^5}{5!} 2 = \frac{4^4 \times 10^{-5}}{15}$$

$$\simeq 1.7 \times 10^{-4} - \text{not small enough!}$$

Try $n = 5$.

$$|R_5| < \frac{(0.4)^6}{6!} 2 = \frac{4^5 \times 10^{-6}}{90}$$

$$\simeq 1.1 \times 10^{-5}.$$

So we must use an approximation of degree at least 5 in this case.

Case (c)

Let $n = 2$ and impose $|R_2| \leqslant 10^{-6}$, say, (quite a fierce condition) and suppose $|y_\theta - \theta h| < 1$ again. Then

$$\left| \frac{h^3}{3!} (1 - \theta h + y_\theta) \right| \leqslant \frac{2h^3}{3!} = \frac{h^3}{3} \leqslant 10^{-6}.$$

$$\therefore \quad h \leqslant \sqrt[3]{3} \times 10^{-2} \simeq 0.014 .$$

Thus using just a quadratic and requiring a high degree of accuracy leads to a small value of h, i.e., a small interval over which the accuracy can be attained.

Notice also that in each of these problems an assumption about $|y_\theta - \theta h|$ has to be made in order to obtain an estimate for the error term. This is sometimes possible, for example from knowledge of the underlying physical problem. But generally it is not possible and we resort to other techniques for error estimation. One of these — step-halving — will be considered in Lesson 21.

Example 20.10, Case (b), shows that for a global Taylor-polynomial to be accurate enough its degree may have to be relatively high. A similar problem in Lesson 12 required a 2000th degree polynomial to obtain quite modest accuracy over a quite small interval! A high degree implies many differentiations of the equation $y^{(1)} = f(x, y)$, which may well be complicated (as in Example 20.8 and Exercise 20.2.) On the other hand a low degree polynomial generally implies a small interval over which the required accuracy can be attained (as in Case (c) of Examples 20.9 and 20.10.) These observations lead naturally to a questioning of the idea of covering with one formula (or, as we say, in one step) the whole of the interval over which the solution is required. Perhaps this interval can be covered by many smaller steps in each of which a simple formula (with low degree n) and high accuracy (small error $|R_n|$) are retained.

In fact this is generally possible, and using this strategy produces a so-called *step-by-step* or *marching* method. We shall consider step-by-step Taylor-polynomial solutions in Lesson 21.

EXERCISE 20.3

Consider the problem

$$\frac{dy}{dx} = y \qquad x \in [0, 2]$$

$$y = 1 \quad \text{when} \quad x = 0$$

and assume that $|y(x)| \leqslant 10$ for $x \in [0, 2]$.

(a) Find the degree of the Taylor polynomial constructed at $x = 0$ which gives values of y correct to 2D for $x \in [0, 2]$.

(b) Find the largest interval contained in $[0, 2]$ for which a quadratic Taylor polynomial constructed at $x = 0$ will give values of y correct to 2D.

20.8   SUMMARY

In this lesson we have defined a number of useful terms related to ODE's and seen that an ODE need not have a solution. The Taylor-polynomial approach has enabled us to produce *formal* expressions to approximate the solution of a first-order initial-value problem in terms of $x - x_0$ (or h), which we called a *global* solution. In general $f(x, y)$ involves y; Example 20.8 showed how nasty the resulting calculation of the higher derivatives can be for a Taylor-polynomial approximate solution. Moreover, investigation of the relations between the degree, the interval length and the accuracy depended on being able to assume (or show) that $y^{(n+1)}(x_\theta)$ is bounded. However, it may prove impossible or very difficult to show that $y^{(n+1)}(x_\theta)$ is bounded and hence to find how the error is behaving. Because of these difficulties we usually employ alternatives to the global approximation. One of these is obtained, essentially, by integrating the ODE numerically from x_0 to $x_0 + h$ in a number of smaller steps instead of in one big step.

This is called step-by-step integration and avoids the calculation of the higher derivatives. However, this does not answer the question of the accuracy of the solution. In the next lesson we consider step-by-step Taylor-polynomial solutions, and also attempt to deal with the accuracy problem.

ANSWERS TO EXERCISES

20.1 $\quad y^{(1)} = 1/(1+y), \qquad y_0^{(1)} = \dfrac{1}{1+1} = \dfrac{1}{2}$

$y^{(2)} = -y^{(1)}/(1+y)^2, \; y_0^{(2)} = -\tfrac{1}{2}/(1+1)^2 = -\tfrac{1}{8}$

$\therefore \quad y(h) \simeq 1 + \dfrac{1}{2}h - \dfrac{1}{8}\dfrac{h^2}{2!} = 1 + \tfrac{1}{2}h - \tfrac{1}{16}h^2$

and

$\qquad y(0.2) \simeq 1 + \dfrac{0.2}{2} - \dfrac{0.04}{16} = \underline{1.0975}$

20.2 $\quad y^{(1)} = 1/(1+y^2), \qquad\qquad y_n^{(1)} = 1/(1+y_n^2)$

$y^{(2)} = -2yy^{(1)}/(1+y^2)^2, \quad y_n^{(2)} = -2y_n/(1+y_n^2)^3$

$y^{(3)} = \big\{(1+y^2)^2(-2yy^{(2)} - 2y^{(1)}y^{(1)})$

$\qquad\quad - (-2yy^{(1)})2(1+y^2)2yy^{(1)}\big\} \div (1+y^2)^4$

$\qquad = 2\big\{4y^2y^{(1)2} - (1+y^2)(yy^{(2)} + y^{(1)2})\big\}/(1+y^2)^3 .$

$\therefore \quad y_n^{(3)} = 2(5y_n^2 - 1)/(1+y_n^2)^5 \quad$ (after some rearrangement!)

The recurrence relation follows.

$\qquad y_1 = 1 + \dfrac{0.2}{2} + \left(\dfrac{0.04}{2}\right)\left(\dfrac{-2}{8}\right) + \left(\dfrac{0.008}{6}\right)\left(\dfrac{2(4)}{32}\right) = \underline{1.0953}$

$\qquad y_2 = 1.0953 + \dfrac{0.2}{2.1998} + \left(\dfrac{0.04}{2}\right)\left(\dfrac{-2.1906}{10.6451}\right)$

$\qquad\qquad + \left(\dfrac{0.008}{6}\right)\left(\dfrac{9.9968}{51.5129}\right) = \underline{1.1824}$

$\qquad y_2 \simeq y(0.4).$

20.3 (a) Since $\quad y^{(1)} = y, \, y^{(n)} = y \quad$ for all $\quad n \in \mathbb{N}$. Hence the Taylor-polynomial solution of degree n, constructed at $x = 0$, is

$\qquad y(h) = y_0\left(1 + h + \dfrac{h^2}{2!} + \dfrac{h^3}{3!} + \ldots + \dfrac{h^n}{n!}\right)$

with error

$\qquad R_n(h) = \dfrac{h^{n+1}}{(n+1)!}y(\xi), \quad 0 < \xi < h$

(because we are only interested in $h > 0$). To obtain 2D accuracy for $h \in [0, 2]$ we need

$$\left| \frac{2^{n+1}}{(n+1)!} y(\xi) \right| \leqslant \tfrac{1}{2} \times 10^{-2}.$$

But $|y(\xi)| \leqslant 10$ so $2^{n+1}/(n+1)! \times 10 \leqslant \tfrac{1}{2} \times 10^{-2}$ will suffice.

Trying some values for n in the RHS leads eventually to $n = 9$ since

$$\frac{2^{10} \times 10}{10!} = 0.0028 \quad \text{whereas} \quad \frac{2^9 \times 10}{9!} = 0.014.$$

(b) Now consider

$$|R_2| = \frac{h^3}{3!} y(\xi) < \frac{10 h^3}{3!}.$$

For 2D accuracy we need $10 h^3 / 3! \leqslant \tfrac{1}{2} \times 10^{-2}$ which implies $h < 0.15$ approx., i.e., for 2D accuracy we are restricted to $x \in [0, 0.15]$.

FURTHER READING

None of the texts we know of treat the global Taylor method in any detail — perhaps rightly! We shall defer recommending any reading until Lesson 21.

LESSON 20 — COMPREHENSION TEST

1. Which of these are first-order ODE's:

(a) $\dfrac{dy}{dx} = 1 + 2x + 3y,$

(b) $y^{(1)} = 1 + 2x + 3y^2,$

(c) $\dfrac{d^2 x}{dt^2} + \dfrac{dx}{dt} + x = \sin 3t,$

(d) $\dfrac{dy}{dx} = 1 \Big/ \dfrac{d^2 y}{dx^2},$

(e) $\dfrac{d^2 z}{dx^2} + 5z = 1,$

(f) $\sqrt{y} \dfrac{dy}{dx} = \sqrt{x}?$

2. Which of these are linear ODE's:

(a) $\dfrac{dy}{dx} = 1 + 2x + 3y,$

(b) $y^{(1)} = 1 + 2x + 3y^2,$

(c) $\dfrac{dy}{dx} = \sqrt{(x/y)},$

(d) $\dfrac{dy}{dx} = \sqrt{x/y},$

(e) $\dfrac{dy}{dx} = y\sqrt{x}$,   (f) $(1 + t^2)\dfrac{dx}{dt} + 2tx = e^{-t}$?

3. Given $y^{(1)} = 1 + 2x + 3y^2$ and $x_0 = 1$, $y_0 = -1$, calculate $y_0^{(1)}, y_0^{(2)}$ and $y_0^{(3)}$. Hence write down the cubic Taylor-polynomial approximation for y, constructed at $x = 1$, and written in terms of $h = x - x_0 = x - 1$.

*4. Given $y^{(1)} = y^2$ and $x_0 = 0$ show that the error in a quadratic Taylor-polynomial approximation to y has the form $h^3[y(\theta h)]^4$, where $0 < \theta < 1$. Assuming that $|y| < 1$ throughout the range of interest show that 5D accuracy will be guaranteed if $|h| < 10^{-2}$.

ANSWERS

1   (a), (b), (f).

2   (a), (e), (f).

3    $y_0^{(1)} = 6$, $y_0^{(2)} = -34$, $y_0^{(3)} = 420$.

∴ $y(h) \simeq -1 + 6h - 17h^2 + 70h^3$.

LESSON 20 – SUPPLEMENTARY EXERCISES

20.4. Obtain the cubic Taylor-polynomial approximation to $y(x)$, constructed at $x = 1$, given that $y^{(1)} = 4x\sqrt{y}$ and $y(1) = 1$. Evaluate the approximation at $x = 1.25(0.25)2.0$.

Verify that $y = x^4$ is the exact solution and compare your numerical solution with the true values.

20.5. Obtain the quadratic Taylor-polynomial approximation y_{n+1} to $y(x_n + h)$, constructed at $x = x_n$, given that $y^{(1)} = 4x\sqrt{y}$ and $y(x_n) = y_n$.

Given further that $x_0 = y_0 = 1$, find y_1 with $h = 0.5$, then y_2 and y_3. Which values of $y(x)$ are these approximating?

20.6. Obtain the cubic Taylor-polynomial approximation to $y(x_0 + h)$, constructed at $x = x_0 = 0$, given that $y^{(1)} = -y^2/(1 + 2x)$ and $y(0) = 2$. Calculate approximations to $y(0.4)$ and $y(0.8)$, comparing them with the true values. (The solution is $y(x) = 2/(1 + \ln(1 + 2x))$.) Similarly calculate approximations to $y(0.04)$ and $y(0.08)$ and compare them with the true values.

*20.7. Given that $y^{(1)} = x^2 - 2y$, $y(0) = -1$, find the mth degree Taylor-polynomial approximation to $y(h)$, constructed at $x = 0$, where $m \geqslant 3$.

(a) Replacing the true remainder R_m by what would be the $(m+1)$th degree term, $h^{m+1} y^{(m+1)}(0)/(m+1)!$, estimate the degree m that would be required to obtain $y(x)$ correct to 6D for $0 \leqslant x \leqslant 1$.

(b) Similarly using the 4th degree term as an approximate R_3, estimate the interval $[0, h]$ over which the cubic approximation to y would be correct to 6D.

Lesson 21 Taylor-Polynomial Methods II

OBJECTIVES

At the end of this lesson you should

(a) appreciate why a global solution may well be inappropriate;

(b) understand and be able to apply the step-by-step method;

(c) understand the need for step-halving and the basic structure of a step-halving algorithm.

CONTENTS

.1 INTRODUCTION

We have seen in Lesson 20 (Section 20.6) how a *global* approximate solution for a first-order initial-value problem may be obtained as a formal series in $x - x_0$ or h. In practice the evaluation of even a few of the higher derivatives can be difficult and prone to mistakes. Moreover the accuracy of the approximation can be poor for a large value of h either because the number of terms in the Taylor-polynomial is not great enough or because h is so large that even an infinite series will not converge to the required value. In consequence we generally prefer an alternative technique known as step-by-step integration.

21.2   STEP-BY-STEP TAYLOR-POLYNOMIAL SOLUTIONS

Suppose that we know that the Taylor-polynomial solution of degree n for y, constructed at x_0 from

$$\frac{dy}{dx} = f(x, y), \quad y = y_0 \quad \text{when} \quad x = x_0$$

is sufficiently accurate for $x \in [x_0, x_0 + h]$.[†]

How can we extend this accuracy to larger intervals, e.g., to $[x_0, x_0 + mh]$ for $m = 2, 3, \ldots$?

The obvious answer is to use the polynomial constructed at x_0 to estimate $y(x_0 + h)$, to construct a new polynomial of degree n at $x = x_1 = x_0 + h$ from which we can estimate $y(x_1 + h) = y(x_0 + 2h)$, and so on. This is the step-by-step, Taylor-polynomial method mentioned in the previous section, and previewed by the last part of Exercise 20.2. In this context h is called the *step-length*.

We can look at this idea diagramatically. Fig. 21.1 supposes that we have a numerical method T (based on Taylor polynomials) which uses the values x_0, y_0 and h to produce an approximate solution $y = y_1$ at $x = x_0 + h$ to the initial-value problem $dy/dx = f(x, y)$ with $y = y_0$ at $x = x_0$.

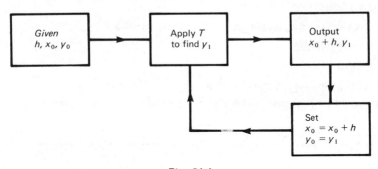

Fig. 21.1

Notice that in this 'computerised' version we do not specifically produce $y_2, y_3, y_4, \ldots$ at $x_2, x_3, x_4, \ldots$. When y_1 is obtained at $x_1 = x_0 + h$ we

---

[†] From now on we shall assume that $h > 0$. If $y(x_0 + h)$ is required for negative values of h this can be achieved by making the change of variable $u = -x$ giving the initial-value problem

$$\frac{dy}{du} = -f(-u, y), \quad y = y_0 \quad \text{when} \quad u = -x_0$$

$x = x_0 + h$ with $h < 0$ then corresponds to $u = u_0 + k$ where $u_0 = -x_0$ and $k = -h > 0$. In other words, decreasing x becomes increasing u.

rename x_1 as x_0 and y_1 as y_0 and start again. So the general step is *always* the obtaining of y_1 from x_0, y_0 and h, rather than the obtaining of y_{k+1} from x_k, y_k and h for $k = 1, 2, 3, \ldots$. Writing the method as in Fig. 21.1 enables a computer program to be written very easily.

As expressed in Fig. 21.1 the process will never stop, but a small alteration will control the number of times T is applied. (Incidentally T does not have to be based on Taylor-polynomials. It could be any step-by-step method.)

In theory we should analyse the error at each step because each new polynomial has a different remainder term. This would mean that the step-length h might vary from one step to the next. Such a process would be too cumbersome and we have an alternative strategy called step-halving which is described in the next section. For the sake of simplicity let us for the moment assume that we know of a constant step-length h which will achieve the required accuracy for each interval $[x_0, x_0 + h]$, $[x_0 + h, x_0 + 2h]$, $[x_0 + 2h, x_0 + 3h]$, $\ldots$, $[x_0 + (m-1)h, x_0 + mh]$ for some positive integer $m > 1$. For example, consider

$$\frac{dy}{dx} = y, \quad y = 1 \quad \text{when} \quad x = 0$$

with the solution required for $x \in [0, 2]$ and suppose that $h = 0.4$ will give the required accuracy. Then the numbers obtained by the step-by-step process and denoted y_1, y_2, y_3, y_4, y_5 are approximations to $y(0.4)$, $y(0.8)$, $y(1.2)$, $y(1.6)$, $y(2.0)$, each one obtained in a single step from the previous one. Fig. 21.2 indicates the results of the step-by-step process — a discrete set of approximate solution values.

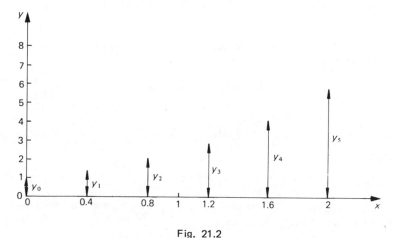

Fig. 21.2

On the other hand the analytical solution to the initial-value problem is the function $y = e^x$, as in Fig. 21.3 — a continuous curve.

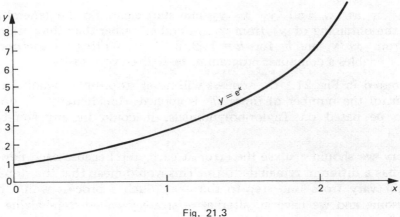

Fig. 21.3

Here are the details of the step-by-step Taylor method for the simple case $n = 1$. The first step is generated by

$$y_1 = y_0 + hy_0^{(1)} = y_0 + hf(x_0, y_0)$$

the second step by

$$y_2 = y_1 + hy_1^{(1)} = y_1 + hf(x_1, y_1)$$

and so on. The general step is given by the *recurrence relation*

$$y_{k+1} = y_k + hf(x_k, y_k), \quad k = 0, 1, 2, \dots . \qquad [21\text{-}1]$$

This rather crude method (we are using linear approximations to y) is historically famous and is known as *Euler's method*. Geometrically it approximates dy/dx at $x = x_k$ by $(y_{k+1} - y_k)/h$, as shown in Fig. 21.4.

EXAMPLE 21.1   For the problem above, i.e., $dy/dx = y$, $y(0) = 1$, and using $h = 0.4$, Euler's method is

$$y_{k+1} = y_k + hy_k = (1 + h)y_k = 1.4y_k.$$

So

$$y(0.4) \simeq y_1 = 1.4y_0 = 1.4(1) = 1.4$$

$$y(0.8) \simeq y_2 = 1.4(1.4) = 1.96$$

$$y(1.2) \simeq y_3 = 1.4(1.96) = 2.744.$$

We can see that, for this simple problem, the numerical solution by this method can be written down immediately:

$$y_k = (1.4)^k$$

so that

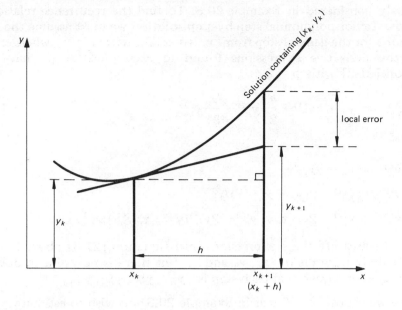

Fig. 21.4

$$y(1.6) \simeq y_4 = (1.4)^4 = 3.8416$$

$$y(2.0) \simeq y_5 = (1.4)^5 = 5.378\,24.$$

These can be compared with the true solution $y = e^x$ which gives

$$y(x_0 + kh) = y(kh) = y(0.4k) = e^{0.4k} = \{(2.718\,28 \ldots)^{0.4}\}^k$$

$$= (1.491\,82 \ldots)^k.$$

Since Euler's method uses only the first degree Taylor polynomial at each step, it is not very accurate. The general recurrence relation for the method is

$$y_{k+1} = y_k + hy_k^{(1)} + \frac{h^2}{2!}y_k^{(2)} + \frac{h^3}{3!}y_k^{(3)} + \ldots + \frac{h^p}{p!}y_k^{(p)}. \qquad [21\text{-}2]$$

We need the expressions for the derivatives $y^{(2)}, \ldots, y^{(p)}$ as with the global method, but they are evaluated at each point $x_0, x_1, x_2, \ldots$ in the step-by-step process. What we hope is that, for the step-by-step method, p can be much smaller for the same accuracy.

Here is a slightly more general problem that shows the amount of calculation typically required to achieve a step-by-step, Taylor-polynomial solution.

EXAMPLE 21.2   Consider the initial-value problem

$$y^{(1)} = y - 2x/y, \quad \text{with} \quad y = 2 \quad \text{when} \quad x = 0$$

previously considered in Example 20.8. To find the recurrence relation for the *cubic* Taylor-polynomial step-by-step solution we must assume the initial conditions for the general step from x_n to x_{n+1}, viz. $y = y_n$ when $x = x_n$. Using the derivative expressions found in Example 20.8 we have from Equation [21-2] with $p = 3$

$$y_{n+1} = y_n + hy_n^{(1)} + \frac{h^2}{2!}y_n^{(2)} + \frac{h^3}{3!}y_n^{(3)} \qquad\qquad [21\text{-}3]$$

where

$$y_n^{(1)} = y_n - 2x_n/y_n$$

$$y_n^{(2)} = y_n^{(1)} - 2(y_n - x_n y_n^{(1)})/y_n^2$$

$$y_n^{(3)} = y_n^{(2)} - 2\{-x_n y_n y_n^{(2)} - 2y_n^{(1)}(y_n - x_n y_n^{(1)})\}/y_n^3.$$

We could substitute these expressions into Equation [21-3] above to obtain y_{n+1} entirely in terms of x_n, y_n and h, but there is no point in doing so. The calculation sequence at each step is: $y_n^{(1)}, y_n^{(2)}, y_n^{(3)}, y_{n+1}$.

Suppose we choose $h = 0.5$ as in Example 20.8 and wish to estimate $y(0.5)$, $y(1.0), y(1.5)$ and $y(2.0)$. Here is a table of results. (The first row is essentially Example 20.8 again.)

n	x_n	y_n	$y_n^{(1)}$	$y_n^{(2)}$	$y_n^{(3)}$	y_{n+1}
0	0	2	2	1	3	3.1875
1	0.5	3.1875	2.8738	2.5292	3.3995	5.0114
2	1.0	5.0114	4.6123	4.5805	5.0038	7.9944
3	1.5	7.9944	7.6191	7.7266	7.8844	12.934
4	2.0	12.934				

The final column is redundant, but helps to show the flow of the calculation. To check that you understand where these numbers come from you should check one row, say that for $n = 2$.

EXERCISE 21.1

Improve the accuracy of the results in Example 21.1 by showing that the *cubic* Taylor-polynomial recurrence relation is

$$y_{k+1} = (1 + h + \tfrac{1}{2}h^2 + \tfrac{1}{6}h^3)y_k$$

so that

$$y_k = (1.4907)^k.$$

EXERCISE 21.2

Given that $y^{(1)} = -2xy$, with $y(0) = 1$, show that the step-by-step Taylor method of order 2 (quadratic approximation) for solving this problem is

$$y_{n+1} = y_n + hy_n^{(1)} + \tfrac{1}{2}h^2 y_n^{(2)}$$
$$= y_n - 2x_n y_n h + (2x_n^2 - 1)y_n h^2$$
$$= y_n \{1 - 2x_n h + (2x_n^2 - 1)h^2\}.$$

Calculate y_1, y_2 and y_3 using $h = 0.2$ and laying out the calculations as follows:

n	x_n	y_n	$y_n^{(1)}$	$y_n^{(2)}$	y_{n+1}
0	0	1	0	-2	0.96
1	0.2	etc.			

Work with 4 decimal places.

*EXERCISE 21.3

Given that $y^{(1)} = x - y$, with $y(1) = 0$, find the Taylor step-by-step relation of order p and write down the remainder term. Estimate the value of p required to obtain $y(2)$ correct to 4D, using $h = 1$. (Assume that $0 < y(\xi) < 2$ for $1 < \xi < 2$.) Similarly estimate the value of h, and hence the number of steps required, to obtain $y(2)$ correct to 4D, using $p = 2$. Carefully compare the amount of work involved in these two approaches. Can you draw general conclusions (i.e., for any first-order initial-value problem)?

Use the recurrence relation with $p = 2$ to estimate $y(1 + h)$ and $y(1 + 2h)$, where h is the step-length chosen to obtain 4D accuracy. (The true solution is $y(x) = x - 1$.)

21.3 STEP-HALVING

As we have seen, deciding on a suitable size for the step-length h may be awkward, difficult or impossible with the information available. This is especially true when the calculations are to be done automatically using a computer. However, there is another way of finding a solution to a given local accuracy which does not need information about the theoretical error, except that it decreases to zero when h does.

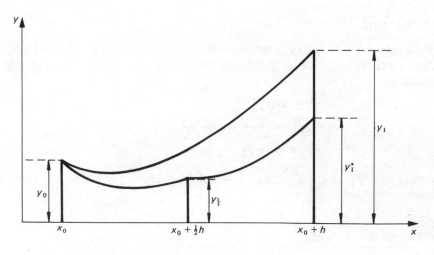

Fig. 21.5

Suppose that x_0, y_0 are fixed and that h is guessed. Suppose also that we have a method that will give a value for y at $x = x_0 + h$, given x_0, y_0 and h. (It could be one based on the Taylor-polynomial approach.) Then consider the following scheme. (Fig. 21.5 illustrates it.)

CALCULATION 1

Starting with x_0, y_0 and h, calculate y_1 (an estimate of y at $x = x_0 + h$).

CALCULATION 2

Starting with x_0, y_0 and $\frac{1}{2}h$, calculate $y_{1/2}$ (an estimate of y at the midpoint of the interval).

CALCULATION 3

Starting with $x_0 + \frac{1}{2}h$, $y_{1/2}$ and $\frac{1}{2}h$, calculate a new $y_1 = y_1^*$ (an estimate for y at $x = x_0 + h$ gained by using two steps of length $\frac{1}{2}h$).

Both y_1 and y_1^* are approximations to y at $x = x_0 + h$. If they agree sufficiently well ($|y_1 - y_1^*| < \epsilon$ say), then we may accept y_1^* (as the step length used is the smaller, implying a smaller error). If they do not agree ($|y_1 - y_1^*| \geqslant \epsilon$) then the step length h is too large and we regard y_1 as out of reach for the moment. It is usual to halve h and use $h_1 = \frac{1}{2}h$ to apply the process again starting at $x = x_0, y = y_0$. We are now aiming to find $y_{1/2}$, and not y_1 for the moment. The flow diagram of Fig. 21.6 may help to make the sequence of calculations clear. (As before the value of h is overwritten by $\frac{1}{2}h$ instead of writing $h_1 = \frac{1}{2}h$.) I denotes an integration rule.

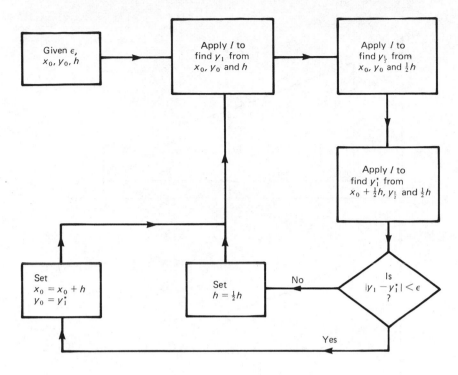

Fig. 21.6

Note that when h is halved what used to be $y_{1/2}$ now becomes y_1, what used to be y_1 becomes y_2, etc. — see Fig. 21.7.

The great advantage of this technique is that the method can be applied on a computer by calls of a procedure in ALGOL or subroutine in FORTRAN. Nothing need be known about error estimates. Other methods (like the Runge–Kutta methods which appear in Lesson 22) can also be used with step-halving.

However it should not be assumed that this technique will necessarily give solutions which are within ϵ of the true solution. The successive solutions merely agree *with each other* (to within ϵ).

These individual errors can accumulate and after many steps produce an unacceptably large error. To see how this might happen suppose y_0 is exact. Then in producing y_1 we introduce a local error of at most ϵ. In producing y_2 from y_1 we introduce a further local error of at most ϵ which *may* reinforce the error in y_1 to give an error of at most 2ϵ in y_2. In general the error bound for y_m will be $m\epsilon$. For example, if $\epsilon = \frac{1}{2} \times 10^{-3}$ then 2000 steps may given an error as large as 1 in y_{2000}. We hope of course, that some of the errors will have opposite signs and cancel, but we cannot rely on that.

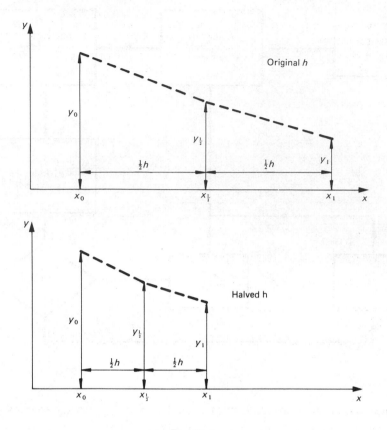

Fig. 21.7

EXERCISE 21.4

Apply the step-halving technique to the problem

$$\frac{dy}{dx} = y, \quad y(0) = 1$$

with Euler's method. Take $h = 0.4$ initially and estimate $y(0.4)$ with the local tolerance $\epsilon = 0.02$.

Is $y(0.4)$ within ϵ of the true solution? Within 2ϵ?

21.4 SUMMARY

We have seen that attempting to find a global solution may involve a great deal of difficult calculation. Even then the resulting Taylor-polynomial solution may be of little use. To overcome this difficulty the total interval over which the solution is required may be divided up into a number of smaller intervals and the Taylor-polynomial method of solution can be applied successively — the so-called step-by-step approach. However, this approach still may not give any idea of how accurate the answers are. The step-halving technique can be applied to obtain, at each stage, what may be called *numerically self-consistent* solutions. Even so, these could conceivably have little to do with the value of the true solution after some large number of steps.

This approach may give the idea that solving initial-value problems in ODE's is full of pit falls. In practice for many ODE's these techniques (step-by-step and step-halving) do give good results, and only fail with the more difficult problems.

ANSWERS TO EXERCISES

21.1   The recurrence relation follows from $y_k^{(n)} = y_k$ since $y^{(1)} = y$ and the solution from

$$1 + 0.4 + \tfrac{1}{2}(0.4)^2 + \tfrac{1}{6}(0.4)^3 = 1.4907 \quad \text{to 4D}.$$

21.2 $y_1 = 0.9600$, $y_2 = 0.8479$, $y_3 = 0.6892$.

(True solution: $y(x) = e^{-x^2}$.)

21.3 $y_{n+1} = y_n + (x_n - y_n)h$

$$+ (1 - x_n + y_n)\left\{\frac{h^2}{2!} - \frac{h^3}{3!} + \frac{h^4}{4!} + \ldots + (-1)^p\,\frac{h^p}{p!}\right\}.$$

Remainder $= (-1)^{p+1}\{1 - \xi + y(\xi)\}\dfrac{h^{p+1}}{(p+1)!}$

where $x_n < \xi < x_{n+1}$.

For $h = 1$, with $1 < \xi < 2$ and assuming $0 < y(\xi) < 2$ we need $(p+1)! > 2 \times 10^4$ MAX $\{1 - \xi + y(\xi)\} \approx 4 \times 10^4$.

∴ $\underline{p > 7 \text{ is needed.}}$

For $p = 2$, we need $h^3 < \tfrac{1}{2} \times 10^{-4} \times 3!/\text{MAX}\,\{1 - \xi + y(\xi)\} \approx \tfrac{3}{2} \times 10^{-4}$.

∴ $\underline{h < 0.05 \text{ approximately,}}$

i.e., it will take 40 steps to reach $x = 2$.

The second approach is longer in this case because there are simple expressions for $y^{(3)}, y^{(4)}$, etc. In general the reverse is true.

Using $h = 0.05$, $y(1.05) \approx 0.05$, $y(1.10) \approx 0.10$.

(Not surprisingly this quadratic approximation finds the linear solution exactly.)

21.4   Euler's formula gives

$$y_{k+1} = y_k + hy_k = (1+h)y_k$$

Step 1 $y_0 = 1$. Therefore $y_1 = 1.4y_0 = 1.4$.

Step 2   Using $\frac{1}{2}h$ we get $y_{1/2} = 1.2$ ($\simeq y(0.2)$).

Again using $\frac{1}{2}h$ we get $y_1^* = 1.2y_{1/2} = 1.2^2 = 1.44$.

Comparing y_1 and y_1^*, $|y_1 - y_1^*| = 0.04$ which is too big ($\epsilon = 0.02$).

So we must go back to $x_0 = 0$, $y_0 = 1$ and change h to 0.2.

Step 1 $y_0 = 1$, $y_1 = 1.2y_0 = 1.2$.

Step 2 $y_{1/2} = 1.1$ ($\simeq y(0.1)$).

$$y_1^* = 1.1y_{1/2} = 1.1^2 = 1.21$$

$|y_1 - y_1^*| = 0.01$ which is less than ϵ.

Hence we accept the value of y_1^* to get $y(0.2) = 1.21$.

Now we start the whole process over again with $x_0 = 0.2$, $y_0 = 1.21$ and $h = 0.2$ (which worked last time).

Step 1 $y_0 = 1.21$. Therefore $y_1 = 1.2y_0 = 1.452$.

Step 2 $y_{1/2} = 1.1y_0 = 1.331$ ($\simeq y(0.3)$)

$$y_1^* = 1.1y_{1/2} = 1.464 \quad \text{to 3D}$$

$|y_1 - y_1^*| = 0.012$ which is less than ϵ.

Hence we accept the value of y_1^* to get $y(0.4) = 1.46$ to 2D.

Now $y = e^x$ is the true solution, and

$$y(0.4) - e^{0.4} \simeq 0.03.$$

Therefore the *cumulative* error in $y(0.4)$ after 2 steps is 0.03, which is less than 2ϵ (as it should be!)

FURTHER READING

For an introduction to differential equations, their numerical solution in step-by-step fashion, and Euler's method in particular, see either Stark, Chapter 7, up to Section 7.7 inclusive, or Dixon, Chapter 9. Stark refers to the step-by-step Taylor method as the extended Euler method.

There is a small amount of *partial differentiation* in Stark's account, as there is in most. If you are familiar with partial differentiation and the idea of a total derivative (or prepared to ignore the bits you don't understand) then Conte (or Conte and deBoor), Section 6.3 is a good reference.

LESSON 21 – COMPREHENSION TEST

1. Explain the difference between global and step-by-step Taylor-polynomial solutions to ODE's.

2. Why is the step-by-step approach necessary?

3. Given $y^{(1)} = y^2$ derive the recurrence relation

$$y_{n+1} = y_n + hy_n^2 + h^2 y_n^3$$

for the step-by-step, quadratic, Taylor-polynomial method. Apply it to estimating $y(0.5)$ and $y(1.0)$ given that $y(0) = 2$.

4. Explain the need for step-halving.

5. Suppose that $h = 0.4$ is used initially in an algorithm using step-halving, that the halving process repeats until $h = 0.1$, and that no further halving is necessary. How many approximate solution values will be calculated in obtaining the approximation to $y(0.4)$? Assume $x_0 = 0$.

ANSWERS

1   A global solution is one Taylor polynomial of sufficient accuracy for all the relevant values of the independent variable.

     A step-by-step solution is a sequence of Taylor polynomials joined end-to-end, each one of sufficient accuracy over its own subinterval only.

2   Because a global solution may have to be of very high degree to obtain the required accuracy over the whole range.

3    $y_n^{(1)} = y_n^2$.

     $y^{(2)} = 2yy^{(1)} = 2y(y^2) = 2y^3$. Therefore $y_n^{(2)} = 2y_n^3$.

     Hence

$$y_{n+1} = y_n + hy_n^{(1)} + \tfrac{1}{2}h^2 y_n^{(2)}$$
$$= y_n + hy_n^2 + h^2 y_n^3$$
$$y(0.5) \simeq 2 + 0.5(2)^2 + 0.25(2)^3 = 6$$
$$y(1.0) \simeq 6 + 0.5(6)^2 + 0.25(6)^3 = 78.$$

4   Estimation of the value of a truncation error is generally too difficult. Step-halving avoids it by using the fact that the truncation error tends to zero with the step-length.

5 Stage	1	2	3	4	5	6
$h, \frac{1}{2}h$	0.4, 0.2	0.2, 0.1	0.1, 0.05	0.1, 0.05	0.1, 0.05	0.1, 0.05
Values of x at which y is evaluated	0.4 0.2 0.4	0.2 0.1 0.2	0.1 0.05 0.1	0.2 0.15 0.2	0.3 0.25 0.3	0.4 0.35 0.4
Result acceptable?	×	×	✓	✓	✓	✓

Total no. = 18

LESSON 21 – SUPPLEMENTARY EXERCISES

21.5. Obtain the recurrence relation for the cubic, Taylor-polynomial, step-by-step method as applied to the problem of Exercise 20.4. Hence estimate $y(x)$ for $x = 1.25(0.25)2.0$ by the step-by-step method, and compare the results with those obtained in Exercise 20.4 using a single (global) cubic, and with the exact values also given in Exercise 20.4.

21.6. Repeat Exercise 21.5 for the cubic approximation of Exercise 20.6, obtaining approximations to $y(0.04)$ and $y(0.08)$ by the step-by-step method and comparing them with the global approximations and the correct values.

21.7. Reconsider Exercise 21.5. Obtain $y(1.25)$ correct to 3D by applying the step-halving process.

21.8. The formula obtained as the first part of Exercise 20.5 is the recurrence relation for the quadratic, Taylor-polynomial, step-by-step method for the problem given there. The second part performs three steps with $h = 0.5$. Use step-halving until the approximations to $y(1.5)$, $y(2.0)$ and $y(2.5)$ are correct to 1D in the sense of numerical self-consistency (i.e., obtain $y(1.5)$ correct to 1D, then $y(2.0)$ correct to 1D assuming the approximation to $y(1.5)$ is exact, and so on). (The exact solution is $y = (3x^2 - 2)^{2/3}$.)

*21.9. In Lesson 9 we described the concept of *ill-conditioning* (where a small change in the data for a problem produces a large change in the answer), as applied to the solution of linear equations. Here is an example of an ill-conditioned initial-value problem.

Verify that $y = e^{1-x}$ is the solution of $y^{(1)} = 10y - 11e^{1-x}$ with $y(0) = e$, but that the solution is $y = e^{1-x} - \epsilon e^{10x}$ if the initial conditon changes to $y(0) = e - \epsilon$. This clearly indicates ill-conditioning for sufficiently large x since ϵe^{10x} will eventually swamp e^{1-x} regardless of how small ϵ is. Demonstrate this by supposing that e is replaced by its 4D value so that

$\epsilon = 2.718\,28 \ldots - 2.7183 \simeq -0.000\,02$, and then drawing accurate graphs of e^{1-x} and $-0.000\,02e^{10x}$ for $0 \leqslant x \leqslant a$, where a is 'a bit larger than one'. (You will see the reason for this unmathematical qualification when you try the question!)

*21.10. We also discussed *induced instability* in Lesson 9. This occurs when the method used produces nonsensical answers. In the case of simultaneous linear equations, large multipliers and small pivots magnify round-off errors. In the case of differential equations, the recurrence relation may not properly represent the problem to be solved.

Verify that $y = k\,e^{-\lambda x}$, $k \in \mathbb{R}$, is the family of solutions of $y^{(1)} = -\lambda y$, where $\lambda > 0$ — i.e., a family of decreasing exponential functions. Show further that the recurrence relation for Euler's method and this differential equation is $y_{n+1} = (1 - \lambda h)y_n$. Hence deduce that induced instability occurs if $h > 1/\lambda$. (If this is easy show that the quadratic Taylor method $y_{n+1} = (1 - \lambda h + \frac{1}{2}\lambda^2 h^2)y_n$ is unstable if $h > 2/\lambda$.)

Lesson 22 **Runge-Kutta Methods**

OBJECTIVES

At the end of this lesson you should

(a)   understand the disadvantages in using Taylor-polynomial approximations for the solution of a first-order initial-value problem;

(b)   understand the concept of Runge—Kutta methods and their advantage over Taylor methods;

(c)   be able to use a given Runge—Kutta method of any order;

(d)   be able to draw a flow-diagram for solving a first-order initial-value problem using a Runge—Kutta method.

CONTENTS

Answers to exercises. Further reading.
Comprehension test and answers. Supplementary exercises.

22.1   INTRODUCTION

In Lesson 20 we found approximate *global* solutions to first-order initial-value problems by constructing Taylor polynomials. For these to be sufficiently accurate over a usefully large interval, they often need to be of high degree. This involves many tedious differentiations of the right-hand side of the equation

$$\frac{dy}{dx} = f(x, y) \qquad \text{[22-1]}$$

in order to find the values of $y_0^{(1)}, y_0^{(2)}, \ldots, y_0^{(p)}$, say.

In Lesson 21 we saw how we could improve this situation by applying a step-by-step philosophy; that is, covering the interval in question by a number of smaller intervals in each of which we use a *local* Taylor approximation. This approach does not remove the disadvantage of the Taylor method, although it does moderate it.

Although the Taylor-polynomial approach has other disadvantages, notably the difficulty of error estimation, it is the calculation of the second and higher derivatives that we would like to avoid. The Runge–Kutta methods we consider in this lesson are designed for that very purpose. The basic idea is a simple one: we are given a differential equation in the form of Equation [22-1] above; can we construct a method which, instead of requiring the values of the higher derivatives of y, requires instead further values of $f(x, y)$, i.e., the first derivative of y? Consider, for example, the second-order Taylor method, based on the recurrence relation.

$$y_{k+1} = y_k + h y_k^{(1)} + \tfrac{1}{2} h^2 y_k^{(2)}$$
$$= y_k + h f(x_k, y_k) + \tfrac{1}{2} h^2 \frac{df}{dx}(x_k, y_k)$$

and having an $O(h^3)$ error. Can we find an alternative, still with an $O(h^3)$ error, but involving a value $f(\hat{x}, \hat{y})$, say, instead of the $(df/dx)(x_k, y_k)$ value (where $\hat{x}$ and $\hat{y}$ are yet to be determined)?

Well, the answer is yes, but it takes some fairly sophisticated mathematics to prove it. Since Runge–Kutta methods are quite widely used we think it is important to include them in this course, but we shall not consider their proper mathematical derivation, contenting ourselves with an intuitive approach to their construction.

22.2 EULER'S METHOD

We met Euler's method in Lesson 21 where it was conceived simply as the step-by-step Taylor method or order one, i.e., the linear method.

Geometrically (Fig. 22.1) Euler's idea is to approximate the value of $y(x_{n+1})$ by replacing the solution curve through the point (x_n, y_n) by its tangent, giving the estimate

$$y_{n+1} = y_n + h y_n^{(1)}$$
$$= y_n + h f(x_n, y_n) \qquad \text{[22-2]}$$

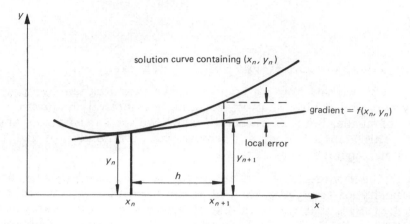

Fig. 22.1

which *is* clearly just the first-degree Taylor-polynomial method, with a local error of $O(h^2)$.

There is another way of looking at Equation [22-2]. The fundamental theorem of calculus tells us that

$$\int_{x_n}^{x_{n+1}} \left(\frac{dy}{dx}\right) dx = y(x_{n+1}) - y(x_n)$$

or

$$y(x_{n+1}) = y(x_n) + \int_{x_n}^{x_{n+1}} \left(\frac{dy}{dx}\right) dx. \qquad [22-3]$$

If we approximate the integral on the right using the rectangular rule then we get Euler's formula again:

$$y(x_{n+1}) \simeq y(x_n) + hy^{(1)}(x_n);$$

i.e., $\quad y_{n+1} = y_n + hy_n^{(1)}.$

In Lesson 23 we shall take a detailed look at using numerical integration formulae to approximate the integral in Equation [22-3], but we can also develop Runge—Kutta methods from the same idea.

22.3 SECOND-ORDER RUNGE—KUTTA METHODS

One of the main facts about Runge—Kutta methods is that there are a whole variety of them of each order — they are not uniquely defined. In this section we shall consider two second-order methods.

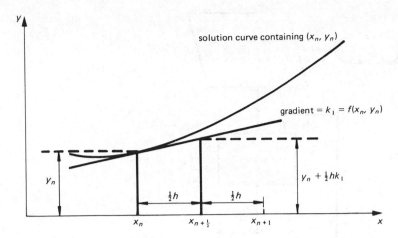

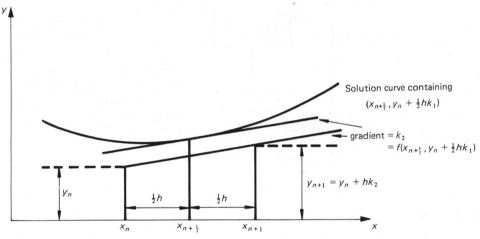

Fig. 22.2

Suppose that we replace the integral in Equation [22-3] by the mid-point rule. We obtain

$$y_{n+1} = y_n + hy^{(1)}(x_n + \tfrac{1}{2}h)$$
$$= y_n + hf(x_n + \tfrac{1}{2}h, y_{n+1/2}). \qquad [22\text{-}4]$$

The problem here is: what value does $y_{n+1/2}$ take? A simple approximation to it could be made by Euler's formula with a step-length of $\tfrac{1}{2}h$:

$$y_{n+1/2} = y_n + (\tfrac{1}{2}h)y_n^{(1)}$$
$$= y_n + \tfrac{1}{2}hf(x_n, y_n). \qquad [22\text{-}5]$$

Putting Equations [22-4] and [22-5] together gives us the *modified Euler method*, which is usually written in three steps thus:

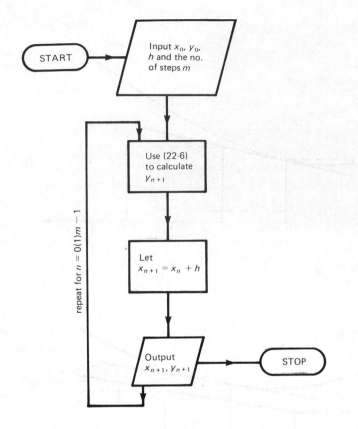

Fig. 22.3

$$k_1 = f(x_n, y_n)$$
$$\left. \begin{array}{l} k_2 = f(x_n + \tfrac{1}{2}h, y_n + \tfrac{1}{2}hk_1) \\[4pt] y_{n+1} = y_n + hk_2. \end{array} \right\} \qquad \text{[22-6]}$$

Fig. 22.2 gives a geometrical interpretation of this method.

k_1 is the gradient of the solution of $y^{(1)} = f(x, y)$ that passes through (x_n, y_n). This is used in Euler's method to estimate $y(x_n + \tfrac{1}{2}h)$ as $y_n + \tfrac{1}{2}hk_1$. k_2 is the gradient of the solution of $y^{(1)} = f(x, y)$ that passes through $(x_n + \tfrac{1}{2}h, y_n + \tfrac{1}{2}hk_1)$ and is used as an average gradient, in Euler fashion, to estimate y_{n+1} as $y_n + hk_2$.

At this stage we remind you of what a recurrence relation such as [22-6] really means. We are given the differential equation $y^{(1)} = f(x, y)$ and the initial condition $y = y_0$ when $x = x_0$. Setting $n = 0$ in Equations [22-6] gives us a 3-step calculation to produce y_1, our approximation to $y(x_0 + h) = y(x_1)$. Then setting $n = 1$ in Equations [22-6] gives us y_2 in identical fashion; and so on. Fig. 22.3 shows an outline flow chart for this process.

Later in this lesson we ask you to draw a similar flow chart, but avoiding the need for the arrays $(x_0, x_1, x_2, \ldots, x_m)$ and $(y_0, y_1, y_2, \ldots, y_m)$.

Returning to the derivation of Equations [22-6] note that we have not shown that the method is equivalent to the Taylor quadratic method, with an error of $O(h^3)$. We have already said that that is beyond the scope of this course. But if we *assume* the equivalence, then we can observe that we have achieved our aim: each step defined by Equations [22-6] contains two evaluations of $f(x, y)$ and no evaluation of its derivatives.

EXAMPLE 22.1   To solve $dy/dx = 2y + x + 1$ with $y(0) = 1$, using $h = 0.1$ and working with 4D, finding $y(0.1), y(0.2), y(0.3)$ and $y(0.4)$ by the modified Euler method. (The exact solution is $y = \frac{1}{4}(7e^{2x} - 2x - 3)$.)

The natural progression of the calculation for this method is: $k_1, x_n + \frac{1}{2}h$, $y_n + \frac{1}{2}hk_1, k_2, y_{n+1}$; hence the following scheme.

n	x_n	y_n	k_1	$x_n + \frac{1}{2}h$	$y_n + \frac{1}{2}hk_1$	k_2	y_{n+1}
0	0	1	3	0.05	1.15	3.35	1.3350
1	0.1	1.3350	3.7700	0.15	1.5235	4.1970	1.7547
2	0.2	1.7547	4.7094	0.25	1.9902	5.2303	2.2777
3	0.3	2.2777	5.8555	0.35	2.5705	6.4909	2.9268

As a matter of interest let us consider the errors in the y_{n+1}.

x	y	$\frac{1}{4}(7e^{2x} - 2x - 3)$	Error
0	1	1	0
0.1	1.3350	1.3375	0.0025
0.2	1.7547	1.7607	0.0060
0.3	2.2777	2.2887	0.0110
0.4	2.9268	2.9447	0.0179

Notice the increasing nature of the errors. We would expect this because the errors are likely to accumulate: the error in y_1 is compounded from the error in y_0 and the local error of the method; the error in y_2 consists of the error in y_1 plus the local error of the method again; and so on.

EXERCISE 22.1

Show that applying the modified Euler method to

$$y^{(1)} = -2xy, \quad y(0) = 1$$

is equivalent to using the recurrence relation

$$y_{n+1} = y_n(1 - 2x_n h + (2x_n^2 - 1)h^2 + x_n h^3).$$

This should agree with the Taylor method of order 2, up to terms in h^2. Check that it does by comparing it with the equivalent result in Exercise 21.2 in Lesson 21.

Then perform the following calculations:

(a) for $h = 0.2$ find y_1, y_2, y_3;

(b) for $h = 0.1$ find y_i for $i = 1(1)6$;

in each case using the modified Euler method and laying out the calculations as shown in Example 22.1.

An alternative second-order Runge–Kutta method can be found by trying to approximate the integral in Equation [22-3] by the trapezium rule:

$$
\begin{aligned}
y_{n+1} &= y_n + \tfrac{1}{2}h(y^{(1)}(x_n) + y^{(1)}(x_n + h)) \\
&= y_n + \tfrac{1}{2}h(y_n^{(1)} + y_{n+1}^{(1)}) \\
&= y_n + \tfrac{1}{2}h(f(x_n, y_n) + f(x_{n+1}, y_{n+1})). \qquad [22\text{-}7]
\end{aligned}
$$

The problem with this formula is that $f(x_{n+1}, y_{n+1})$ requires the value of y_{n+1} which we don't know. In fact it's y_{n+1} that we are trying to calculate! We can overcome this by estimating y_{n+1} by Euler's method (just as we estimated $y_{n+1/2}$ above), i.e., by letting $y_{n+1} = y_n + hy_n^{(1)}$ in the right-hand side. This gives a method usually referred to as the simple (second-order) Runge–Kutta method:

$$
\left.
\begin{aligned}
k_1 &= f(x_n, y_n) \\
k_2 &= f(x_n + h, y_n + hk_1) \\
y_{n+1} &= y_n + \tfrac{1}{2}h(k_1 + k_2).
\end{aligned}
\right\} \qquad [22\text{-}8]
$$

Note that we have again achieved what we set out to do. This method (like the modified Euler method) involves two evaluations of $f(x, y)$, as opposed to one such evaluation and an evaluation of df/dx in the equivalent Taylor method. Once again we shall not actually *prove* that Equations [22-8] are equivalent to the quadratic Taylor method, but it can certainly be done.

EXAMPLE 22.2   Applying Equations [22-8] to $dy/dx = 2y + x + 1$ with $y(0) = 1$ under the same conditions as in Example 22.1 we obtain

n	x_n	y_n	k_1	$x_n + h$	$y_n + hk_1$	k_2	y_{n+1}
0	0	1	3	0.1	1.3	3.7	1.3350
1	0.1	1.3350	3.7700	0.2	1.7120	4.6240	1.7547
2	0.2	1.7547	4.7094	0.3	2.2256	5.7512	2.2777
3	0.3	2.2777	5.8555	0.4	2.8633	7.1266	2.9268

We see that, *for this particular problem*, the results from the modified Euler and second-order Runge–Kutta methods are identical. In fact they both reduce exactly to the second-order Taylor recurrence relation

$$y_{n+1} = y_n + h(2y_n + x_n + 1) + \tfrac{1}{2}h^2(4y_n + 2x_n + 3)$$

as you can check for yourself by substituting $f(x, y) = 2y + x + 1$ into Equations [22-6] and [22-8].

EXERCISE 22.2

Given that $y^{(1)} = -2xy$, $y(0) = 1$, show that the step-by-step simple Runge–Kutta method of order 2 for this problem is equivalent to

$$y_{n+1} = y_n(1 - 2x_n h + (2x_n^2 - 1)h^2 + 2x_n h^3).$$

How does this compare with the Taylor method of order 2? (See Lesson 21, Exercise 21.2.) With the modified Euler method? (See Exercise 22.1.)

Perform the same calculations as in Exercise 22.1, i.e.,

(a) 3 steps with $h = 0.2$;

(b) 6 steps with $h = 0.1$;

using this 2nd-order Runge–Kutta method and laying out the calculations as in Example 22.2.

Compare the results with those of Exercise 22.1.

22.4 OTHER RUNGE–KUTTA METHODS

We have considered the well-known second-order methods first because they are relatively simple. In general they are not sufficiently accurate for practical purposes, in other words they require too small a step-length for reasonable accuracy and are therefore inefficient in calculation time.

On the other hand higher-order Runge–Kutta methods are more complicated (just as Simpson's rule, for example, is more complicated than the trapezium rule). For no particular reason the best compromise between complexity and accuracy, and hence the most efficient methods, seem to be those of fourth order. These are even more difficult to derive and analyse than the second-order variety so we shall content ourselves with stating the most well-known example.

Remember that we are claiming that the formulae will be equivalent to the quartic Taylor method

$$y_{n+1} = y_n + hy_n^{(1)} + \tfrac{1}{2}h^2 y_n^{(2)} + \tfrac{1}{6}h^3 y_n^{(3)} + \tfrac{1}{24}h^4 y_n^{(4)}$$

with four evaluations of $f(x, y)$ in some sense replacing the evaluation of the four derivatives of y. Here then is the *classical fourth-order Runge–Kutta method*.

$$
\left.
\begin{aligned}
k_1 &= f(x_n, y_n) \\
k_2 &= f(x_n + \tfrac{1}{2}h, y_n + \tfrac{1}{2}hk_1) \\
k_3 &= f(x_n + \tfrac{1}{2}h, y_n + \tfrac{1}{2}hk_2) \\
k_4 &= f(x_n + h, y_n + hk_3) \\
y_{n+1} &= y_n + \tfrac{1}{6}h(k_1 + 2k_2 + 2k_3 + k_4).
\end{aligned}
\right\}
\qquad [22\text{-}9]
$$

(You might observe that this results from applying a kind of distorted Simpson's rule to the integral in Equation [22-3].)

EXAMPLE 22.3   Applying Equations [22-9] to $dy/dx = 2y + x + 1$ with $y(0) = 1$ and $h = 0.1$, working to 6D this time because we expect greater accuracy with the fourth-order method, we obtain

n	0	1	2	3
x_n	0	0.1	0.2	0.3
y_n	1	1.337 450	1.760 681	2.288 686
k_1	3	3.774 900	4.721 363	5.877 372
$x_n + \tfrac{1}{2}h$	0.05	0.15	0.25	0.35
$y_n + \tfrac{1}{2}hk_1$	1.15	1.526 195	1.996 749	2.582 554
k_2	3.35	4.202 390	5.243 498	6.515 109
$y_n + \tfrac{1}{2}hk_2$	1.1675	1.547 570	2.022 856	2.614 441
k_3	3.385	4.245 139	5.295 712	6.578 882
$x_n + h$	0.1	0.2	0.3	0.4
$y_n + hk_3$	1.3385	1.761 964	2.290 252	2.946 574
k_4	3.777	4.723 928	5.880 504	7.293 148
y_{n+1}	1.337 450	1.760 681	2.288 686	2.944 661
Correct solution (to 6D)	1.337 455	1.760 693	2.288 708	2.944 697
Error	0.000 005	0.000 012	0.000 022	0.000 036

where the correct solution is calculated from $\tfrac{1}{4}(7e^{2x} - 2x - 3)$ as before. We observe, as expected, greater accuracy than the second-order methods obtain. The error still increases with the numbers of steps of course.

EXERCISE 22.3

Apply the classical 4th-order Runge—Kutta method, with $h = 0.2$ and working with 4D, to find y_1, y_2, y_3 for the problem of Exercise 22.1.

Compare the accuracy of the solutions and the effectiveness of the method in this exercise with those in Exercises 22.1 and 22.2 and with the true solutions to 4D given below. (In fact $y(x) = \exp(-x^2)$.)

$$y(0.2) = 0.9608$$
$$y(0.4) = 0.8521$$
$$y(0.6) = 0.6977$$

.5 PROPERTIES OF RUNGE—KUTTA (R—K) METHODS

Basically the advantage of the Runge—Kutta methods is that they are simple to use and can be programmed for a computer very easily. The calculations are tedious of course, but that is true of all practical methods (unless you like that sort of thing!). In practice we use computers to apply such methods and only do the calculations by hand to gain understanding of the process. Not only are R—K methods simpler to program than Taylor methods; they are generally more economical. The time taken to evaluate $f(x, y)$ 4 times (as for the 4th order R—K method) is usually much shorter than that taken to calculate the corresponding high-order derivatives for the Taylor-polynomial method.

There are methods (see Lessons 23 and 24) which are more economical than R—K methods, but they usually have other disadvantages such as poor stability properties. R—K methods can be made unstable, but the instability depends upon the step-length h and can be eliminated simply by making h sufficiently small.

A major disadvantage of R—K methods is that, like the Taylor methods, it is difficult to estimate the errors in the solution values. In fact, *unlike* the Taylor methods, it is difficult even to write down a formula for the local error of an R—K method. This means that in practice we usually resort to the interval-halving technique described in Lesson 21, seeking numerically self-consistent solutions.

To conclude this lesson here are two exercises designed to make sure that you understand the structure of a R—K calculation.

EXERCISE 22.4

Draw a flow diagram for the classical 4th-order Runge—Kutta method with the following specification:

input x_0, y_0, h, m (number of steps required)

output $y_1, y_2, \ldots, y_m$

(You need not use arrays. Interpret the recurrence relation $y_{n+1} = y_n + \ldots$ as $y_{\text{new}} = y_{\text{old}} + \ldots$ so that at any one time only two successive y values are stored. You must then print out each y_{new} as it is obtained and before it is overwritten.)

*EXERCISE 22.5

Adapt your answer to Exercise 22.4 to include the interval-halving technique. (Input a tolerance ϵ and successively halve the step-length until successive values of y_i agree to within ϵ.)

22.6   POSTSCRIPT

The titles given to numerical methods for solving ODE's are somewhat confusing. The method that we have called the simple (second-order) Runge–Kutta method is sometimes called the modified Euler method! The method that we have called the modified Euler method is sometimes classed as a predictor–corrector method (see Lesson 24) — which it is — rather than an R–K method — which it also is! (A rose by any other name . . . .) This simply reflects the fact that you can classify the methods in more than one way, and that there is overlap between different groups. The best way to think of R–K methods is as methods directly equivalent to the corresponding Taylor methods, but avoiding the calculation of higher derivatives, and set out in the form

$$\left.\begin{aligned}
k_1 &= f(x_n, y_n) \\
k_2 &= f(x_n + \ldots, y_n + \ldots) \\
&\cdots \\
y_{n+1} &= y_n + h(\alpha_1 k_1 + \alpha_2 k_2 + \ldots).
\end{aligned}\right\}$$

ANSWERS TO EXERCISES

22.1   The difference between modified Euler and Taylor is just the $x_n h^3 y_n$ term, which is of the order of the local error.

$h = 0.2$

n	x_n	y_n	k_1	$x_1 + \frac{1}{2}h$	$y_n + \frac{1}{2}hk_1$	k_2	y_{n+1}
0	0	1	0	0.1	1	-0.2	0.96
1	0.2	0.96	-0.384	0.3	0.9216	-0.5530	0.8494
2	0.4	0.8494	-0.6795	0.5	0.7814	-0.7814	0.6931
3	0.6	0.6931					

$h = 0.1$

n	x_n	y_n	k_1	$x_1 + \frac{1}{2}h$	$y_n + \frac{1}{2}hk_1$	k_2	y_{n+1}
0	0	1	0	0.05	1	-0.1	0.99
1	0.1	0.99	-0.198	0.15	0.9801	-0.2940	0.9606
2	0.2	0.9606	-0.3842	0.25	0.9414	-0.4707	0.9135
3	0.3	0.9135	-0.5481	0.35	0.8861	-0.6203	0.8515
4	0.4	0.8515	-0.6812	0.45	0.8174	-0.7357	0.7779
5	0.5	0.7779	-0.7779	0.55	0.7390	-0.8129	0.6966
6	0.6	0.6966					

22.2   The difference now is $2x_n h^3 y_n$, i.e., twice the corresponding term in the modified Euler method.

$h = 0.2$

n	x_n	y_n	k_1	$x_n + h$	$y_n + hk_1$	k_2	y_{n+1}
0	0	1	0	0.2	1	-0.4	0.96
1	0.2	0.96	-0.384	0.4	0.8832	-0.7066	0.8509
2	0.4	0.8509	-0.6807	0.6	0.7148	-0.8578	0.6970
3	0.6	0.6970					

$h = 0.1$

n	x_n	y_n	k_1	$x_n + h$	$y_n + hk_1$	k_2	y_{n+1}
0	0	1	0	0.1	1	-0.2	0.99
1	0.1	0.99	-0.198	0.2	0.9702	-0.3881	0.9607
2	0.2	0.9607	-0.3843	0.3	0.9223	-0.5534	0.9138
3	0.3	0.9138	-0.5483	0.4	0.8590	-0.6872	0.8520
4	0.4	0.8520	-0.6816	0.5	0.7838	-0.7838	0.7787
5	0.5	0.7787	-0.7787	0.6	0.7008	-0.8410	0.6977
6	0.6	0.6977					

n	0	1	2	3
x_n	0	0.2	0.4	0.6
y_n	1	0.9608	0.8522	0.6977
k_1	0	-0.3843	-0.6818	
$x_n + \tfrac{1}{2}h$	0.1	0.3	0.5	
$y_n + \tfrac{1}{2}hk_1$	1	0.9224	0.7840	
k_2	-0.2	-0.5534	-0.7840	
$y_n + \tfrac{1}{2}hk_2$	0.98	0.9055	0.7738	
k_3	-0.196	-0.5433	-0.7738	
$x_n + h$	0.2	0.4	0.6	
$y_n + hk_3$	0.9608	0.8521	0.6974	
k_4	-0.3843	-0.6817	-0.8369	
y_{n+1}	0.9608	0.8522	0.6977	

FURTHER READING

We shall not recommend any further reading on Runge—Kutta methods because they are usually described with their derivation, which, as we have already said, requires mathematical techniques above the level assumed in this course.

However, if you really want to see this derivation for yourself, then you can find it in most books on numerical mathematics, and in particular in

(a) Conte and deBoor, Section 6.5,

(b) Conte, Section 6.5,

(c) Phillips and Taylor, Section 11.3.

LESSON 22 – COMPREHENSION TEST

1. True or false?

(a) Runge—Kutta methods are simpler to apply than the Taylor-polynomial method of the same order.

(b) Runge—Kutta methods do not need step-halving.

(c) Runge—Kutta methods of order k have truncation error of the form $h^{k+1}y^{(k+1)}(\xi)/(k+1)!$

2. In what sense is a kth-order Runge—Kutta method equivalent to the kth degree Taylor-polynomial method?

3. List the three obvious mistakes in this statement of a second-order Runge—Kutta method:

$$k_1 = f(x_n, y_n)$$
$$k_2 = f(x_n + \tfrac{3}{4}h, y_n + \tfrac{3}{4}k_1)$$
$$y_{n+1} = \tfrac{1}{3}h(k_1 + k_2).$$

4. Perform one step of this second-order Runge—Kutta method:

$$k_1 = f(x_n, y_n)$$
$$k_2 = f(x_n + \tfrac{2}{3}h, y_n + \tfrac{2}{3}hk_1)$$
$$y_{n+1} = y_n + \tfrac{1}{4}h(k_1 + 3k_2)$$

as applied to $y^{(1)} = x/y$, $y(1) = 1$, with $h = 1$ (i.e., estimate $y(2)$). Use exact arithmetic.

ANSWERS

1 (a) True.

(b) False.

(c) False — the expressions are much more complicated than the Taylor remainder, and we have not stated them.

2 Their truncation errors are both of $O(h^{k+1})$.

3 (a) The increment $\tfrac{3}{4}k_1$ to y_n in k_2 should involve h; e.g., $\tfrac{3}{4}hk_1$,

(b) $y_{n+1} = y_n + \dots$,

(c) $\tfrac{1}{3}h(k_1 + k_2)$ is dimensionally incorrect. The sum of the coefficients of k_1 and k_2, times the leading fraction, must equal unity, e.g., $\tfrac{1}{3}h(k_1 + 2k_2)$.

4 $k_1 = x_n/y_n = 1/1 = 1.$

$k_2 = (x_n + \tfrac{2}{3}h)/(y_n + \tfrac{2}{3}hk_1) = \tfrac{5}{3}/\tfrac{5}{3} = 1.$

$y_{n+1} = 1 + \tfrac{1}{4}(1 + 3) = 2.$

LESSON 22 – SUPPLEMENTARY EXERCISES

22.6. Apply both of the second-order Runge—Kutta methods in Equations [22-6] and [22-8] to $y^{(1)} = 4x/\sqrt{y}$, $y(1) = 1$, obtaining estimates of $y(1.5)$, $y(2.0)$ and $y(2.5)$ using $h = 0.5$. Compare the results with the equivalent quadratic Taylor-polynomial values of Exercise 20.5. Any comments on the step-length? (See also Exercise 21.8.)

22.7. Apply the third-order Runge–Kutta method

$$k_1 = f(x_n, y_n)$$
$$k_2 = f(x_n + \tfrac{1}{3}h, y_n + \tfrac{1}{3}hk_1)$$
$$k_3 = f(x_n + \tfrac{2}{3}h, y_n + \tfrac{2}{3}hk_2)$$
$$y_{n+1} = y_n + \tfrac{1}{4}h(k_1 + 3k_3)$$

;o $y^{(1)} = 4x\sqrt{y}$, $y(1) = 1$, performing 4 steps with $h = 0.25$. Use 4D working (cf. Exercises 20.4 and 21.5).

Apply step-halving with this problem and method to obtain $y(1.25)$ correct to 3D (cf. Exercise 21.7).

22.8. Apply the third-order Runge–Kutta method

$$k_1 = f(x_n, y_n)$$
$$k_2 = f(x_n + \tfrac{1}{2}h, y_n + \tfrac{1}{2}hk_1)$$
$$k_3 = f(x_n + h, y_n + h(2k_2 - k_1))$$
$$y_{n+1} = y_n + \tfrac{1}{6}h(k_1 + 4k_2 + k_3)$$

to $y^{(1)} = -y^2/(1 + 2x)$, $y(0) = 2$, obtaining approximations to $y(0.04)$ and $y(0.08)$ using $h = 0.04$. Work to 5D (cf. Exercises 20.6 and 21.6).

22.9. Apply the Kutta–Merson (fourth-order) method

$$k_1 = f(x_n, y_n)$$
$$k_2 = f(x_n + \tfrac{1}{3}h, y_n + \tfrac{1}{3}hk_1)$$
$$k_3 = f(x_n + \tfrac{1}{3}h, y_n + \tfrac{1}{6}h(k_1 + k_2))$$
$$k_4 = f(x_n + \tfrac{1}{2}h, y_n + \tfrac{1}{8}h(k_1 + 3k_3))$$
$$k_5 = f(x_n + h, y_n + \tfrac{1}{2}h(k_1 - 3k_3 + 4k_4))$$
$$y_{n+1} = y_n + \tfrac{1}{6}h(k_1 + 4k_4 + k_5)$$

to the problem of Exercise 22.7. Work now to 6D and compare the results with the previously obtained third-order results and the exact values given by $y = x^4$.

Solve the same problem using the classical fourth-order method in Equations [22-9] and compare the results.

22.10. Solve $y^{(1)} = x^2 - 2y$, $y(0) = -1$, for $x = 0.02(0.02)0.1$, using the third-order method of Exercise 22.7 and the Kutta–Merson method of Exercise 22.9. Work to 7D (cf. Exercise 20.7(b)). What can you conclude about the solution values? (Note that this is an alternative to step-halving, i.e., increasing order implies reducing errors, therefore increase the order until consistency is obtained. This is a perfectly feasible method on a computer.)

Lesson 23 **Using Integration Rules**

OBJECTIVES

At the end of this lesson you should

(a) understand how the integration of $y^{(1)} = f(x, y)$ can be approximated using quadrature rules;

(b) know the difference between one-step and multi-step methods and understand how multi-step methods may be started;

(c) appreciate the difference between an explicit (open) method and an implicit (closed) method;

(d) be able to apply formulae based on integration rules to the solution of first-order initial-value problems;

(e) know the relative accuracies of the popular formulae.

CONTENTS

23.1 THE INTEGRATION OF $y^{(1)} = f(x, y)$

If the ODE

$$\frac{dy}{dx} = y^{(1)} = f(x, y)$$

is integrated between $x = x_n$ and $x = x_{n+1}$ we get

$$\int_{x_n}^{x_{n+1}} \left(\frac{dy}{dx}\right) dx = \int_{x_n}^{x_{n+1}} f(x, y)\, dx;$$

i.e., $\quad y(x_{n+1}) - y(x_n) = \int_{x_n}^{x_{n+1}} f(x, y)\, dx;$

i.e., $\quad y(x_{n+1}) = y(x_n) + \int_{x_n}^{x_{n+1}} f(x, y)\, dx.$ [23-1]

In Equation [23-1] $y(x_n)$ and $y(x_{n+1})$ are the exact values of y at $x = x_n$ and $x = x_{n+1}$. In practice we will want a recurrence relation for computing values of $y_1, y_2, y_3, \ldots$ so we replace $y(x_n)$ and $y(x_{n+1})$ by their respective approximations, y_n and y_{n+1}.

Of course $\int_{x_n}^{x_{n+1}} f(x, y)\, dx$ contains y, but we know that y is a function of x so that this integral has a meaning.

The recurrence relation will now take the form

$$y_{n+1} = y_n + \int_{x_n}^{x_{n+1}} f(x, y)\, dx.$$ [23-2]

We have encountered the idea in Lesson 22 that we might construct methods equivalent to Taylor-polynomial methods by approximating the integral in Equation [23-2]. We called the resulting recurrence relations Runge–Kutta formulae although we did not actually prove that the Runge–Kutta formulae and the Taylor-polynomial methods were equivalent.

In this lesson we will also approximate the integral, using for instance the integration rules of Lesson 17, to produce different kinds of recurrence formulae.

23.2 ONE-STEP METHODS

In Lessons 21 and 22 we have seen examples of methods where, given $y = y_n$ at $x = x_n$ we have been able to calculate an estimate y_{n+1} of the value of y at $x = x_{n+1}$. In a sense we have stepped from x_n to x_{n+1}, i.e., we have performed an integration over one step and only needed information at the beginning of the step. Both Taylor-polynomial and Runge–Kutta methods are examples of *one-step methods*.

In this section we consider the one-step formulae obtained by approximating $\int_{x_n}^{x_{n+1}} f(x, y)\, dx$ using the rectangular and trapezium rules.

23.2.1 USING THE RECTANGULAR RULE

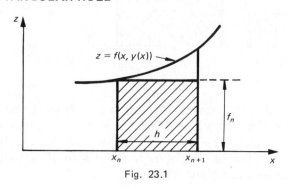

Fig. 23.1

The rectangular rule (see also Lesson 17) approximates the area under f between $x = x_n$ and $x = x_{n+1}$ by

$$f_n(x_{n+1} - x_n) = hf_n$$

where $x_{n+1} - x_n = h$ and $f_n = f(x_n, y_n)$. (See Fig. 23.1.)

Therefore we have for our recurrence relation

$$y_{n+1} = y_n + hf_n \qquad\qquad [23\text{-}3]$$

giving an approximation for $y(x_{n+1})$. (Euler's formula again!) Here notice that all that is needed to find y_{n+1} using this formula is knowledge of y_n at $x = x_n$. The formula gives y_{n+1} *explicitly*, i.e., we can calculate y_{n+1} directly because y_{n+1} does not appear on both sides of the equation. Such a formula is called an *explicit* or *open* formula.

23.2.2 USING THE TRAPEZIUM RULE

The trapezium rule (see also Lesson 17) approximates the area under f between $x = x_n$ and $x = x_{n+1}$ by (see Fig. 23.2)

$$\tfrac{1}{2}(x_{n+1} - x_n)(f_n + f_{n+1}) = \tfrac{1}{2}h(f_n + f_{n+1}).$$

Therefore we have for our recurrence relation

$$y_{n+1} = y_n + \tfrac{1}{2}h(f_n + f_{n+1}). \qquad\qquad [23\text{-}4]$$

Now $f_{n+1} = f(x_{n+1}, y_{n+1})$, and y_{n+1} therefore appears on *both* sides of the equation. So this formula does not give y_{n+1} explicitly. It is an example

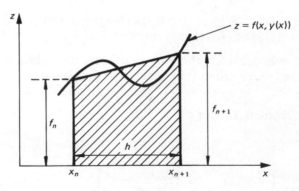

Fig. 23.2

of an *implicit* or *closed* formula. Finding y_{n+1} in such a situation will only be straightforward if $f(x, y)$ is *linear* in y, i.e., if the ODE is linear, as shown in the following examples. (In Lesson 22 we avoided the problem by estimating y_{n+1} for use in the right hand side of Equation [23-4], and we return to that idea in Lesson 24.)

EXAMPLE 23.1   Consider the first-order initial-value problem

$$y^{(1)} = x - y, \quad y(0) = 1.$$

Using the trapezium rule

$$y_{n+1} = y_n + \tfrac{1}{2}h(f_n + f_{n+1})$$
$$= y_n + \tfrac{1}{2}h(x_n - y_n + x_{n+1} - y_{n+1})$$
$$\therefore \quad (1 + \tfrac{1}{2}h)y_{n+1} = (1 - \tfrac{1}{2}h)y_n + \tfrac{1}{2}h(2x_n + h)$$

since $x_{n+1} = x_n + h$, and $f(x, y) = x - y$.

Here we can see that, because f is linear in y, the formula can be re-arranged so that it becomes explicit for y_{n+1}.

If the ODE is non-linear then the implicit formula cannot be used directly, as the following example shows.

EXAMPLE 23.2   Consider $y^{(1)} = x - y^2$, $y(0) = 1$.

Again applying the trapezium rule

$$y_{n+1} = y_n + \tfrac{1}{2}h(f_n + f_{n+1})$$
$$= y_n + \tfrac{1}{2}h(x_n - y_n^2 + x_{n+1} - y_{n+1}^2).$$

Here we have a quadratic in y_{n+1} and so the formula cannot easily be made explicit for y_{n+1}.

EXERCISE 23.1

For the first-order initial-value problem

$$y^{(1)} = x^2 - 3y, \quad y(1) = 2$$

find an explicit recurrence relation for y_{n+1} in terms of x_n and y_n by applying

(a) the rectangular rule (Euler's formula);

(b) the trapezium rule.

In a numerical application which do you think would be the more effective and in what sense? Give reasons for your answer.

Estimate $y(1.1), y(1.2)$ and $y(1.3)$ using the formulae you have found in (a) and in (b), working with 4D. Compare your solutions with the true solution

$$y = \frac{49}{27} e^{3-3x} + \frac{2}{27} - \frac{2x}{9} + \frac{x^2}{3} .$$

23.3 MULTI-STEP METHODS

23.3.1 ADAMS–BASHFORTH FORMULAE

In both the rectangular rule and the trapezium rule the approximation is based on the idea of replacing $f(x, y)$ by another function — a straight line —

(a) through the point (x_n, y_n) and parallel to the x-axis in the rectangular rule; and

(b) through the points (x_n, y_n) and (x_{n+1}, y_{n+1}) in the trapezium rule.

In both cases the approximating function is an *interpolatory* function. In (a) (above) the interpolation only depends on one point and in (b) it depends on two points. Simple formulae of not very high accuracy result.

One way to increase the accuracy might be to use more points, i.e., to interpolate using (x_{n-1}, y_{n-1}), (x_{n-2}, y_{n-2}), etc. Such an extension will naturally give rise to formulae which involve $x_n, x_{n-1}, x_{n-2}, \ldots$ and thus range over more than one step. Such formulae are called *multistep formulae*. If f_{n+1} appears in the approximation to $\int_{x_n}^{x_{n+1}} f(x, y) \, dx$ any such formula will be closed; if not, it will be open.

EXAMPLE 23.3   If we choose to replace $f(x, y)$ by a linear expression based on the points (x_{n-1}, y_{n-1}) and (x_n, y_n) then the function $f(x, y(x))$ of x is replaced by the function of s

$$f_s = f_n + s(f_n - f_{n-1})$$

where $x = x_n + sh$. (Here if $s = 0$, $f = f_n$ and if $s = -1$, $f = f_{n-1}$, as required. In fact we are using a linear, backward-difference, interpolating polynomial — see Section 16.6.)

Geometrically we are replacing the curve $z = f(x, y(x))$ by the line ABC in Fig. 23.3.

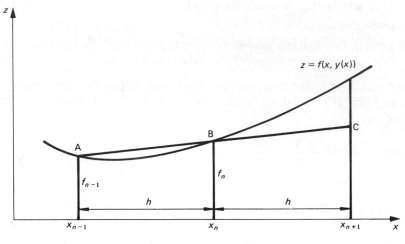

Fig. 23.3

Then $\int_{x_n}^{x_{n+1}} f(x, y)\, \mathrm{d}x$ is replaced by

$$\int_{x_n}^{x_{n+1}} (f_n + s(f_n - f_{n-1}))\, \mathrm{d}x = \int_0^1 (f_n + s(f_n - f_{n-1}))h\, \mathrm{d}s$$

$$= hf_n + \tfrac{1}{2}h(f_n - f_{n-1})$$

$$= \tfrac{1}{2}h(3f_n - f_{n-1}).$$

The complete formula then reads

$$y_{n+1} = y_n + \tfrac{1}{2}h(3f_n - f_{n-1}). \qquad\qquad [23\text{-}5]$$

This is a two-step open formula because

(a) it requires information at x_{n-1} and x_n to obtain y_{n+1} — hence two-step, and

(b) it gives y_{n+1} directly, needing only values of y and f at $x = x_n$ and $x = x_{n-1}$ (which are presumed available. We will deal later with how we get such information!) — hence open.

This type of formula (which depends on polynomial approximation using backward differences) is usually called an Adams–Bashforth formula.

EXERCISE 23.2

Find the Adams–Bashforth formula which depends on f_n, f_{n-1} and f_{n-2}. The relevant quadratic interpolating polynomial is

$$f_s = f_n + s(f_n - f_{n-1}) + \tfrac{1}{2}s(s+1)(f_n - 2f_{n-1} + f_{n-2}).$$

* EXERCISE 23.3

Find the Adams–Bashforth formula which depends on f_n, f_{n-1}, f_{n-2} and f_{n-3}. The relevant cubic interpolating polynomial is

$$f_s = f_n + s(f_n - f_{n-1}) + \tfrac{1}{2}s(s+1)(f_n - 2f_{n-1} + f_{n-2})$$
$$+ \tfrac{1}{6}s(s+1)(s+2)(f_n - 3f_{n-1} + 3f_{n-2} - f_{n-3}).$$

23.3.2   NYSTRÖM FORMULAE

So far we have integrated

$$\frac{dy}{dx} = f(x, y)$$

over the interval (x_n, x_{n+1}) to get

$$y(x_{n+1}) = y(x_n) + \int_{x_n}^{x_{n+1}} f(x, y)\, dx$$

and then approximated the integral in different ways.

Nyström's idea was to integrate over more than one interval. We could have integrated, for instance, over the interval (x_{n-1}, x_{n+1}) to get

$$y(x_{n+1}) = y(x_{n-1}) + \int_{x_{n-1}}^{x_{n+1}} f(x, y)\, dx$$

and then approximated the integral. This would give a two-step example of a Nyström method. In general, in a Nyström method, we could integrate over (x_{n-r}, x_{n+1}) $(r \in \mathbb{N})$ and get an $(r+1)$-step method.

EXAMPLE 23.4   Starting from the formula

$$y_{n+1} = y_{n-1} + \int_{x_{n-1}}^{x_{n+1}} f(x, y)\, dx$$

(on replacing $y(x_{n+1})$ and $y(x_{n-1})$ by y_{n+1} and y_{n-1} respectively), if we approximate the integral in this formula using the mid-point rule we get

$$y_{n+1} = y_{n-1} + 2hf_n \qquad\qquad [23\text{-}6]$$

which is an open two-step formula of the Nyström type.

EXERCISE 23.4

Use Simpson's Rule with

$$y_{n+1} = y_{n-1} + \int_{x_{n-1}}^{x_{n+1}} f(x, y) \, dx$$

to find an appropriate closed two-step formula of the Nyström type.

23.4 APPLYING MULTI-STEP METHODS

23.4.1 OPEN FORMULAE

Let us take the simple example of the formula

$$y_{n+1} = y_n + \tfrac{1}{2}h(3f_n - f_{n-1}).$$

Zero is the lowest possible subscript, by convention, so $n \geqslant 1$ is implied. So we have, to start with,

$$y_2 = y_1 + \tfrac{1}{2}h(3f_1 - f_0).$$

If we know y_1 and y_0 we can calculate $f_1 = f(x_1, y_1)$ and $f_0 = f(x_0, y_0)$ and hence y_2, and subsequently $f_2 = f(x_2, y_2)$. Similarly y_3, and so f_3, can be calculated, and in like manner as many steps as we please.

Thus, once the process has been started, the recurrence relation enables us to carry on finding as many approximate y values at equally-spaced intervals as we need.

The only problem then is that we need both y_0 *and* y_1 to start this two-step process when we almost always are given only the initial value $y = y_0$ when $x = x_0$. Unless the extra information is available from some other source (the underlying physical problem for example) the usual technique is to produce all the extra starting values required by using a one-step method — Runge–Kutta is the most popular — of the same order of accuracy as the integration-rule method being considered. So, if a k-step mth-order integration rule is to be used, then $k - 1$ steps of an mth-order one-step method will be required before the integration rule can be started.

EXERCISE 23.5

What values of y need to be known to start the application of these formulae?

(a) $y_{n+1} = y_n + \dfrac{h}{12}(23f_n - 16f_{n-1} + 5f_{n-2})$

(b) $y_{n+1} = y_n + \dfrac{h}{24}(55f_n - 59f_{n-1} + 37f_{n-2} - 9f_{n-3})$

(c) $y_{n+1} = y_{n-1} + 2hf_n$.

EXERCISE 23.6

Use $y_{n+1} = y_n + \frac{1}{2}h(3f_n - f_{n-1})$ to solve the problem

$$y^{(1)} = -2xy, \quad y(0) = 1$$

using $h = 0.2$ to find y_2, y_3, y_4. You may assume that $y_1 = y(0.2) = 0.9600$, as obtained by a second-order Runge–Kutta method in Lesson 22. Work with 4D and use the calculation scheme

$h = 0.2$	n	x_n	y_n	$f_n = -2x_n y_n$	y_{n+1}
	0	0	1	0	0.9600
	1	0.2	0.9600	-0.3840	0.8448
	2	0.4	0.8448	etc.	

Compare your results with those obtained by the Taylor-polynomial method (Exercise 21.2 of Lesson 21) and Runge–Kutta methods (Exercises 22.1 and 22.2).

.4.2   CLOSED FORMULAE

The only closed multi-step method we have considered is that resulting from Simpson's rule and produced in Exercise 23.4, viz.

$$y_{n+1} = y_{n-1} + \tfrac{1}{3}h(f_{n-1} + 4f_n + f_{n+1}). \qquad [23\text{-}7]$$

It is closed because f_{n+1} will depend in general on y_{n+1}, and is two-step because information is required at x_{n-1} and x_n before y_{n+1} can be calculated.

As with closed one-step formulae, such as that based on the trapezium rule

$$y_{n+1} = y_n + \tfrac{1}{2}h(f_n + f_{n+1})$$

we cannot use Equation [23-7] above directly unless the ODE is linear, and its application will therefore be left, along with the trapezium rule, until Lesson 24.

EXERCISE 23.7

Show that applying Simpson's rule to the linear differential equation $y^{(1)} = x - y$ leads to the recurrence relation

$$(1 + \tfrac{1}{3}h)y_{n+1} = (1 - \tfrac{1}{3}h)y_{n-1} - \tfrac{4}{3}hy_n + 2h(x_{n-1} + h).$$

23.5 TRUNCATION ERRORS AND ERROR ACCUMULATION

The remarks that have been made previously about estimating the truncation errors in Taylor-polynomial and Runge–Kutta methods apply equally well to integration-rule methods. However, if step-halving is to be used with a multi-step method, to check the consistency of solutions, the strategy of the calculations does become rather complicated. In consequence the details will be omitted here. Error terms can however be found (see Lesson 18 for the well-known ones), but they are usually used to gain qualitative rather than quantitative information about the accuracy of the solution.

For example, it can be shown that

$$y(x_{n+1}) = y(x_n) + \tfrac{1}{2}h(3f_n - f_{n-1}) + \tfrac{5}{12}h^3 y^{(3)}(\xi_1)$$ [23-8]

and we already know from Lesson 18 that

$$y(x_{n+1}) = y(x_n) + \tfrac{1}{2}h(f_{n+1} + f_n) - \tfrac{1}{12}h^3 y^{(3)}(\xi_2)$$ [23-9]

where $x_{n-1} < \xi_1 < x_{n+1}$ and $x_n < \xi_2 < x_{n+1}$.

This illustrates what might be expected — that closed rules are generally more accurate than open ones of the same order, as the open ones are extrapolatory in nature (they estimate y_{n+1} without using any information at x_{n+1}), whereas the closed ones are interpolatory (they include the value of $f(x, y)$ at x_{n+1} in the formula for calculating y_{n+1}).

In Lesson 24 we shall discuss combining open and closed formulae to obtain methods having both the applicability of the open formulae *and* the accuracy of the closed formulae.

The local truncation error for a given formula is only a small part of the error story of course. We also have at each step (indeed at each arithmetic operation) a potential round-off error, and all the errors at each step are propagated through to the following steps. As we've already said, it is extremely difficult to analyse the effect of all these error sources, and generally speaking we don't try. We can however consider ways of reducing the errors.

The round-off error will be worse the more steps we take, so we should choose a formula which implies a good compromise between accuracy (more accurate ⇒ less steps ⇒ less round-off error) and complexity (more complex ⇒ more round-off error). We should carry as many significant figures as possible in the calculation (in a computer we could use 'double precision' if available).

We can use interval-halving (bearing in mind the potential increase in round-off error) to improve the solution at each step.

Lastly we should be aware that, despite all precautions, it is possible for some of our methods to give rubbish for answers. If errors introduced accumulate until they swamp the exact solution then we have a case of *instability*. It is possible for the initial-value problem itself to be unstable (so-called *inherent instability*) so that small changes in the initial condition, say, produce large changes in the solution even if we solve the problems exactly. More likely (we hope!) a solution process goes wrong because of the method used (so-called *induced* instability). Here we state without analysis some basic facts:

(a) Taylor and Runge–Kutta methods (including Euler's) can be unstable if the step-length is too large, as can the lower-order Adams–Bashforth methods.

(b) The trapezium-rule method is unconditionally stable.

(c) The method based on Simpson's rule can be unstable if used to find a solution which decreases rapidly.

.6 MILNE'S FORMULA

In Lesson 24 we shall need one formula of the type produced in this lesson which we have yet to consider. We need an explicit formula of the same order of local error as Simpson's rule, i.e., $O(h^5)$, but of a simpler form than the fourth-order Adams–Bashforth. To obtain this with an open formula it has to be a many-step formula. The one that is most widely known is due to Milne and is the four-step formula

$$y_{n+1} = y_{n-3} + \frac{4h}{3}(2f_n - f_{n-1} + 2f_{n-2}). \qquad [23\text{-}10]$$

As a further illustration of the difference between the accuracies of open and closed formulae we state that the local error in Milne's formula has the form $\frac{14}{45}h^5 y^{(5)}(\xi_1)$, as compared to Simpson's $-\frac{1}{90}h^5 y^{(5)}(\xi_2)$.

---

*

EXERCISE 23.8

Derive Milne's formula by approximating $f(x, y)$ in

$$y_{n+1} = y_{n-3} + \int_{x_{n-3}}^{x_{n+1}} f(x, y)\, dx$$

by the quadratic interpolating polynomial matching f at $x = x_{n-2}, x_{n-1}$ and x_n. (*Hint*: if you use the forward-difference interpolating polynomial it's easier if you make the problem specific by setting $n = 2$.)

23.7 SUMMARY

Integration rules applied to the integral in

$$y(x_{n+1}) - y(x_{n-r}) = \int_{x_{n-r}}^{x_{n+1}} f(x, y)\, dx, \quad r = 0, 1, 2, \ldots$$

lead to various recurrence relations for solving $y^{(1)} = f(x, y)$ in step-by-step fashion. A particular formula is characterised by being open or closed; one-step, two-step, etc.; and by the order of the error. Here is a list of the formulae we have encountered in this lesson. (A–B means Adams–Bashforth.)

Name	Type	Step	Error order	
Euler	Open	One	$O(h^2)$	$y_{n+1} = y_n + hf_n$
Trapezium	Closed	One	$O(h^3)$	$y_{n+1} = y_n + \tfrac{1}{2}h(f_n + f_{n+1})$
A–B	Open	Two	$O(h^3)$	$y_{n+1} = y_n + \tfrac{1}{2}h(3f_n - f_{n-1})$
A–B	Open	Three	$O(h^4)$	$y_{n+1} = y_n + \dfrac{h}{12}(23f_n - 16f_{n-1} + 5f_{n-2})$
A–B	Open	Four	$O(h^5)$	$y_{n+1} = y_n + \dfrac{h}{24}(55f_n - 59f_{n-1} + 37f_{n-2} - 9f_{n-3})$
Midpoint	Open	Two	$O(h^3)$	$y_{n+1} = y_{n-1} + 2hf_n$
Simpson	Closed	Two	$O(h^5)$	$y_{n+1} = y_{n-1} + \dfrac{h}{3}(f_{n-1} + 4f_n + f_{n+1})$
Milne	Open	Four	$O(h^5)$	$y_{n+1} = y_{n-3} + \dfrac{4h}{3}(2f_n - f_{n-1} + 2f_{n-2})$

ANSWERS TO EXERCISES

23.1 (a) $y_{n+1} = y_n + hf_n$

$$= y_n + h(x_n^2 - 3y_n).$$

$$y_{n+1} = y_n(1 - 3h) + hx_n^2.$$

(b) $y_{n+1} = y_n + \tfrac{1}{2}h(f_n + f_{n+1})$

$$= y_n + \tfrac{1}{2}h(x_n^2 - 3y_n + x_{n+1}^2 - 3y_{n+1}).$$

$$\therefore \quad y_{n+1}\left(1 + \frac{3h}{2}\right) = y_n\left(1 - \frac{3h}{2}\right) + \tfrac{1}{2}h(2x_n^2 + 2x_n h + h^2).$$

$$\therefore \quad y_{n+1} = y_n\left(\frac{2 - 3h}{2 + 3h}\right) + \frac{h(2x_n^2 + 2x_n h + h^2)}{2 + 3h}.$$

(b) should be better as the trapezium rule approximates integrals better than the rectangular rule. The order of the error in (b) is $O(h^3)$ and in (a) is $O(h^2)$, so one should be able to use larger steps for the same accuracy. (See Lesson 18 for the truncation errors of integration rules.)

Using $y(1) = 2$, $h = 0.1$:

(a) $[1 - 3h = 0.7]$

n	x_n	y_n	x_n^2	y_{n+1}
0	1	2	1	1.5
1	1.1	1.5	1.21	1.171
2	1.2	1.171	1.44	0.9637
3	1.3	0.9637		

(b) $\left[\dfrac{2 - 3h}{2 + 3h} = \dfrac{1.7}{2.3} = 0.7391 \qquad h/(2 + 3h) = 0.1/2.3 = 0.0435 \right]$

n	x_n	y_n	$0.7391y_n$	$2x_n^2$	$2x_n h$	y_{n+1}
0	1	2	1.4782	2	0.2	1.5743
1	1.1	1.5743	1.1636	2.42	0.22	1.2789
2	1.2	1.2789	0.9452	2.88	0.24	1.0814
3	1.3	1.0814				

If $y = \dfrac{49e^{3-3x}}{27} + \dfrac{2}{27} - \dfrac{2x}{9} + \dfrac{x^2}{3}$

$\qquad y(1.1) = 1.5774$

$\qquad y(1.2) = 1.2835$

$\qquad y(1.3) = 1.0863.$

Method (a) is poor.

Method (b) nearly gives 2D accuracy.

23.2 $\quad y_{n+1} = y_n + \dfrac{h}{12} (23f_n - 16f_{n-1} + 5f_{n-2}).$

23.3 $\quad y_{n+1} = y_n + \dfrac{h}{24} (55f_n - 59f_{n-1} + 37f_{n-2} - 9f_{n-3}).$

23.4 Using Simpson's Rule gives

$$\int_{x_{n-1}}^{x_{n+1}} f(x, y)\, dx \simeq \frac{h}{3}\left(f_{n+1} + 4f_n + f_{n-1}\right)$$

so $\quad y_{n+1} = y_{n-1} + \dfrac{h}{3}\left(f_{n+1} + 4f_n + f_{n-1}\right)$

which is a closed two-step method.

23.5 (a) y_0, y_1, y_2.

(b) y_0, y_1, y_2, y_3.

(c) y_0, y_1.

23.6

$h = 0.2$	n	x_n	y_n	$f_n = -2x_n y_n$	y_{n+1}
	0	0	1	0	0.9600
	1	0.2	0.9600	-0.3840	0.8448
	2	0.4	0.8448	-0.6758	0.6805
	3	0.6	0.6805	-0.8166	0.5031
	4	0.8	0.5031		

23.8 Consider

$$y_3 = y_{-1} + \int_{x_{-1}}^{x_3} f(x, y)\, dx$$

and replace $f(x, y)$ by $f_s = f_0 + s\Delta f_0 + \frac{1}{2}s(s-1)\Delta^2 f_0$ where $s = (x - x_0)/h$. Then the integral is replaced by

$$h\left[sf_0 + \tfrac{1}{2}s^2\Delta f_0 + \left(\frac{s^3}{6} - \frac{s^2}{4}\right)\Delta^2 f_0\right]_{-1}^{3} = \frac{4h}{3}\left(2f_0 - f_1 + 2f_2\right)$$

after manipulation.

FURTHER READING

We have yet to discover a text which treats this subject as we would like. The normal treatment is usually tied up with predictor–corrector methods, which we consider in the next lesson. We therefore delay recommending any further reading until the end of that lesson.

LESSON 23 – COMPREHENSION TEST

1. Derive the general recurrence relation

$$y_{n+1} = y_n + \int_{x_n}^{x_{n+1}} f(x, y)\, dx.$$

2. Which of these are one-step methods?
(a) Taylor-polynomial methods.
(b) Runge–Kutta methods.
(c) Adams–Bashforth methods.
(d) The method based on the trapezium rule.
(e) The method based on the mid-point rule.
(f) The method based on Simpson's rule.
(g) The method based on Milne's formula.

3. Which of (c) through (g) in Question 2 are based on explicit (open) formulae?

4. How are the starting values for a multi-step method normally obtained?

5. Explain why a closed formula is not directly applicable in general.

6. If an open and a closed formula of the same order have truncation errors $Ah^p y^{(p)}(\xi_1)$ and $Bh^p y^{(p)}(\xi_2)$ respectively, is $A > B$ or is $B > A$?

ANSWERS

1 See Section 23.1.

2 (a), (b) and (d).

3 (c), (e) and (g).

4 A one-step method of the same order is used, typically a Runge–Kutta method.

5 Because it is an implicit equation for y_{n+1}, requiring an iterative method for its solution (see Lesson 24).

6 $A > B$. (The closed formula is more accurate.)

LESSON 23 – SUPPLEMENTARY EXERCISES

23.9. Obtain the recurrence relations for the methods based on (a) Simpson's rule and (b) Milne's formula, as applied to the linear ODE $y^{(1)} = x^2 - 2y$. Given that $y(0) = -1$ use the recurrence relations to calculate

y_1 through y_5 with $h = 0.02$. Compare the results with each other and the true values given by $y = \frac{1}{4}(2x^2 - 2x + 1 - 5e^{-2x})$. (Use this formula to calculate any starting values required.) Work throughout to 7D. (See also Exercise 22.10.)

23.10. Show that applying Simpson's rule to the nonlinear ODE $y^{(1)} = 4x\sqrt{y}$ does not produce an explicit recurrence relation.

23.11. Obtain the closed, 3-step, Nyström formula

$$y_{n+1} = y_n + \tfrac{3}{8}h(f_{n+1} + 3f_n + 3f_{n-1} + f_{n-2})$$

based on the $\frac{3}{8}$ths rule

$$\int_{x_0}^{x_3} g(x)\, dx \simeq \tfrac{3}{8}h(g_0 + 3g_1 + 3g_2 + g_3).$$

Apply it to the problem of Exercise 23.9.

*23.12. The Adams–Bashforth formulae are open, being based on replacing f in $\int_{x_n}^{x_{n+1}} f$ by its interpolating polynomial for the points $x_{n-r}, x_{n-r+1}, \ldots, x_n$ for $r \in \{0, 1, 2, 3, \ldots\}$. Corresponding closed formulae of the same orders can be obtained by replacing f by its interpolating polynomial for the points $x_{n-r+1}, x_{n-r+2}, \ldots, x_{n+1}$. Show that the method of this kind which uses linear interpolation is just the trapezium rule, and obtain these methods based on quadratic and cubic interpolation:

$$y_{n+1} = y_n + \frac{h}{12}(5f_{n+1} + 8f_n - f_{n-1})$$

$$y_{n+1} = y_n + \frac{h}{24}(9f_{n+1} + 19f_n - 5f_{n-1} + f_{n-2}).$$

(See Example 23.3 and Exercises 23.2 and 23.3.)

Lesson 24 **Predictor-Corrector Methods**

OBJECTIVES

At the end of this lesson you should

(a)   understand what a predictor formula is;

(b)   understand what a corrector formula is;

(c)   know how to use a predictor—corrector pair of formulae, whether one-step or multi-step;

(d)   have a qualitative appreciation of the errors involved;

(e)   be able to compare the advantages and disadvantages of the Runge—Kutta methods and the predictor—corrector methods.

CONTENTS

Answers to exercises. Further reading.
Comprehension test and answers. Supplementary exercises.

24.1   INTRODUCTION

In Lesson 23 we saw how integration formulae could be used to approximate for instance

$$\int_{x_n}^{x_{n+1}} f(x, y)\, \mathrm{d}x$$

in the formula

$$y_{n+1} = y_n + \int_{x_n}^{x_{n+1}} f(x, y)\, dx.$$

In particular we gained the closed formula (from the trapezium rule)

$$y_{n+1} = y_n + \tfrac{1}{2}h(f_n + f_{n+1})$$

or

$$y_{n+1} = y_n + \tfrac{1}{2}h\{f(x_n, y_n) + f(x_{n+1}, y_{n+1})\}.$$

Unless $f(x, y)$ is linear in y this formula is not explicit nor can it be re-arranged as an explicit formula, i.e., we do not have a direct way of calcu-lating y_{n+1}.

However, if we had an approximation for y_{n+1}, (let us call it y_{n+1}^*) we could use this value in the expression $f(x_{n+1}, y_{n+1})$ on the right-hand side and so find (hopefully) a better approximation for y_{n+1} (let us call it y_{n+1}^{**}).

Thus

$$y_{n+1}^{**} = y_n + \tfrac{1}{2}h\{f(x_n, y_n) + f(x_{n+1}, y_{n+1}^*)\}. \qquad \text{[24-1]}$$

To recap, if we can guess y_{n+1}^* we can find y_{n+1}^{**}. So all we need is a value for y_{n+1}^*, and an obvious possibility for finding a reasonable value is by using an *explicit* formula. We could, for example, use Euler's formula. (Recall that Euler's formula has an error $O(h^2)$ and the trapezium-rule formula has an error $O(h^3)$).

Hence y_{n+1}^* can be found from

$$y_{n+1}^* = y_n + hf(x_n, y_n). \qquad \text{[24-2]}$$

A diagram may help us to see the position (Fig. 24.1).

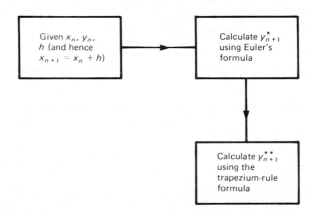

Fig. 24.1

But now we have two estimates of y_{n+1}, i.e., y^*_{n+1} and y^{**}_{n+1}, and we can compare them to find out if they are close enough. If ϵ is the local accuracy required we can accept y^{**}_{n+1} as being the required value of y_{n+1} (to within ϵ), in something like the self-consistent sense described in Lesson 21, Section 21.3, if $|y^{**}_{n+1} - y^*_{n+1}| < \epsilon$. y^{**}_{n+1} will be better than y^*_{n+1} because it arises from a formula with a smaller truncation error.

Note that this method is essentially the same as the second-order Runge–Kutta method of Lesson 22. After the next example we shall develop the essential difference between the Runge–Kutta and predictor–corrector implementations.

EXAMPLE 24.1   Consider $dy/dx = x - y$ with $y(0) = 1$. Use the Euler and trapezium-rule formulae [24-1] and [24-2] with $h = 0.1$, and let $\epsilon = 0.1$.

With these values we can work safely with 3D provided that we only do a small number of steps. In fact we shall calculate just y_1, y_2 and y_3.

n	x_n	y_n	$f_n = x_n - y_n$	y^*_{n+1}	$f_{n+1} = x_{n+1} - y^*_{n+1}$	y^{**}_{n+1}
0	0	1	-1	0.9	$0.1 - 0.9 = -0.8$	0.910
1	0.1	0.910	-0.810	0.829	$0.2 - 0.829 = -0.629$	0.838
2	0.2	0.838	-0.638	0.774	$0.3 - 0.774 = -0.474$	0.782
3	0.3	0.782				

Clearly we can proceed as far as we like! As it happens ϵ is fairly large and y^*_{n+1} and y^{**}_{n+1} agree to within ϵ at each stage.

If $|y^{**}_{n+1} - y^*_{n+1}| \geqslant \epsilon$, i.e., if our test fails, we must clearly find another estimate of y_{n+1} which is better than y^{**}_{n+1}. The obvious thing to try is a further application of the trapezium-rule formula using y^{**}_{n+1} instead of y^*_{n+1}. This gives

$$y^{***}_{n+1} = y_n + \tfrac{1}{2}h\{f(x_n, y_n) + f(x_{n+1}, y^{**}_{n+1})\}.$$

The values y^{***}_{n+1} and y^{**}_{n+1} must now be compared. Clearly if we proceed with a notation like this we could have rather a lot of stars on the page! Fig. 24.2 shows how y^*_{n+1} can (in computer terminology) be over-written so that we need only two estimates for y_{n+1} at any stage, i.e., y^*_{n+1} and y^{**}_{n+1}.

At A the old value of y^*_{n+1} is replaced by the value of y^{**}_{n+1}, and on returning to B, a new value for y^{**}_{n+1} is calculated. This flow diagram gives the method for calculating y_{n+1} from y_n by a one-step method and is clearly an example of an iterative method of the kind considered in Lesson 5.

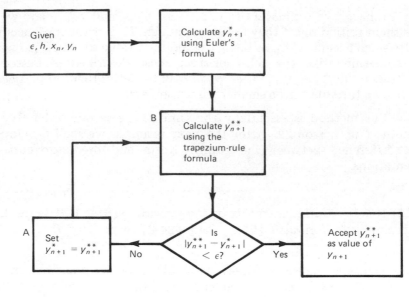

Fig. 24.2

In this discussion the Euler formula has been used to *predict* a value for y_{n+1} and the trapezium-rule formula has been used (possibly many times) to *correct* this value until convergence is obtained. In this context the Euler formula is called a *predictor* formula and the trapezium-rule formula a *corrector* formula. Put together they are called a *predictor—corrector pair*. We shall consider other such pairs later in this lesson.

EXAMPLE 24.2 Using the problem of Example 24.1 we shall look for more accurate solutions from the same formulae and the same step-length by decreasing the local error criterion to $\epsilon = 0.001$. We expect that this will mean iterating with the corrector (the trapezium-rule formula) as described above. To match the reduced ϵ we work to 4D.

n	x_n	y_n	$f_n = x_n - y_n$	y_{n+1}^*	$f_{n+1} = x_{n+1} - y_{n+1}^*$	y_{n+1}^{**}
0	0	1	-1	0.9	-0.8	0.9100
				0.9100	-0.8100	0.9095
1	0.1	0.9095	-0.8095	0.8286	-0.6286	0.8376
				0.8376	-0.6376	0.8371
2	0.2	0.8371	-0.6371	0.7698	-0.4698	0.7818
				0.7818	-0.4818	0.7812
3	0.3	0.7812				

We see that the corrector needed to be applied twice at each step, the correction stopping because

$$|0.9095 - 0.9100| < \epsilon = 0.001$$
and
$$|0.8371 - 0.8376| < \epsilon$$
and
$$|0.7812 - 0.7818| < \epsilon.$$

In the above example we have laid out the calculations in an obvious but inefficient form. As y^*_{n+1} changes, only those terms in the corrector affected by y^*_{n+1} need to be recalculated. It is worthwhile to treat each problem separately by dividing the actual corrector for that problem into those terms depending on y^*_{n+1} and those not, as follows.

EXAMPLE 24.3   For the problem and formulae of Example 24.2
$$y^{**}_{n+1} = y_n + 0.05(x_n - y_n + x_{n+1} - y^*_{n+1})$$
$$= y_n + 0.05(2x_n + 0.1 - y_n) - 0.05y^*_{n+1}.$$

Separating out the terms not involving y^*_{n+1}, let
$$g_n = 0.95y_n + 0.1x_n + 0.005$$
so that
$$y^{**}_{n+1} = g_n - 0.05y^*_{n+1}.$$

At each correction g_n is unchanged, giving us the new, more efficient scheme

n	x_n	y_n	f_n	g_n	y^*_{n+1}	y^{**}_{n+1}
0	0	1	-1	0.955	0.9	0.9100
					0.9100	0.9095
1	0.1	0.9095	-0.8095	0.8790	0.8286	0.8376
					0.8376	0.8371
2	0.2	0.8371	-0.6371	0.8202	0.7698	0.7817
					0.7817	0.7811
3	0.3	0.7811				

We see that the different form of the calculation produces a slight change in the last result, due simply to the different effects of rounding-off. We see also that we could now dispense with the last column and simply produce successive corrections in one column.

To ensure that you see the difference between the way the calculations are performed you should check the working of one step in each of Examples 24.2 and 24.3.

24.2   THE GENERAL FORM OF A PREDICTOR–CORRECTOR METHOD

We can generalise the approach we have just described. Fig. 24.3 does this in a flow diagram which is simply an adaptation of Fig. 24.2.

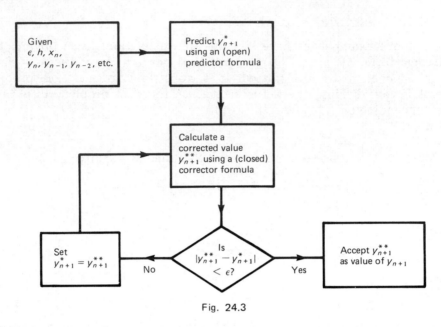

Fig. 24.3

In the 'given' box we may have to have y_{n-1}, y_{n-2}, etc., as well as y_n because the methods of prediction and correction may be based on multistep formulae.

Note that the flow diagram given here is just for the one step from x_n to x_{n+1}. A complete predictor-corrector algorithm would have an outer loop in which n increased in steps of one from its lowest possible value (taking account of the fact that some starting values may need to be found first) up to the maximum required value.

Fig. 24.4 shows an outline of the complete process.

This flow diagram assumes that the x_n and y_n values would be held in arrays in a computer program. This is not strictly necessary but it does make the flow diagram easier to construct.

24.3   EXAMPLES OF PREDICTOR–CORRECTOR PAIRS

Here are some of the more well-known examples of predictor—corrector pairs.

(a) The *Euler-trapezium* pair is, as before,

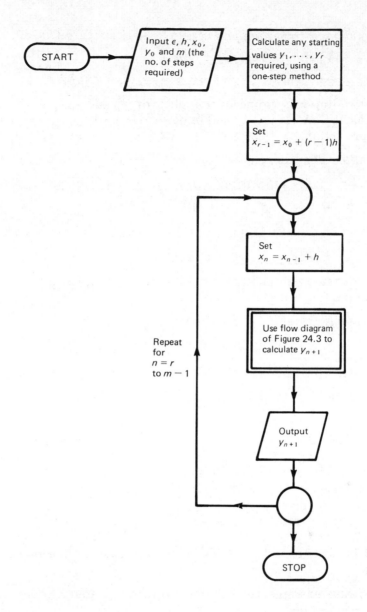

Fig. 24.4

$$y_{n+1} = y_n + hf(x_n, y_n) + O(h^2)$$
$$y_{n+1} = y_n + \tfrac{1}{2}h\{f(x_n, y_n) + f(x_{n+1}, y_{n+1})\} + O(h^3).$$ [24-3]

This is a one-step pair and needs only the value of y_n in the 'given' box in the flow diagram above.

(b) The *mid-point-trapezium pair* is

$$y_{n+1} = y_{n-1} + 2hf(x_n, y_n) + O(h^3)$$

$$y_{n+1} = y_n + \tfrac{1}{2}h\{f(x_n, y_n) + f(x_{n+1}, y_{n+1})\} + O(h^3). \qquad [24\text{-}4]$$

This is a two-step pair and needs the values of y_n and y_{n-1} as starting values for the calculation of y_{n+1}, and in particular needs y_0 and y_1 before even the first prediction.

(c) The *Milne–Simpson pair* is

$$y_{n+1} = y_{n-3} + (4h/3)\{2f(x_n, y_n) - f(x_{n-1}, y_{n-1})$$

$$+ \, 2f(x_{n-2}, y_{n-2})\} + O(h^5)$$

$$y_{n+1} = y_{n-1} + (h/3)\{f(x_{n+1}, y_{n+1}) + 4f(x_n, y_n)$$

$$+ \, f(x_{n-1}, y_{n-1})\} + O(h^5). \qquad [24\text{-}5]$$

This is a four-step pair and needs the values of $y_{n-3}, y_{n-2}, y_{n-1},$ and y_n as starting values for the calculation of y_{n+1}.

(d) The *Adams–Moulton* (three-step) *pair* is

$$y_{n+1} = y_n + (h/12)\{23f(x_n, y_n) - 16f(x_{n-1}, y_{n-1})$$

$$+ \, 5f(x_{n-2}, y_{n-2})\} + O(h^4)$$

$$y_{n+1} = y_n + (h/12)\{5f(x_{n+1}, y_{n+1}) + 8f(x_n, y_n)$$

$$- \, f(x_{n-1}, y_{n-1})\} + O(h^4). \qquad [24\text{-}6]$$

This pair needs the values of y_n, y_{n-1} and y_{n-2} as starting values for the calculation of y_{n+1}. The Adams–Moulton formulae will probably be the least familiar. The predictor is the Adams–Bashforth third-order formula obtained in Lesson 23 by integrating over $[x_n, x_{n+1}]$ a quadratic interpolating polynomial for $f(x, y(x))$ agreeing with f at x_{n-2}, x_{n-1} and x_n. The corrector is obtained similarly, but to make it implicit the polynomial agrees with f at x_{n-1}, x_n and x_{n+1}. Similar Adams–Moulton pairs can be constructed of other orders, the fourth-order pair being quite popular.

The following example shows how powerful a fourth-order predictor–corrector pair can be.

EXAMPLE 24.4   Continuing our study of $dy/dx = x - y,$ $y(0) = 1$, from the examples of Section 24.1, we shall decrease ϵ further to 0.000 01 (and work to 6D in consequence), using the Milne–Simpson pair given in (c) above. (Since this pair are $O(h^5)$, whereas the Euler-trapezium pair are $O(h^3)$ at best, we might expect to be able to reduce ϵ by a factor of 0.01 (since $h = 0.1$) and still require only two corrections at each step.)

For our particular problem

$$y^*_{n+1} = y_{n-3} + (4h/3)(2f_{n-2} - f_{n-1} + 2f_n)$$

$$y^{**}_{n+1} = y_{n-1} + (h/3)(f_{n-1} + 4f_n + f(x_{n+1}, y^*_{n+1}))$$

i.e. $\quad y^*_{n+1} = y_{n-3} + (0.4/3)(2f_{n-2} - f_{n-1} + 2f_n)$

$$y^{**}_{n+1} = \{y_{n-1} + (0.1/3)(f_{n-1} + 4f_n + x_{n+1})\} - 0.1y^*_{n+1}/3$$

$$= g_n - 0.1y^*_{n+1}/3, \quad \text{say.}$$

To be able to compare our results with those obtained previously we shall suppose that a fourth-order one-step method has been applied to the problem with $h = -0.1$ (i.e., stepping 'backwards') giving

$$y(-0.1) = 1.110\,342$$

$$y(-0.2) = 1.242\,806$$

$$y(-0.3) = 1.399\,718.$$

With these and $y(0) = 1$ as starting values we can now estimate $y(0.1)$, $y(0.2)$ and $y(0.3)$ as before.

n	x_n	y_n	f_n	g_n	y_{n+1}
-3	-0.3	1.399 718	—	—	—
-2	-0.2	1.242 806	$-1.442\,806$	—	—
-1	-0.1	1.110 342	$-1.210\,342$	—	—
0	0	1	-1	0.939 997	0.909 682 (P)
					0.909 674 (C)
1	0.1	0.909 674	$-0.809\,674$	0.865 377	0.837 468 (P)
					0.837 461 (C)
2	0.2	0.837 461	$-0.637\,461$	0.807 691	0.781 642 (P)
					0.781 636 (C)

We see that only one correction is needed for $\epsilon = 0.000\,01$, probably due in part to the error in Simpson's rule having a coefficient of $1/90$ compared with the trapezium rule's $1/12$. In fact the corrected values are in error by at most one in the sixth decimal place. (The exact solution is $y = 2e^{-x} + x - 1$.)

Having seen the power of this method we might be tempted to increase the step-length considerably, especially if what we actually need is an estimate of $y(50)$ for example. Let us try $h = 1$, still asking for $\epsilon = 0.000\,01$. This time we shall need

$$y(-1) = 3.436\,564$$

$$y(-2) = 11.778\,112$$

$$y(-3) = 36.171\,074$$

and suitable adjustments to the formulae for y^*_{n+1} and y^{**}_{n+1} to take account of $h = 1$.

n	x_n	y_n	f_n	g_n	y_{n+1}
-3	-3	36.171 074	—	—	—
-2	-2	11.778 112	$-$ 13.778 112	—	—
-1	-1	3.436 564	$-$ 4.436 564	—	—
0	0	1	$-$ 1	0.957 709	2.678 194 (P)
					0.064 978 (C)
					0.936 050 (C)
					0.645 692 (C)
					0.742 478 (C)
					0.710 216 (C)
					0.720 970 (C)
					0.717 386 (C)

This is obviously converging — but very slowly. It looks as though we have gone too far in increasing h. In fact with $h = 1$ the truncation error is approximately 0.02. Let us reduce our sights and try $h = 0.5$, starting from

$$y(-0.5) = 1.797\,443$$

$$y(-1.0) = 3.436\,564$$

$$y(-1.5) = 6.463\,378$$

and with suitable adjustments again to the formulae for y^*_{n+1} and y^{**}_{n+1}.

n	x_n	y_n	f_n	g_n	y_{n+1}
-3	-1.5	6.463 378	—	—	—
-2	-1.0	3.436 564	$-$ 4.436 564	—	—
-1	-0.5	1.797 443	$-$ 2.297 443	—	—
0	0	1	$-$ 1	0.831 202	0.746 255 (P)
					0.706 826 (C)
					0.713 398 (C)
					0.712 302 (C)
					0.712 485 (C)
					0.712 454 (C)
					0.712 460 (C)
1	0.5	0.712 460			

If we are looking for the largest step-length that will give a 'reasonable' convergence rate then it appears that $h = 0.5$ is fairly close to it, and requires, for this problem, of the order of six corrections per step for $\epsilon = 0.000\,01$.

Finally in this section, a cautionary note: the ϵ-criterion indicates the accuracy to which the implicit (corrector) equation for y_{n+1} is solved. It does not mean that y_{n+1} is necessarily correct to within $\pm \epsilon$. To assess the accuracy of the values obtained we would normally have to resort again to step-halving and checking, just as we would with Taylor-polynomial and Runge–Kutta methods. Unfortunately such checking is much more complicated when applied to predictor–corrector methods. We shall not consider it in this course.

If it *is* possible to estimate the truncation error of the corrector at each step then ϵ should be set at a value a little less than the truncation error so that the maximum accuracy is obtained with the minimum of corrections. In the last part of the example above (with $h = 0.5$) the truncation error was of the order of 0.0006 (using our knowledge of the exact solution) so $\epsilon = 0.000\,01$ could reasonably have been ten times larger.

EXERCISE 24.1

Consider $y^{(1)} = -2xy$, $y(0) = 1$. Estimate $y(0.4)$ and $y(0.6)$, using $h = 0.2$ and $\epsilon = 0.0005$ with the mid-point-trapezium pair of rules; i.e.,

$$y_{n+1}^* = y_{n-1} + 2hf(x_n, y_n)$$
$$y_{n+1}^{**} = y_n + \tfrac{1}{2}h\{f(x_n, y_n) + f(x_{n+1}, y_{n+1}^*)\}.$$

To start the calculation assume that $y(0.2) = 0.960\,79$ correct to 5D. Work to 5D throughout. (*Hint*: Don't forget to split y_{n+1}^{**} into those terms that involve y_{n+1}^* and those that do not.)

EXERCISE 24.2

Apply the Adams–Moulton three-step pair of Equations [24-6] to the problem of Example 24.4, i.e., $y^{(1)} = x - y$, $y(0) = 1$, with $h = 0.1$ and $\epsilon = 0.000\,01$, and assuming

$y(-0.1) = 1.110\,342$

$y(-0.2) = 1.242\,806$

$y(-0.3) = 1.399\,718$.

Obtain approximations to $y(0.1)$, $y(0.2)$ and $y(0.3)$, working to 6D throughout.

24.4 CHOICE AND APPLICATION OF PREDICTOR–CORRECTOR PAIRS

There are clearly a great number of possible ways of constructing predictor–corrector pairs. In practice the number of pairs used is quite small. It is intuitively fairly obvious that the overall accuracy of a pair is essentially the accuracy of the corrector; the predictor merely starts the iterative process for solving the implicit (corrector) equation for y_{n+1}.

To limit the corrections to a reasonable number requires a reasonable predicted value, but not an exceptionally good one. The generally-accepted view is that the order of the predictor should be either the same as that of the corrector (as in mid-point-trapezium) or one less (as in Euler-trapezium).

It is also widely accepted that fourth-order methods provide the best compromise between accuracy and economy of calculation, although such methods are more likely to be unstable than low-order methods. Thus, in the past, Milne–Simpson and fourth-order Adams–Moulton have been much used and investigated, and they provide good results for all but the most difficult problems.

Having chosen a method, the user has to decide on a step-length to meet an accuracy requirement. We have seen in Example 24.4 how this can be an important decision. In the simpler computer programs for predictor–corrector methods an upper limit is set to the number of corrections allowed; if the number is exceeded then non-convergence is assumed and the program stops. In Exercise 24.2 below we ask you to draw a flow diagram for such a program. If the limit were reached the user would have to choose a smaller step-length and try again. This is clearly far from ideal and the more powerful algorithms have an automatic step-halving and step-doubling mechanism which tries to maintain the number of corrections at some optimum level and also provides the step-halving required to check the solution at each stage. You might like to consider what would be involved in constructing such an algorithm. Needless to say we shall not attempt it here!

EXERCISE 24.3

Combine the flow diagrams of Sections 24.2 and introduce a limit on the number of corrections (call it L, say), to produce a convergence-checking predictor–corrector flow diagram.

24.5 PREDICTOR–CORRECTOR METHODS AND RUNGE–KUTTA METHODS – ADVANTAGES AND DISADVANTAGES

RUNGE–KUTTA METHODS

(a) They are self-starting, i.e., only x_n and y_n need to be known to calculate y_{n+1}.

(b) They are explicit formulae, i.e., no iterative processes are involved.

(c) Programming Runge—Kutta methods for a computer is relatively easy.

(d) Error estimation is difficult and step-halving and checking must be applied (see Lesson 21, Section 21.3).

(e) High-order formulae are available. They may allow large values of h but may involve much calculation.

(f) Many evaluations of $f(x, y)$ are usually required, implying a large amount of calculation time.

PREDICTOR—CORRECTOR METHODS

(a) They may be self-starting but will then have low-order truncation errors which may imply very small values of h for even reasonable accuracy. However, the commonly used methods require more than one starting value.

(b) These methods are implicit and require an iterative solution. To avoid an excessive number of iterations h may have to be fairly small.

(c) Error terms are available but may yield only qualitative information. In consequence to achieve a specific accuracy in the solution step-halving and checking will usually be necessary.

(d) High-order formulae are available. These may allow fairly large values of h with few iterations and few consequent evaluations of $f(x, y)$, but may introduce instability.

(e) Programming Predictor—Corrector Methods is not easy because of the necessity of step-halving and checking. If h must be halved new intermediate values of y $(y_n, y_{n-2}, $ etc.) will be necessary for restarting the integration rules. These can be calculated by Runge—Kutta methods — but the resulting computer program will be quite complicated.

(f) If h can be taken at a reasonably large value and does not have to be changed these methods can prove fast and accurate, involving as they do fewer evaluations of $f(x, y)$ than the corresponding Runge—Kutta methods.

A very short summary would say that Runge—Kutta methods are slow, but are generally simple and can be made stable, whereas predictor—corrector methods are generally fast but may be complicated, and possibly unstable.

ANSWERS TO EXERCISES

24.1 $y_{n+1}^* = y_{n-1} + 0.4f_n$

$y_{n+1}^{**} = y_n + 0.1(f_n - 2x_{n+1}y_{n+1}^*)$

$\qquad = \{y_n + 0.1f_n\} - 0.2x_{n+1}y_{n+1}^*$

$\qquad = g_n - 0.2x_{n+1}y_{n+1}^*, \quad$ say

where

$$g_n = y_n + 0.1f_n.$$

n	x_n	y_n	y_{n-1}	f_n	g_n	y_{n+1}
0	0	1	—	—	—	—
1	0.2	0.960 79	1	$-0.384\,32$	0.922 36	0.846 27 (P)
						0.854 66 (C)
						0.853 99 (C)
						0.854 04 (C)
2	0.4	0.854 04	0.960 79	$-0.683\,23$	0.785 72	0.687 50 (P)
						0.703 22 (C)
						0.701 33 (C)
						0.701 56 (C)
3	0.6	0.701 56				

The true solutions are (from $y = e^{-x^2}$)

$$y(0.4) = 0.8521 \quad \text{to 4D}$$
$$y(0.6) = 0.6977 \quad \text{to 4D}$$

clearly showing that solving the corrector equation to 3D does *not* mean the solution estimates are correct to 3D. The errors in the estimated values are of course those of the trapezium rule.

24.2 $$y^*_{n+1} = y_n + \frac{h}{12}\{23f_n - 16f_{n-1} + 5f_{n-2}\}$$

$$y^{**}_{n+1} = y_n + \frac{h}{12}\{5f(x_{n+1}, y^*_{n+1}) + 8f_n - f_{n-1}\}$$

$$= y_n + \frac{h}{12}\{5(x_{n+1} - y^*_{n+1}) + 8f_n - f_{n-1}\}$$

$$= g_n - \frac{5h}{12}y^*_{n+1}$$

where

$$g_n = y_n + \frac{h}{12}\{5x_{n+1} + 8f_n - f_{n-1}\}.$$

n	x_n	y_n	f_n	g_n	y_{n+1}
-3	-0.3	1.399 718	—	—	—
-2	-0.2	1.242 806	$-1.442\,806$	—	—
-1	-0.1	1.110 342	$-1.210\,342$	—	--

n	x_n	y_n	f_n	g_n	y_{n+1}
0	0	1	-1	0.947 586	0.909 596 (P)
					0.909 686 (C)
					0.909 683 (C)
1	0.1	0.909 683	$-0.809\,683$	0.872 371	0.837 396 (P)
					0.837 479 (C)
					0.837 476 (C)
2	0.2	0.837 476	$-0.637\,476$	0.814 225	0.781 584 (P)
					0.781 659 (C)
					0.781 656 (C)
3	0.3	0.781 656			

The true solutions (from $y = 2e^{-x} + x - 1$) are 0.909 675, 0.837 462 and 0.781 636, correct to 6D.

FURTHER READING

For a fairly complete and readable account covering Lessons 20 to 24 see either of Stark, Chapter 7 (except Sections 7.13 and 7.16) or Williams, Chapter 5.

Be prepared, however, to skip sections which use mathematics beyond your experience.

Most accounts of the numerical solution of ODE's involve partial differentiation. If you are familar with this then either of Phillips and Taylor, Sections 11.3, 11.7 and 11.8, or Conte, Sections 6.3, 6.5, 6.6, 6.7 and 6.8 may be suitable, but they are at a somewhat higher mathematical level than this course.

LESSON 24 – COMPREHENSION TEST

1. In a predictor—corrector pair of formulae which one is open and which one closed?

2. If a large number of corrections is required at each step of a predictor—corrector solution process should we make the step-length larger or smaller?

3. If the corrector equation is solved for y_{n+1} correct to 6D (say), what does this tell us about the accuracy of the final value of y_{n+1} as an approximation to $y(x_{n+1})$?

4. (a) In constructing a predictor—corrector pair what is the rule normally used to decide the relative orders of the errors in the two formulae?

(b) Which of the two errors dominates the error in the approximate solution values?

5. In general, in comparison with Runge—Kutta methods, are predictor—corrector methods:

(a) faster?   (b) simpler?   (c) more stable?

ANSWERS

1   The predictor is open and the corrector is closed.

2   Smaller.

3   Nothing. The accuracy of the final y_{n+1} is determined by the truncation error of the corrector, provided the working accuracy is greater (i.e., the round-off errors smaller).

4   (a) The predictor should be of the same order as, or one order less than the corrector.

   (b) The corrector error.

5   (a) Yes,   (b) no,   (c) no.

LESSON 24 – SUPPLEMENTARY EXERCISES

24.4. Apply the Euler-trapezium pair of Equations [24-3] to the problem of Exercise 20.4, i.e., $y^{(1)} = 4x\sqrt{y}$ and $y(1) = 1$, obtaining values of y at $x = 1.25(0.25)2.0$. Solve the corrector equation to 2D at each step. (Use $\epsilon = 0.001$.) Do you get the solution values correct to 2D? Why not? (The solution is $y = x^4$; cf. Exercises 20.4, 21.5, 22.7 and 22.9.)

24.5. Apply the fourth-order Adams—Moulton pair

$$y_{n+1}^{*} = y_n + \frac{h}{24}(55f_n - 59f_{n-1} + 37f_{n-2} - 9f_{n-3})$$

$$y_{n+1}^{**} = y_n + \frac{h}{24}(9f(x_{n+1}, y_{n+1}^{*}) + 19f_n - 5f_{n-1} + f_{n-2})$$

to the problem of Exercise 20.6, i.e., $y^{(1)} = -y^2/(1 + 2x)$ with $y(0) = 2$, performing 2 steps with $h = 0.04$ and working to 6D. Solve the corrector equation to 4D at each step. (Use $\epsilon = 10^{-5}$.) Obtain the necessary starting values $y(-0.04), y(-0.08), \ldots$ by any fourth-order one-step method (cf. Exercises 20.6, 21.6 and 22.8).

*24.6. When the predictor and corrector are of the same order the truncation-error expressions can often be used to improve the first correction to such an extent that no further corrections are necessary (indeed they would make the result worse in general).

Reconsider Exercise 24.1. The error in y^*_{n+1} (the mid-point rule) is

$$y_{n+1} - y^*_{n+1} = \tfrac{1}{3}h^3 y^{(3)}(\xi_1)$$

and in y^{**}_{n+1} (the trapezium rule) is

$$y_{n+1} - y^{**}_{n+1} = -\tfrac{1}{12}h^3 y^{(3)}(\xi_2)$$

where ξ_1 and ξ_2 lie in $[x_n, x_{n+1}]$ but are otherwise unknown. If we make the familiar assumption that $y^{(3)}$ does not vary much in $[x_n, x_{n+1}]$, then

$$y_{n+1} - y^{**}_{n+1} \simeq -\tfrac{1}{4}(y_{n+1} - y^*_{n+1}).$$

Hence

$$y_{n+1} \simeq \tfrac{1}{5}(4y^{**}_{n+1} + y^*_{n+1}).$$

If our assumption is correct this extrapolated value will be an order of accuracy better than the final y^{**}_{n+1} (the solution of the corrector equation).

Apply the formula obtained above to the results of Exercise 24.1. (The true solution-values are given in the Answers to Exercises section).

Given that the truncation errors for the formulae in Exercise 24.5 are of the form

$$\tfrac{251}{720}h^5 y^{(5)}(\xi_3) \quad \text{and} \quad -\tfrac{19}{720}h^5 y^{(5)}(\xi_4)$$

respectively, find an approximate expression for y_{n+1} in terms of y^*_{n+1} and y^{**}_{n+1} (as above). Apply it to the results of Exercise 24.5. (The true solution is $y = 2/(1 + \ln(1 + 2x))$.)

Appendix 1 The Triangle Inequality

FOR REAL NUMBERS

Let $a, b \in \mathbb{R}$; then

$$|a + b| \leqslant |a| + |b|. \qquad\qquad\qquad\qquad\qquad\qquad\text{[A1-1]}$$

It is not difficult to see that this is true. If a and b have the same sign then $|a + b| = |a| + |b|$.

E.g., $|5 + 7| = |12| = 12$

$\qquad |5| + |7| = 5 + 7 = 12$

and $|-5 - 7| = |-12| = 12$

$\qquad |-5| + |-7| = 5 + 7 = 12.$

On the other hand, if a and b have opposite signs then $|a + b| < |a| + |b|$.

E.g., $|-5 + 7| = |2| = 2$

$\qquad |-5| + |7| = 5 + 7 = 12.$

Lastly, if one or both of a and b are zero, then obviously $|a + b| = |a| + |b|$.

E.g. $|a + 0| = |a|$

$\qquad |a| + |0| = |a|.$

Hence, if a and b are *any* real numbers,

$$|a + b| \leqslant |a| + |b|.$$

Note that we have not given a formal proof. Illustrations such as $|5 + 7| \leqslant |5| + |7|$ do not constitute part of a proof.

FOR COMPLEX NUMBERS

The inequality is also true for complex numbers, when it says, in effect, that the sum of the lengths of two sides of a triangle ($|a| + |b|$) is greater than, or equal to, the length of the third side ($|a + b|$).

Hence the name 'triangle inequality'. What does the triangle look like when $|a + b| = |a| + |b|$?

Appendix 2 The First Mean Value Theorem (MVT)

STATEMENT OF THE THEOREM

If $f(x)$ is continuous in $[a, b]$ and differentiable in (a, b) then there exists a number ξ in (a, b), (i.e., $a < \xi < b$) such that

$$f'(\xi) = \frac{f(b) - f(a)}{b - a} ;$$

i.e. $f(b) = f(a) + (b - a)f'(\xi)$.

Note that the theorem only says that ξ exists. It does not say how it might be found.

EXPLANATION

We can show what the theorem means geometrically. The tangent to the curve at $x = \xi$ is parallel to the chord AB joining the points $(a, f(a))$, $(b, f(b))$ — when f satisfies the conditions of the theorem. (See Fig. A2.1.)

The proof of the theorem does not concern us; it can be found in elementary books on analysis. However it is worth considering the statement of the theorem for a moment.

By '$f(x)$ is continuous in $[a, b]$' we mean that the graph of f has no breaks in it between (and including) the points $(a, f(a))$ and $(b, f(b))$. By '$f(x)$ is differentiable in (a, b)' we mean that the derivative of f exists (i.e., its graph has a well-defined tangent) at all points between $x = a$ and $x = b$, but not necessarily at $x = a$ or $x = b$.

If you wish you can leave the understanding of this theorem at this geometrical, intuitive level. If you wish to go further, here are some examples which may illuminate the theorem.

EXAMPLE A2.1   Consider $f(x) = 4x - x^2$, $a = 0$ and $b = 3$. f is continuous on $[0, 3]$ and differentiable on $(0, 3)$ so there exists a number ξ such that

$$f(3) - f(0) = (3 - 0)f'(\xi)$$

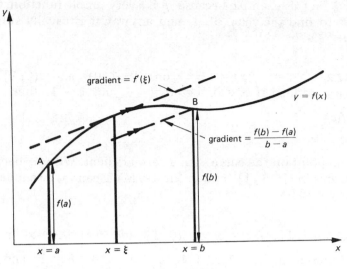

gradient = $f'(\xi)$

B

$y = f(x)$

gradient = $\dfrac{f(b) - f(a)}{b - a}$

$f(b)$

A

$f(a)$

$x = a$ $x = \xi$ $x = b$ x

Fig. A2.1

and where $0 < \xi < 3$. But we know that

$$f(3) - f(0) = 4(3) - 3^2 = 3$$

∴ $f'(\xi) = 1$

i.e. $4 - 2\xi = 1$

i.e. $\underline{\xi = 3/2.}$

Here is a diagram to show you what is going on in this case.

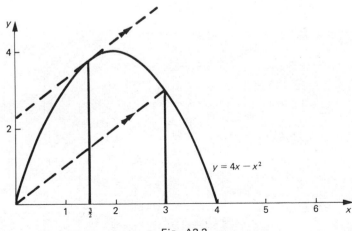

$y = 4x - x^2$

Fig. A2.2

We can find ξ in this example because f is a very simple function. Generally it is difficult to find the value of ξ and anyway it is usually sufficient to know that $a < \xi < b$.

EXAMPLE A2.2   Let $f(x) = 1/x^2$ on $[-1, 0)$ and $(0, 1]$. Notice that $f(x)$ is not defined at $x = 0$. Let $a = -1$ and $b = 1$; then

$$\frac{f(b) - f(a)}{b - a} = \frac{1 - 1}{2} = 0.$$

But there is no point on the curve with a zero gradient. Why? — because $f(x)$ is not continuous in $[-1, 1]$ and so the 1st MVT cannot be applied. Sketch the curve for yourself!

EXAMPLE A2.3   Let $f(x) = x$ on $[0, 1]$ and $f(x) = 2 - x$ on $(1, 2]$.

$f(x)$ is continuous in $[0, 2]$. Let $a = 0$ and $b = 2$ and calculate $\dfrac{f(b) - f(a)}{b - a}$ which equals $\dfrac{0 - 0}{2 - 0} = 0$.

Again there is no point at which $f'(x) = 0$. Why? — because $f(x)$, although continuous in $[0, 2]$, is not differentiable at $x = 1$ and so the 1st MVT cannot be applied. Sketch the curve for yourself.

Appendix 3 Differences of a Polynomial

The following theorem explains why the nth differences of a table of values of an nth degree polynomial are constant, and hence all higher differences zero. You should at least read and try to understand the statements of the theorem and its corollary.

THEOREM

If p_n is a polynomial of degree n then Δp_n is a polynomial of degree $n-1$.

COROLLARY

$\Delta^n p_n$ is a constant function (a polynomial of degree zero) and $\Delta^k p_n$ is the zero function for $k > n$.

PROOF OF THE THEOREM

Let $p_n(x) = a_n x^n + a_{n-1} x^{n-1} + \ldots + a_1 x + a_0 \ (a_n \neq 0)$. Then

$$\Delta p_n(x) = p_n(x+h) - p_n(x)$$

$$= a_n(x+h)^n + a_{n-1}(x+h)^{n-1} + \ldots + a_1(x+h) + a_0 - a_n x^n$$
$$- a_{n-1}x^{n-1} - \ldots - a_1 x - a_0$$

$$= a_n\{(x+h)^n - x^n\} + a_{n-1}\{(x+h)^{n-1} - x^{n-1}\} + \ldots$$
$$+ a_1\{(x+h) - x\} + a_0\{1 - 1\}$$

$$= a_n\left\{x^n + nhx^{n-1} + \frac{n(n-1)}{2}h^2 x^{n-2} + \ldots + h^n - x^n\right\}$$

$$+ a_{n-1}\{x^{n-1} + (n-1)hx^{n-2} + \ldots + h^{n-1} - x^{n-1}\} + \ldots + a_1 h$$

$$= a_n\{nhx^{n-1} + \tfrac{1}{2}n(n-1)h^2 x^{n-2} + \ldots + h^n\}$$

$$+ a_{n-1}\{(n-1)hx^{n-2} + \ldots + h^{n-1}\} + \ldots + a_1 h$$

which is a polynomial of degree $n-1$, as required. (The leading term is $a_n nhx^{n-1}$.)

The proof of the corollary is easy. Just apply the theorem n times, i.e., $\Delta^2 p_n$ is of degree $n - 2$, $\Delta^3 p_n$ is of degree $n - 3$, etc., until $\Delta^n p_n$ is of degree $n - n = 0$. Obviously higher differences (of this constant function) are zero.

*EXERCISE

Prove that $\Delta^n p_n(x) = n! \, a_n h^n$ (*Hint*: $\Delta p_n(x) = n a_n h x^{n-1}$ + a polynomial of degree $n - 2$. Apply this fact to get the leading terms of $\Delta^2 p_n$, $\Delta^3 p_n$, etc., e.g.

$$\Delta^2 p_n(x) = (n - 1)\{n a_n h\} h x^{n-2} + \text{a polynomial of degree } n - 3.)$$

Appendix 4 The Mean Value Theorem for Integrals

If (a) $f(x)$ is continuous on $[a, b]$, (b) $\int_a^b g(x)\,\mathrm{d}x$ exists, (c) $g(x) \geqslant 0$ for all $x \in [a, b]$ or $g(x) \leqslant 0$ for all $x \in [a, b]$,

then there exists a number $\xi \in (a, b)$ such that

$$\int_a^b f(x)\,g(x)\,\mathrm{d}x \;=\; f(\xi) \int_a^b g(x)\,\mathrm{d}x.$$

The proof of this theorem is to be found in books on analysis. In particular (taking $g(x) = 1$) it says that

$$\int_a^b f(x)\,\mathrm{d}x \;=\; (b-a)f(\xi).$$

This can be demonstrated geometrically as in Fig. A4.1.

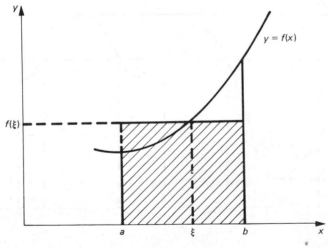

Fig. A4.1

It says that there exists a value $\xi \in (a, b)$ such that the area under the graph $y = f(x)$ between $x = a$ and $x = b$ equals the shaded area. Introducing the $g(x)$ factor simply modifies the shape of $y = f(x)$ *provided g does not change sign in* $[a, b]$ (see (c) above).

Appendix 5 The Intermediate Value Theorem

The result that we require is equivalent to: if g is a continuous function on $[x_0, x_0 + nh]$ and $x_0 < \xi_1 < \xi_2 < \ldots < \xi_n < x_0 + nh$, then there exists a $\xi \in (x_0, x_0 + nh)$ such that

$$g(\xi_1) + g(\xi_2) + \ldots + g(\xi_n) = ng(\xi).$$

This result depends on the following theorem of analysis, called the Intermediate Value Theorem, which again can be found in books on that subject.

If $f(x)$ is continuous in $[a, b]$, and, if c lies between the maximum value of $f(x)$ and the minimum value of $f(x)$ where $x \in [a, b]$, then there exists a number ξ in (a, b) such that $f(\xi) = c$.

Fig. A5.1 may help you to see what is going on.

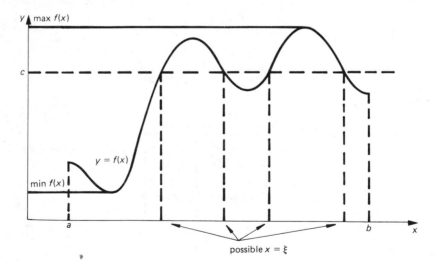

Fig. A5.1

In this case there are four possible choices for ξ.

Now you may ask what this has to do with the result we want. Well, if we take $c = (1/n)(g(\xi_1) + \ldots + g(\xi_n))$ c is the average of $g(\xi_1), \ldots, g(\xi_n)$

and so lies between the greatest of the $g(\xi_i)$ and the least of the $g(\xi_i)$. (In Fig. A5.2 $g(\xi_3)$ is the least and $g(\xi_{n-1})$ is the greatest of the $g(\xi_i)$). Hence c certainly lies between the maximum and the minimum values of $g(x)$ for $x \in [\xi_1, \xi_n]$. Therefore, since g is continuous on $[\xi_1, \xi_n]$, there exists a ξ in (ξ_1, ξ_n) such that, as required,

$$c = g(\xi).$$

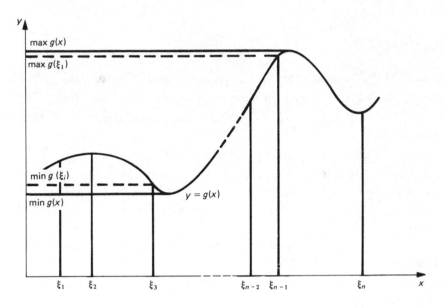

Fig. A5.2

Since

$$x_0 < \xi_1 < \xi_2 < \ldots < \xi_n < x_0 + nh$$

and $\xi_1 < \xi < \xi_n$, we certainly have, as required, $x_0 < \xi < x_0 + nh$.

Mathematicians may find this interesting if a little hard. It does show how mathematical analysis is necessary to obtain results. For non-mathematicians, if you can't follow it, or you are not particularly interested, don't worry. Perhaps it is not for you.

Bibliography

BIBLIOGRAPHY

Cohn, P. M., *Linear Equations*. Routledge and Kegan Paul (1958).

Conte, S. D., *Elementary Numerical Analysis*. McGraw-Hill (1965).

Conte, S. D. and de Boor, C., *Elementary Numerical Analysis*. McGraw-Hill (1972).

Dixon, C., *Numerical Analysis*. Blackie and Son Chambers (1974).

Goult, R. J., Hoskins, R. F., Milner, J. A., and Pratt, M. J., *Computational Methods in Linear Algebra*. Stanley Thornes (1974).

Henrici, P., *Elements of Numerical Analysis*. John Wiley and Sons (1964).

Hildebrand, F. B., *Introduction to Numerical Analysis*. McGraw-Hill (1956).

Hosking, R. J., Joyce, D. C., and Turner, J. C., *First Steps in Numerical Analysis*. Hodder and Stoughton (1978).

Noble, B., *Numerical Methods Volume 1*. Oliver and Boyd (1964).

Pennington, R. H., *Introductory Computer Methods and Numerical Analysis*. Collier-MacMillan (1970).

Phillips, G. M. and Taylor, P. J., *Theory and Application of Numerical Analysis*. Academic Press (1973).

Ribbans, J., *Basic Numerical Analysis Book I*. Intertext Books (1969).

Ribbans, J., *Basic Numerical Analysis Book II*. Intertext Books (1970).

Stark, P. A., *Introduction to Numerical Methods*. Collier-Macmillan (1970).

Walsh, J., *Numerical Analysis — An Introduction*. Academic Press (1966).

Wilkes, M. V., *A Short Introduction to Numerical Analysis*. Cambridge Univerity Press (1966).

Williams, P. W., *Numerical Computation*. Nelson (1972).

Index